T0226137

Lecture Notes in Computer Science   3266

Commenced Publication in 1973
Founding and Former Series Editors:
Gerhard Goos, Juris Hartmanis, and Jan van Leeuwen

Editorial Board

David Hutchison
   Lancaster University, UK
Takeo Kanade
   Carnegie Mellon University, Pittsburgh, PA, USA
Josef Kittler
   University of Surrey, Guildford, UK
Jon M. Kleinberg
   Cornell University, Ithaca, NY, USA
Friedemann Mattern
   ETH Zurich, Switzerland
John C. Mitchell
   Stanford University, CA, USA
Moni Naor
   Weizmann Institute of Science, Rehovot, Israel
Oscar Nierstrasz
   University of Bern, Switzerland
C. Pandu Rangan
   Indian Institute of Technology, Madras, India
Bernhard Steffen
   University of Dortmund, Germany
Madhu Sudan
   Massachusetts Institute of Technology, MA, USA
Demetri Terzopoulos
   New York University, NY, USA
Doug Tygar
   University of California, Berkeley, CA, USA
Moshe Y. Vardi
   Rice University, Houston, TX, USA
Gerhard Weikum
   Max-Planck Institute of Computer Science, Saarbruecken, Germany

Josep Solé-Pareta   Michael Smirnov
Piet Van Mieghem   Jordi Domingo-Pascual
Edmundo Monteiro   Peter Reichl
Burkhard Stiller   Richard J. Gibbens (Eds.)

Quality of Service in the Emerging Networking Panorama

Fifth International Workshop on Quality of Future Internet Services,
QofIS 2004 and First Workshop on Quality of Service Routing,
WQoSR 2004 and Fourth International Workshop
on Internet Charging and QoS Technology, ICQT 2004
Barcelona, Catalonia, Spain, September 29 - October 1, 2004
Proceedings

 Springer

Volume Editors

Josep Solé-Pareta
Jordi Domingo-Pascual
Universitat Politècnica de Catalunya, Departament d'Arquitectura de Computadors
Jordi Girona, 1-3, Mòdul D6 (Campus Nord), 08034 Barcelona, Catalonia, Spain
E-mail: {pareta,jordi}@ac.upc.es

Michael Smirnov
Fraunhofer FOKUS, Kaiserin-Augusta-Allee 31, 10589 Berlin, Germany
E-mail: smirnow@fokus.fraunhofer.de

Piet Van Mieghem
Delft University of Technology, Faculty of Information Technology and Systems
P.O. Box 5031, 2600 GA Delft, The Netherlands
E-mail: P.VanMieghem@ewi.tudelft.nl

Edmundo Monteiro
University of Coimbra, Department Computer Engineering
POLO II, Pinhal de Marrocos, 3030 Coimbra, Portugal
E-mail: edmundo@dei.uc.pt

Peter Reichl
ForschungszentrumWien, FTW, Donau-City-Str. 1, 1220 Wien, Austria
E-mail: reichl@ftw.at

Burkhard Stiller
Computer Engineering and Networks Laboratory TIK, ETH Zürich
Gloriastrasse 35, CH-8092 Zürich, Switzerland
and University of Federal Armed Forces Munich, Information Systems Laboratory
Werner-Heisenberg-Weg 39, 85577 Neubiberg, Germany
E-mail: stiller@tik.ee.ethz.ch

Richard J. Gibbens
University of Cambridge, Computer Laboratory
15 JJ Thomson Avenue, Cambridge CB 3 0FD, UK
E-mail: Richard.Gibbens@cl.cam.ac.uk

Library of Congress Control Number: 2004113086

CR Subject Classification (1998): C.2, H.4, H.3, J.1
ISSN 0302-9743
ISBN 3-540-23238-9 Springer Berlin Heidelberg New York

This work is subject to copyright. All rights are reserved, whether the whole or part of the material is
concerned, specifically the rights of translation, reprinting, re-use of illustrations, recitation, broadcasting,
reproduction on microfilms or in any other way, and storage in data banks. Duplication of this publication
or parts thereof is permitted only under the provisions of the German Copyright Law of September 9, 1965,
in its current version, and permission for use must always be obtained from Springer. Violations are liable
to prosecution under the German Copyright Law.

Springer is a part of Springer Science+Business Media

springeronline.com

© Springer-Verlag Berlin Heidelberg 2004
Printed in Germany

Typesetting: Camera-ready by author, data conversion by PTP-Berlin, Protago-TeX-Production GmbH
Printed on acid-free paper   SPIN: 11326953   06/3142   5 4 3 2 1 0

Preface

This volume of the Lecture Notes in Computer Science series contains the set of papers accepted for presentation at the 5th International Workshop on Quality of future Internet Services (QofIS 2004) and at the two one-day workshops co-located with QofIS 2004, namely the 1st International Workshop on QoS Routing (WQoSR 2004) and the 4th International Workshop on Internet Charging and QoS Technology (ICQT 2004).

QofIS 2004, the fifth international event, was organized under the umbrella of the E-NEXT Network of Excellence on "Emerging Networking Experiments and Technologies", which started its activities in January 2004. QofIS 2004 took place on September 29–30, 2004 at the Telefónica premises in Barcelona, and was arranged by the Universitat Politècnica de Catalunya (UPC). QofIS 2004 in Barcelona followed the highly successful workshops in Stockholm in 2003, Zürich in 2002, Coimbra in 2001, and Berlin in 2000. The purpose of QofIS 2004, as of all QofIS events, was to present and discuss design and implementation techniques for providing quality of service in the Internet.

The impact of emerging terminals, mobility and embedded systems is creating a new environment where networks are ambient. New challenges are opened by this new space where networks of interest ranging from personal networks to large-scale application networks need to be designed and often integrated. Protocol mechanisms for supporting quality of service at the different layers of the networks need to be assessed and eventually redesigned in such environments. In this context, the focus of the QofIS 2004 workshop was on the provisioning of *Quality of Service in the Emerging Networking Panorama*, assessed by results of experiments carried out in simulation platforms and test-beds, and given the progressive irruption of optical technologies.

QofIS 2004 contributed to this LNCS volume with 22 research papers selected from the 91 submissions received by the workshop, which address specific problems of quality-of-service provisioning in the fields of Internet applications, such as P2P and VoIP; service differentiation and congestion control; traffic engineering and routing; wireless LAN, ad hoc and sensor networks; and mobility in general. According to this, the workshop was organized in five sessions and featured two invited talks, by Prof. Ian F. Akyildiz of the Georgia Institute of Technology (USA) and Prof. David Hutchison of the University of Lancaster (UK), and was closed with a panel session, "New Face for QoS", featuring QofIS 2004 invited speakers and leaders of FP6 Networks of Excellence in networking and QoS provisioning.

August, 2004

QofIS'04

Josep Solé-Pareta
Michael Smirnov

The 1st Workshop on Quality of Service Routing was motivated by the growing number of contributions on this topic within the papers submitted to previous QofIS editions. We thought it would be worth having a set of sessions focused on QoS routing aspects and we decided to organize it as a co-located workshop within QofIS.

Quality of service routing poses several challenges that must be addressed to enable the support of advanced services in the Internet, both at intra- and inter-domain levels. The challenges of QoS routing are related to the distribution of routing information and to path selection and setup throughout the network. Extensive research has been carried out on QoS routing in the past few years. New frontiers are opening up to QoS routing such as the introduction of QoS routing in ad hoc, wireless multihop, sensor and self-organized networks, in content delivery networks and in optical technologies.

The purpose of this workshop was to summarize current research in QoS routing, describing experimental and theoretical studies, and to point out new research directions leading to *Smart Routing*. We were glad to see that 28 papers were submitted to the workshop, with authors from 15 different countries and covering a wide range of topics focused on QoS routing. After the reviewing process, 8 papers were selected, those that are included in this LNCS volume. The final WQoSR 2004 program was structured in two technical sessions, respectively devoted to Algorithms and Scalability issues and to Novel Ideas and Protocol Enhancements, and an invited talk given by Prof. Ariel Orda from the Technion at Haifa (Israel).

We hope the reading of these selected papers might be appealing and stimulating for the research community and that this workshop will be continued in the future.

August, 2004

Piet Van Mieghem
Jordi Domingo-Pascual
Edmundo Monteiro

The International Workshop on Internet Charging and QoS Technology (ICQT 2004) was the fourth event in a series of very successful annual workshops on network economics and Internet charging mechanisms. After establishing ICQT in 2001 in Vienna, Austria, the workshop was co-located once before with QofIS in 2002 in Zürich, Switzerland. The 2003 workshop took place in Munich together with NGC 2003. In 2004, ICQT was again co-located with QofIS and provided further vivid proof of the stimulating interdisciplinary combination of economics and networking technology, which has made these workshops a success story.

As in previous years, ICQT 2004 received more than 20 submissions from 14 different countries. Our enthusiastic Technical Program Committee managed to provide between 3 and 5 reviews per paper. Eventually, 8 papers were selected for the final program and arranged to form sessions on Auctions and Game Theory,

Charging in Mobile Networks, and QoS Provisioning and Monitoring. Together with the traditional invited lecture, this program presented a broad view on current research work in the interesting area where economy meets technology, where theory meets application, and where *"QoS has its price"*, as is stated in the title of ICQT 2004.

August, 2004

Peter Reichl
Burkhard Stiller
Richard Gibbens

Acknowledgments

It is our pleasure to acknowledge the excellent work done by the members of the QofIS 2004, WQoSR 2004 and ICQT 2004 program committees, and by the reviewers assigned by them, in helping to select the papers for this LNCS volume from the respective submissions. It was work done in addition to the daily line of business and we were fortunate to have had such a committed and careful group of experts help us.

The arrangements for the workshop were handled with wonderful dedication by the local organization committee, with the inestimable help of Anna Cutillas and Lluïsa Romanillos of the *Servei de Relacions Institucionals i Internacionals* of the UPC.

The QofIS 2004 website was built and maintained by the staff of the *Laboratori de Càlcul de la Facultat d'Informàtica de Barcelona*, also of UPC, led by Rosa Maria Martín. In particular we thank Xavier Rica (our webmaster) for his dedication and care in fixing all the details of the QofIS 2004 website.

Finally, the administrative issues were carried out by the secretariat of the Computer Architecture Department, our special thanks to Mercè Calvet and Juani Luna.

Organization

QofIS 2004 was organized by the Advanced Broadband Communications Centre (CCABA) of UPC, the Technical University of Catalonia (Spain). The CCABA (http://www.ccaba.upc.es) is a multidisciplinary laboratory composed of researchers from both the Computer Architecture Dept. and the Signal Theory and Communications Dept. of the UPC.

General Chair

Josep Solé-Pareta — Universitat Politècnica de Catalunya, Spain

Steering Committee

Jon Crowcroft — University of Cambridge, UK
James Roberts — France Telecom R&D, France
Michael Smirnov — Fraunhofer FOKUS, Germany
Fernando Boavida — University of Coimbra, Portugal
Burkhard Stiller — UniBw Munich, Germany and ETH Zürich, Switzerland
Gunnar Karlsson — Royal Institute of Technology, Sweden
Josep Solé-Pareta — Universitat Politècnica de Catalunya, Spain

Program Co-chairs QofIS 2004

Josep Solé-Pareta — Universitat Politècnica de Catalunya, Spain
Michael Smirnov — Fraunhofer FOKUS, Germany

Program Co-chairs WQoSR 2004

Piet Van Mieghem — Delft University of Technology, The Netherlands
Jordi Domingo-Pascual — Universitat Politècnica de Catalunya, Spain
Edmundo Monteiro — University of Coimbra, Portugal

Program Co-chairs ICQT 2004

Peter Reichl — FTW Vienna, Austria
Burkhard Stiller — UniBw Munich, Germany and ETH Zürich, Switzerland
Richard Gibbens — Cambridge University, UK

Program Committee QofIS 2004

Ian F. Akyildiz	Georgia Institute of Technology, USA
Arturo Azcorra	Universidad Carlos III de Madrid, Spain
Bengt Ahlgren	Swedish Institute of Computer Science, Sweden
Mario Baldi	Politecnico di Torino, Italy
Roberto Battiti	Università degli Studi di Trento, Italy
J.L. van den Berg	TNO Telecom, The Netherlands
Chris Blondia	University of Antwerp, Belgium
Fernando Boavida	Universidade de Coimbra, Portugal
Olivier Bonaventure	Université Catholique de Louvain, Belgium
David Bonyuet	Delta Search Labs, USA
Torsten Braun	University of Bern, Switzerland
Franco Callegati	Università di Bologna, Italy
Georg Carle	Universität Tübingen, Germany
Jon Crowcroft	University of Cambridge, UK
Michel Diaz	Centre National de la Recherche Scientifique, France
Jordi Domingo-Pascual	Universitat Politècnica de Catalunya, Spain
Peder J. Emstad	Norwegian University of Science and Technology, Norway
Ramon Fabregat	Universitat de Girona, Spain
Serge Fdida	Université Pierre et Marie Curie, France
Gísli Hjálmtýsson	Reykjavik University, Iceland
David Hutchison	University of Lancaster, UK
Andrzej Jajszczyk	AGH University of Science and Technology, Poland
Gunnar Karlsson	Royal Institute of Technology, Sweden
Yevgeni Koucheryavy	Tampere University of Technology, Finland
Artur Lason	AGH University of Science and Technology, Poland
Thomas Lindh	Royal Institute of Technology, Sweden
Marko Luoma	Helsinki University of Technology, Finland
Josep Lluís Marzo	Universitat de Girona, Spain
Xavier Masip-Bruin	Universitat Politècnica de Catalunya, Spain
Hermann de Meer	University of Passau, Germany
Edmundo Monteiro	Universidade de Coimbra, Portugal
Jaudelice C. de Oliveira	Drexel University, USA
Giovanni Pacifici	IBM T.J. Watson Research Center, USA
George Pavlou	University of Surrey, UK
Guido H. Petit	Alcatel, Belgium
Thomas Plagemann	University of Oslo, Norway
Mihai Popa	Unaffiliated, Romania
Carla Raffaelli	Università di Bologna, Italy
Simon Pietro Romano	Università di Napoli Federico II, Italy

Sergi Sánchez-López   Universitat Politècnica de Catalunya, Spain
Caterina Maria Scoglio   Gerogia Institute of Technology, USA
Dimitrios N. Serpanos   University of Patras, Greece
Vasilios Siris   ICS-FORTH and University of Crete, Greece
Michael Smirnov   Fraunhofer FOKUS, Germany
Josep Solé-Pareta   Universitat Politècnica de Catalunya, Spain
Ioannis Stavrakakis   University of Athens, Greece
Burkhard Stiller   UniBw Munich, Germany and ETH Zürich,
                                             Switzerland
Piet Van Mieghem   Delft University of Technology,
                                             The Netherlands
Giorgio Ventre   Università di Napoli Federico II, Italy
Adam Wolisz   Technical University Berlin, Germany
Lars Wolf   Technical University at Braunschweig,
                                             Germany

Program Committee WQoSR 2004

Fernando Boavida   Universidade de Coimbra, Portugal
Olivier Bonaventure   Université Catholique de Louvain, Belgium
Jon Crowcroft   University of Cambridge, UK
J.J. Garcia-Luna-Aceves   University of California at Santa Cruz, USA
Raimo Kantola   Helsinki University of Technology, Finland
Fernando Kuipers   Delft University of Technology,
                                             The Netherlands
Xavier Masip-Bruin   Universitat Politècnica de Catalunya, Spain
Klara Nahrstedt   University of Illinois at Urbana-Champaign,
                                             USA
Ariel Orda   Technion, Israel
Michael Smirnov   Fraunhofer FOKUS, Germany
Josep Solé-Pareta   Universitat Politècnica de Catalunya, Spain
Alexander Sprintson   California Institute of Technology, USA
Giorgio Ventre   Università di Napoli, Federico II, Italy

Program Committee ICQT 2004

Jörn Altmann   International University Bruchsal, Germany
Ragnar Andreassen   Telenor, Norway
Sandford Bessler   FTW Vienna, Austria
Torsten Braun   University of Bern, Switzerland
Bob Briscoe   BTexact Technologies, UK
Roland Büschkes   T-Mobile International, Germany
Costas Courcoubetis   Athens Univ. of Economics and Business,
                                             Greece
Chris Edwards   Lancaster University, UK

Martin Karsten   University of Waterloo, Canada
Peter Key   Microsoft Research Cambridge, UK
Claudia Linnhoff-Popien   LMU Munich, Germany
Simon Leinen   SWITCH, Zürich, Switzerland
Peter Marbach   University of Toronto, Canada
Robin Mason   University of Southampton, UK
Lee McKnight   Tufts University, USA
Jose Ignacio Moreno   Universidad Carlos III Madrid, Spain
Andrew Odlyzko   University of Minnesota, USA
Huw Oliver   Lancaster University, UK
Maximilian Ott   Semandex Networks, USA
Kihong Park   Purdue University, USA
Guido Petit   Alcatel, Belgium
Douglas Reeves   North Carolina State University, USA
Björn Rupp   Arthur D. Little, Germany
Jens Schmitt   Universität Kaiserslautern, Germany
Vasilios Siris   ICS Forth, Greece
Otto Spaniol   RWTH Aachen, Germany
Bruno Tuffin   INRIA Rennes, France

Local Organization (UPC, Spain)

Xavier Masip-Bruin (Co-chair)   Albert Cabellos
Sergi Sánchez-López (Co-chair)   René Serral
Davide Careglio   Carles Kishimoto
Salvatore Spadaro
Pere Barlet

Reviewers

Anas Abou El Kalam   Gyorgy Dan   Salvatore Iacono
Rui L.A. Aguiar   Luis De La Cruz   Luigi Iannelli
Tricha Anjali   Cédric De Launois   Eva Ibarrola
Panayiotis Antoniadis   Danny De Vleeschauwer   Arnaud Jacquet
Yuri Babich   Manos Dramitinos   Jingwen Jin
Albert Banchs   Wojciech Dziunikowski   Jouni Karvo
Florian Baumgartner   Armando Ferro   Karl-Heinz Krempels
Bela Berde   Olivier Fourmaux   Jérôme Lacan
Marilia Curado   Pierre François   Nikolaos Laoutaris
Maria-Dolores Cano   Mario Freire   Pasi Lassila
Selin Kardelen Cerav   Giulio Galante   Bo Li
Walter Cerroni   Vehbi Cagri Gungor   Fidel Liberal
Chao Chen   Jarmo Harju   Yunfeng Lin
Antonio Cuevas Casado   Tom Hofkens   Renato Lo Cigno

Emmanuel Lochin
King-Shan Lui
Iannone Luigi
Michel R.H. Mandjes
Josep Mangues-Bafalluy
Eva Marin-Tordera
Marco Mellia
Tommaso Melodia
Paulo Mendes
Dmitri Moltchanov
Fernando Moreira
Giovanni Muretto
Ruy Oliveira
Evgueni Ossipov
Joao Orvalho
Philippe Owesarski
Thanasis Papaioannou
Manuel Pedro
Sandra Peeters
Cristel Pelsser

Rajendra Persaud
Markus Peuhkuri
Reneé Pilz
Dario Pompili
Jan Potemans
Berthold Rathke
Luis Rodrigues
Ricardo Romeral
Jacek Rzasa
Alexandre Santos
Susana Sargento
Michael Savoric
Matthias Scheidegger
Morten Schläger
Pablo Serrano
Marina Settembre
Jorge Silva
Lukasz Sliwczynski
Rute Sofia
Sergios Soursos

Kathleen Spaey
Promethee Spathis
Rafal Stankiewicz
George Stamoulis
Andrzej Szymanski
Apostolos Traganitis
Hung Tran
Manuel Urueña
Luca Valcarenghi
Gianluca Varenni
Teresa Vazao
Monica Visintin
Michael Voorhaen
Jun Wan
Martin Wenig
Attila Weyland
Li Xiao
Marcelo Yannuzzi
Paolo Zaffoni

Sponsoring Institutions

Table of Contents

Traffic Engineering and Routing

Enforcing Mobility

Quality of Service Routing

Algorithms and Scalability Issues

Novel Ideas and Protocol Enhancements

Internet Charging and QoS Technology

Auctions and Game Theory

Charging in Mobile Networks

QoS Provisioning and Monitoring

Author Index 389

Performance Analysis of
Peer-to-Peer Networks for File Distribution

Ernst W. Biersack[1], Pablo Rodriguez[2], and Pascal Felber[1]

[1] Institut EURECOM, France
{erbi,felber}@eurecom.fr
[2] Microsoft Research, UK
pablo@microsoft.com

Abstract. Peer-to-peer networks have been commonly used for tasks such as file sharing or file distribution. We study a class of cooperative file distribution systems where a file is broken up into many chunks that can be downloaded independently. The different peers cooperate by mutually exchanging the different chunks of the file, each peer being client and server at the same time. While such systems are already in widespread use, little is known about their performance and scaling behavior. We develop analytic models that provide insights into how long it takes to deliver a file to N clients. Our results indicate that the service capacity of these systems grows exponentially with the number of chunks a file consists of.

1 Introduction

Peer-to-peer systems, in which peer computers form a cooperative network and share their resources (storage, CPU, bandwidth), have attracted a lot of interest lately. They provide a great potential for building cooperative networks that are self-organizing, efficient, and scalable.

Research in peer-to-peer networks has so far mainly focused on content storage and lookup, but fewer efforts have been spent on content distribution. By capitalizing on the *bandwidth* of peer nodes, cooperative architectures offer great potential for addressing some of the most challenging issues of today's Internet: the cost-effective distribution of bandwidth-intensive content to thousands of simultaneous users both Internet-wide and in private networks.

Cooperative content distribution networks are inherently *self-scalable*, in that the bandwidth capacity of the system increases as more peers arrive: each new peer requests service from, but also provides service to, the other peers. The network can thus spontaneously adapt to the demand by taking advantage of the resources provided by every peer.

We present a deterministic analysis that provides insights into how different approaches for distributing a file to a large number of clients compare. We consider the simple case of N peers that arrive simultaneously and request to download the same file. Initially, the file exists in a single copy stored at a node called *source* or *server*. We assume that the file is broken up into *chunks* and

J. Solé-Pareta et al. (Eds.): QofIS 2004, LNCS 3266, pp. 1–10, 2004.
© Springer-Verlag Berlin Heidelberg 2004

that peers cooperate, i.e., a peer that has completely received a chunk will offer to upload this chunk to other peers. The time it takes to download the file to all peers will depend on *how* the chunks are exchanged among the peers, which is referred to as peer organization strategy.

To get some insights into the performance of different peer organization strategies, we analytically study three different distribution models:

- A linear chain architecture, referred to as *Linear*, where the peers are organized in a chain with the server uploading the chunks to peer P_1, which in turn uploads the chunks to P_2 and so on.
- A tree architecture, referred to as *Tree*^{k}, where the peers are organized in a tree with an outdegree k. All the peers that are not leaves in the tree will upload the chunks to k peers.
- A forest of trees consisting of k different trees, referred to as *PTree*^{k}, which partitions the file into k parts and constructs k spanning trees to distribute the k parts to all peers.

We analyze the performance of these three architectures and derive an upper bound on the number of peers served within an interval of time t. We consider a scenario where each peer has equal upload and download rates of b. The upload rate of the server is also b. We focus on the distribution of a single file that is partitioned into C chunks. The time needed to download the complete file at rate b is referred to as *one round* or 1 unit of time. Thus, the time needed to download a single chunk at rate b is $1/C$. For the sake of simplicity, we completely ignore the bandwidth fluctuation in the network or node failures. We assume that the only constraint is the upload/download capacity of peers.

Several systems have been recently proposed to leverage peer-to-peer architectures for application-layer multicast. Most of these systems target streaming media (e.g., [1,2,3,4]) but some also consider bulk file transfers (e.g., [5,6]). Experimental evaluation and measurements have been conducted on real-world several peer-to-peer systems to observe their properties and behavior [7,8,9,10] but, to the best of our knowledge, there has been scarcely any analytical study of distribution architectures for file distribution. We are only aware of one other paper that evaluates the performance and scalability of peer-to-peer systems by modeling the propagation of the file as a branching process [11]. However, no particular distribution architecture is assumed. The results of this paper indicate that the number of clients that complete the download grows exponentially in time and are in accordance with our results.

The rest of the paper is organized as follows. Section 2 introduces the *Linear* architecture. In Section 3 we study *Tree*^{k} and we evaluate *PTree*^{k} in Section 4. We then presents a comparative analysis of the three distribution models in Section 5 and conclude the paper in Section 6.

2   *Linear*: A Linear Chain Architecture

In this section, we study the evolution over time of the number of served peers for the *Linear* architecture. We make the following assumptions:

- The server serves sequentially and infinitely the file at rate b. At any point in time, the server uploads the file to a single peer.
- Each peer starts serving the file once it receives the first chunk.

We consider the case where each peer uploads the whole file at rate b to exactly one other peer before it disconnects. Thus, each peer contributes the same amount of data to the system as it receives from the system. At time 0, the server starts serving a first peer. At time $1/C$, the first peer has completely received the first chunk and starts serving a second peer. Likewise, once the second peer has received the first chunk at time $2/C$, it starts serving a third peer and so on. As a result, peers are connected in a chain with each peer receiving chunks from the previous one and serving the next one. The length (i.e., the number of peers) of the chain increases by one peer each $1/C$ unit of time. At time 1, the server finishes uploading the file to the first peer. If there are still peers left that have not even received a single chunk, the server starts a new chain that increases also by one peer each $1/C$ unit of time. The same process repeats at each round, as shown in Figure 1 (the black circle represents the server, the black squares are peers that start downloading the file, and the lines connecting the peers correspond to active connections). This makes $(t+1)$ chains within t rounds. The number of served peers at time t over all those chains includes only the peers that have joined the network on or before time $t-1$. Clients that arrive after time $t-1$ will take one unit of time to download the file and will be done after time t. Given a chain initiated at time 0, its length at time t is $(1 + t \cdot C)$ and the number of served peers in that chain is $1 + (t-1)C$ peers. Over all chains, the number of served peers within t rounds is given by

$$N_{Linear}(C,t) = \sum_{i=1}^{t}(1 + (i-1)C) = t + \frac{C \cdot t(t-1)}{2} \approx C \cdot t^2 \qquad (1)$$

We see that the number of peers served grows linearly with the number of chunks C and quadratically with the number of rounds t. From Equation (1) we derive the formula for the time needed to completely serve N peers as

$$T_{Linear}(C,N) = \frac{(C-2) + \sqrt{(C-2)^2 + 8 \cdot N \cdot C}}{2 \cdot C} \approx \frac{1}{2} + \sqrt{\frac{1}{4} + \frac{2 \cdot N}{C}} \qquad (2)$$

If N/C denotes the node to chunk ratio, we can distinguish the following cases:

1. $T_{Linear}(C,N) \approx \frac{1}{2} + \sqrt{\frac{1}{4}} = 1$, for $\frac{N}{C} \ll 1$
2. $T_{Linear}(C,N) \approx \frac{1}{2} + \sqrt{2} \approx 2$, for $\frac{N}{C} = 1$
3. $T_{Linear}(C,N) \approx \sqrt{\frac{N}{C}}$, for $\frac{N}{C} \gg 1$

Figure 2 plots $N_{Linear}(C,t)$ as a function of the number of rounds for different values of the number of chunks C. It appears clearly that, for a given number of peers N, the smaller the node to chunk ratio N/C, the shorter the time to serve

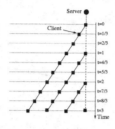

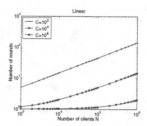

Fig. 1. Evolution of the *Linear* architecture with time ($C = 3$).

Fig. 2. $T_{Linear}(C, N)$ as a function of N and C.

these N peers. In fact, for $N/C \ll 1$ all peers will be active uploading chunks for most of the time and T_{Linear} will be approximately one round. On the other hand, for $N/C > 1$ only C out of the N peers will be uploading at any point in time, while the other $N - C$ peers have either already forwarded the entire file or not yet received a single chunk.

3   *Tree^k*: A Tree Distribution Architecture

As we have just seen, for $N/C > 1$ the linear chain fails to keep all the peers working most of the time. To alleviate this problem we now consider *Tree^k*, a tree architecture with outdegree k where the number of "hops" from the server to the last peer is $\log_k N$, as compared to N for the linear chain. We make the following assumptions:

- The server serves k peers in parallel, each at rate b/k.
- Each peer downloads the whole file at rate b/k.
- A peer that is interior (i.e., non leaf) node of the distribution tree starts uploading the file to k other peers, each rate b/k, soon as it has received the first chunk. This means that interior nodes upload an amount of data equivalent to k times the size of the file, while leaf nodes do not upload the file at all.

Given a download rate of b/k, a peer needs k/C units of time to receive a single chunk. Note that *Tree^{k=1}* is equivalent to *Linear*. We first explain the evolution for $k = 2$. At time 0, the server serves 2 peers each at rate $b/2$. Each of those peers starts serving 2 new peers $2/C$ units later, which will need to wait another $2/C$ units before they have completely received a chunk and can in turn serve other peers. The two peers served by the server become each a root of a tree with an outdegree 2. The height of each tree increases by one level every $2/C$ units of time (see Figure 3).

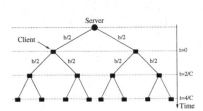

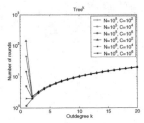

Fig. 3. Evolution of the $Tree^{k=2}$ architecture with time.

Fig. 4. $T_{Tree}(C, k, N)$ as a function of k, N, and C.

In the $Tree^k$ architecture, k identical trees are initiated by the server at time 0, each of which will include N/k peers. The time needed to serve N peers is

$$T_{Tree}(C, k, N) = k + \lfloor \log_k(\frac{N}{k}) \rfloor \cdot \frac{k}{C} \tag{3}$$

where k/C represents the delay induced by each level in the tree. Leaf peers start receiving the first chunk $\lfloor \log_k(N/k) \rfloor \cdot k/C$ units of time after the root peer and complete the download k units of time later.

We derive from Equation (3) the number of peers served within t rounds as

$$N_{Tree}(C, k, t) \approx k^{(\frac{t}{k}-1)C+1} = k^{(t-k)\frac{C}{k}+1} \tag{4}$$

It follows from Equation (3) and (4) that the performance of file distribution directly depends on the degree k of the tree. We can compute the optimal value of k by taking the derivative of T_{tree} with respect to k. This gives

$$k_{opt} = e^{\frac{\blacksquare \log N + \sqrt{(\log N)^2 + 4 \cdot (C \blacksquare 1) \cdot \log N}}{2 \cdot (C \blacksquare 1)}} \qquad \text{given that} \quad N < k \cdot \frac{k^{(C+1)} - 1}{k-1} \tag{5}$$

The optimal outdegree k_{opt} depends on the peer to chunk ratio N/C. For $N/C \leq 1$, the optimal outdegree is 1, i.e., a linear chain, since the linear chain assures that the peers are uploading most of the time at their full bandwidth capacity. For $N/C > 1$, an increase in N/C leads to an increase in the optimal outdegree as the linear chain becomes less and less effective (remember that only C out of the N peers are uploading simultaneously).

In practice, the outdegree can only take integer values and we see from Figure 4 that for $N/C > 1$ the binary tree yields lower download times that the linear chain. The binary tree is also the optimal tree. Remember that in T_{tree} the outdegree k appears as an additive constant that is typically much larger than the other term ($\lfloor \log_k(N/k) \rfloor \cdot k/C$)

While the binary tree improves the download time as compared to the linear chain when $N/C > 1$, it suffers from two important shortcomings. First, although the maximum upload and download rate is b, the peers in a binary tree download only at rate $b/2$. As a consequence, the download time is at least *twice* the time it takes if the file were downloaded at the maximum possible download rate.

Second, in a binary tree of height h, there are 2^h leaf nodes and $2^h - 1$ interior nodes. Since only the interior nodes upload chunks to other peers, this means that half of the peers will not upload even a single block. As the outdegree k of the tree increases, the percentage of peers uploading data keeps decreasing, with only about one out of k peers uploading data. Also, the peers that upload must upload the entire file k times.

4   *PTree^k*: An Architecture Based on Parallel Trees

The overall performance of the tree architecture would be significantly improved if we could capitalize on the unused upload capacity of the leaves to utilize the $b - b/k$ unused download capacity at each of the peers. It is not possible, however, for a leaf to serve other peers upwards its tree because it only holds chunks that its ancestors already have. Given a tree architecture with k trees rooted at the server, the basic intuition underlying the *PTree^k* architecture is to "connect" the leaves of one of the trees to peers of the other $k - 1$ trees to ultimately produce k spanning trees, and have the server send distinct chunks to each of these trees.

More specifically, the *PTree^k* architecture organizes the peers in k different trees such that each peer is an interior peer in at most one tree and a leaf peer in the remaining $k - 1$ trees. The file is then partitioned into k parts, where each part is distributed on a different tree: tree T^k for part P^k. All k parts have the same size in terms of number of bytes. If the entire file is divided into C chunks, each of the k parts will comprise C/k disjoint chunks.[1] Such a distribution architecture was first proposed under the name of SplitStream [3] to increase the resilience against churn (i.e., peers failing or leaving prematurely) in a video streaming application.

In *PTree^k*, a peer receives the k parts in parallel from k different peers, each part at rate b/k, while the peer helps distributing at most one part of the file to k other peers. Therefore, the total amount of data a peer uploads corresponds exactly to the amount contained in the file, regardless the outdegree k of the trees.

Figure 5 depicts the basic idea of *PTree$^{k=2}$*, where k denotes the outdegree of the tree. Each peer, except for peer 4, is an interior peer in one tree and a leaf peer in another tree. It is easy to show that, independent of the outdegree k, there will always be one peer in *PTree^k* that is leaf in all k trees. Each tree includes all N peers. A tree with outdegree k has $1 + \lfloor log_k N \rfloor$ levels and a height of $\lfloor log_k N \rfloor$. Since peers transmit at a rate b/k, each level in the tree induces a delay of k/C units of time.

Consider a leaf peer C_0 in tree T^k that is located $\lfloor log_k N \rfloor$ levels down from the root of T^k. C_0 starts receiving part P^k at time $\lfloor log_k N \rfloor \cdot k/C$ and the time

---

[1] For the sake of simplicity, we assume that the number of chunks C is a multiple of the number of parts k.

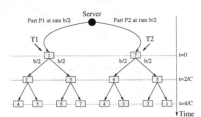

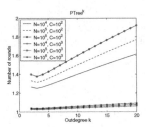

Fig. 5. Evolution of the $PTree^{k=2}$ architecture with time.

Fig. 6. $T_{PTree}(C, k, N)$ as a function of k, N, and C.

to receive P^k entirely, once reception has started, is 1. Therefore, C_0 will have completely received part P^k at time $1 + \lfloor log_k N \rfloor \cdot k/C$.

A peer has completed its download after it has received the last byte of each of the k parts of the file. A $PTree^k$ peer is a leaf node in $k-1$ trees and an interior node in one tree, and it receives all k parts in parallel. This means that all peers complete their download at the same time $1 + \lfloor log_k N \rfloor \cdot k/C$. We therefore have

$$T_{PTree}(C, k, N) = 1 + \lfloor log_k N \rfloor \cdot \frac{k}{C} \qquad (6)$$

We derive from Equation (6) the number of peers served within t rounds as

$$N_{PTree}(C, k, t) \approx k^{(t-1)\frac{C}{k}} \qquad (7)$$

Similarly to the tree distribution architecture in Section 3, there is an optimal value k for $PTree^k$ that minimizes the service time. Intuitively, a very deep tree should be quite inefficient in engaging peers early since leaves are quite far from the source. In fact, $PTree^{k=1}$ is equivalent to *Linear*, which is very inefficient in engaging peers for $N/C > 1$. On the other hand, when the outdegree of the tree is large, leaf peers are only a few hops from the source and can be engaged fast. However, this intuition is not completely correct: flat trees with large outdegrees suffer from the problem that, as the outdegree k increases, the rate b/k at which each chunk is transmitted from one level to the next one decreases linearly with k. This rate reduction can negate the benefits of having many peers reachable within few hops.

We can compute the optimal tree outdegree that provides the best $PTree^k$ performance by taking the derivative of Equation (6) with respect to k and equating the result to zero. We find T_{PTree} to be minimal for $k = e$, independant of the peer to chunk ratio N/C.

Figure 6 depicts the performance of $PTree^k$ as a function of the outdegree. We see that the optimal $PTree^k$ performance is obtained for trees with an outdegree $k = 3$. However, the performance for $k = 2$ and $k = 4$ is almost the same as for $k = 3$. As the outdegree increases the performance of PTree degrades: for $N/C \approx 1$ the degradation is very small while for $N/C \gg 1$ it is quite pronounced.

By striping content across multiple trees, $PTree^k$ can ensure that the departure of one peer causes only a minimal disruption to the system, reducing the peer's throughput only by b/k. Given that the overhead caused by churn can be minimized by striping content across a higher number of trees, one can consider slightly higher outdegrees than the optimal value (e.g., 5) to minimize the impact of churn at the expense of a minimal increase in transfer time.

5   Comparative Analysis

In this section, we compare the performance of the *Linear*, *Tree^k* and *PTree^k* architecture. We first investigate how the time needed to serve N peers varies as a function of the number of peers N and the number of chunks C.

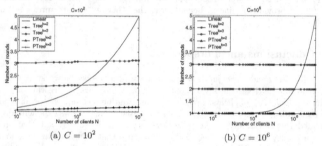

(a) $C = 10^2$ (b) $C = 10^6$

Fig. 7. Performance of *Linear*, *Tree$^{k=\{2,3\}}$* and *PTree$^{k=\{2,3\}}$* as a function of N.

From Figure 7, we see that independent of the number of nodes and chunks, *PTree^k* is able to offer download times close to 1. On the other hand, as already pointed out, the download times for *Tree^k* are always larger than k units of time (see Equation (3)). *PTree^k* with optimal outdegree $k = 3$ provides clear benefits over *Linear* for $N/C \gg 1$ since peers are engaged much faster into uploading chunks than with a linear chain.

When the propagation delay of the first chunk is very small compared to the transmission time of the file, the peers stay engaged most of the time in the linear chain and the benefit of *PTree^k* diminishes. This is the case when the number of chunks is very large ($C \rightarrow \infty$), the number of peers is small, or the transmission rate is very high. The pivotal point where *PTree^k* starts to significantly outperform *Linear* is around $N/C > 10^{-1}$ (see Figure 7(b)).

The comparison of the different approaches would be incomplete if we did not address aspects such as robustness and ease of deployment. Cooperative file distribution relies on the collaboration of individual computers that are not subject to any centralized control or administration. Therefore, the failure or departure of some computers during the data exchange are most likely to occur and should also be taken into account when comparing approaches. For the linear chain and the tree, the departure of the node will disconnect all the nodes "downstream" from the data forwarding. With $PTree^k$, the impact of the departure of a node effects only affects one out of k trees, which makes parallel trees the most robust of the three approaches.

The *Linear* and $PTree^k$ architectures both assume that the upload and download rates are the same. In practice, this is often not the case (e.g., the upload rate of ADSL lines is only a fraction of the download rate). In such cases, the performance will be limited by the upload rate and some of the download bandwidth capacity will remain unused. The $Tree^k$ architecture assumes that nodes can *upload* data at a rate k times higher than the download rate, which is exactly the opposite to what ADSL offers. The tree approach is therefore particularly ill-suited for such environments.

From these results we can conclude that, for most typical file-transfer scenarios, $PTree^k$ provides significant benefits over the other two architectures studied in this paper.

6   Conclusions and Perspectives

The *self-scaling* and *self-organizing* properties of peer-to-peer networks offer the technical capabilities to quickly and efficiently distribute large or critical content to huge populations of clients. Cooperative distribution techniques capitalize on the bandwidth of every peer to offer a service capacity that grows exponentially, provided the blocks among the peers are exchanged in such a way that the peers are busy most of the time. The architecture that best achieves this goal among those studied in the paper, independently of the peer to chunk ratio N/C, is $PTree^k$. For both $Tree^k$ and $PTree^k$ there is an optimal outdegree that minimizes the download time.

Our analysis provided some important insights as to how to choose certain key parameters such as C and k. First, the file should be partitioned into a large number of chunks C, since the performance scales exponentially with C (but not too many as each chunk adds some coordination and connection overhead). Second, each peer should limit the number k of simultaneous uploads to other peers. We saw that for $PTree^k$ a good value for k is between 3 and 5.

The results of our study also guide the design of cooperative peer-to-peer file sharing applications that do not organize the peers in a such a static way as do the linear chain or tree(s) but use a *mesh* instead (e.g., BitTorrent [6]). Here, a peer must decide how many peers to serve simultaneously (the outdegree k) and what chunks to serve next (the "chunk selection strategy"). For each chunk, a peer selects the peer it wants to upload that chunk to (the "peer selection strategy").

Consider a peer selection strategy that gives preference to the peers that are closest to completion (among those that have the fewest incoming connections), and a chunk selection strategy that favors the chunks that are least widely held in the system. Assume that each peer only accepts a single inbound connection. With 1 outbound connection per peer, we trivially obtain a linear chain; with 2 outbound connections, we obtain a binary tree $Tree^{k=2}$; and so on. Failures are a handled gracefully as the parent of a failed peer automatically reconnects to the next peer in the chain or tree.

If we now allow each peer to have k inbound and k outbound connections, we obtain a configuration equivalent to $PTree^k$. Indeed, the source will fork k trees to which it will send distinct chunks (remember that we give preference to the rarest chunks). The leaves of the trees, which have free outbound capacity, will connect to the peers of the other trees to eventually create k parallel spanning trees. Such mesh-based systems, whose topology dynamically evolves according to predefined peer and chunk selection strategies, offer service times as low as the ones of $PTree^k$ and adjust dynamically to bandwidth fluctuations, bandwidth heterogeneity, and node failures [12].

References

1. Chu, Y.H., Rao, S., Zhang, H.: A case for end system multicast. In: Proceedings of ACM SIGMETRICS. (2000)
2. Jannotti, J., Gifford, D., Johnson, K.: Overcast: Reliable multicasting with an overlay network. In: Proceedings of the Symposium on Operating Systems Design and Implementation (OSDI). (2000)
3. Castro, M., Druschel, P., Kermarrec, A.M., Nandi, A., Rowstron, A., Singh, A.: SplitStream: High-bandwidth multicast in a cooperative environment. In: Proceedings of the ACM Symposium on Operating Systems Principles (SOSP). (2003)
4. Tran, D., Hua, K., Do, T.: ZIGZAG: An efficient peer-to-peer scheme for media streaming. In: Proceedings of IEEE INFOCOM, San Francisco, CA, USA (2003)
5. Sherwood, R., Braud, R., Bhattacharjee, B.: Slurpie: A cooperative bulk data transfer protocol. In: Proceedings of IEEE INFOCOM. (2004)
6. Cohen, B.: Incentives to build robustness in bittorrent. Technical report, http://bitconjurer.org/BitTorrent/bittorrentecon.pdf (2003)
7. Saroiu, S., Gummadi, K., Gribble, S.: A measurement study of peer-to-peer file sharing systems. In: Proceedings of Multimedia Computing and Networking (MMCN). (2002)
8. Ripeanu, M., Foster, I., Iamnitchi, A.: Mapping the Gnutella network: Properties of large-scale peer-to-peer systems and implications for system design. IEEE Internet Computing Journal **6** (2002)
9. Gummadi, K., Dunn, R., Saroiu, S., Gribble, S., Levy, H., Zahorjan, J.: Measurement, modeling, and analysis of a peer-to-peer file-sharing workload. In: Proceedings of the ACM Symposium on Operating Systems Principles (SOSP). (2003)
10. Izal, M., Urvoy-Keller, G., Biersack, E., Felber, P., Hamra, A.A., Garces-Erice, L.: Dissecting BitTorrent: Five months in a torrent's lifetime. In: Proceedings of the 5th Passive and Active Measurement Workshop (PAM). (2004)
11. Yang, X., de Veciana, G.: Service capacity of peer-to-peer networks. In: Proceedings of IEEE INFOCOM. (2004)
12. Felber, P., Biersack, E.: Self-scaling networks for content distribution. In: Proceedings of the International Workshop on Self-* Properties in Complex Information Systems. (2004)

MULTI+: Building Topology-Aware Overlay Multicast Trees

Luis Garcés-Erice, Ernst W. Biersack, and Pascal A. Felber

Institut EURECOM
06904 Sophia Antipolis, France
{garces|erbi|felber}@eurecom.fr

Abstract. TOPLUS is a lookup service for structured peer-to-peer networks that is based on the hierarchical grouping of peers according to network IP prefixes. In this paper we present MULTI+, an application-level multicast protocol for content distribution over a peer-to-peer (P2P) TOPLUS-based network. We use the characteristics of TOPLUS to design a protocol that allows for every peer to connect to an available peer that is close. MULTI+ trees also reduce the amount of redundant flows leaving and entering each network, making more efficient bandwidth usage.

1 Introduction

IP Multicast seems (or, at least, was designed) to be the ideal solution for content distribution over the Internet: (1) it can serve content to an unlimited number of destinations, and (2) it is bandwidth-wise economic. These two characteristics are strongly correlated. IP Multicast saves bandwidth because a single data flow can feed many recipients. The data flow is only split at routers where destinations for the data are found in more than one outgoing port. Thus n clients do not need n independent data flows, which allows for IP Multicast's scalability. However, IP Multicast was never widely deployed in the Internet: security reasons, its open-loop nature, made IP multicast remain as a limited use tool for other protocols in LANs. The core Internet lacks of an infrastructure with the characteristics of IP Multicast.

Lately, with the advent of broadband links like ADSL and the generalization of LANs at the workplace, the *edges* of the Internet started to increase their bandwidth. Together with the ever-cheaper and yet more powerful equipment (computational power, storage capacity), they give the millions of hosts connected to the Internet the possibility of implementing themselves services that augment at the application level the capabilities of the network: the Peer-to-Peer (P2P) systems. Various application-level multicast implementations have been proposed [1,2,3,4,5], most of which are directly implemented on top of P2P infrastructures (Chord [6], CAN [7] or Pastry [8]). The good scalability of the underlying P2P networks give these application-level multicast one of the properties of the original IP Multicast service, that of serving content to a virtually unlimited number of clients (peers). However, these P2P networks are generally conceived as an application layer system completely isolated from the underlying IP network.

Thus, the P2P multicast systems that we know may fail at the second goal of IP Multicast: a LAN hosting a number of peers in a P2P multicast tree may find its outbound

J. Solé-Pareta et al. (Eds.): QofIS 2004, LNCS 3266, pp. 11–20, 2004.
© Springer-Verlag Berlin Heidelberg 2004

link saturated by *identical* data flowing to and from its local peers, unless those peers are somehow *aware* of the fact that they are sharing the same network. This is a critical issue for ISPs due to P2P file-sharing applications and flat-rate commercial offers that allow a home computer to be downloading content 24 hours a day. This problem also affects application-level multicast sessions if peers do not take care of the network topology. We have based our P2P multicast protocol on TOPLUS because of its inherent topology-awareness. We consider that there is a large population of peers, which justifies the utilization of Multicast, that many Multicast groups may coexist without interfering each other, and that each peer must accept to cooperate with others in low-bandwidth maintenance tasks, but they are not forced to transmit content that does not interest them.

Related Work. Some examples of overlay networks which introduce topology-awareness in their design are SkipNet [9], Coral[10], Pastry [11], CAN [12]. Application-level Multicast has given some interesting results like the NICE project [1] or End System Multicast [2]. Other application-level multicast implementations use overlay networks as we do to create Multicast trees: Bayeux [3], CAN [4] Pastry (Scribe) [5]. However, these approaches are not designed to optimize some metric like delay or bandwidth utilization. There is an interesting comparative in [13]. Content distribution overlay examples are SplitStream [14] and [15]. Recently, the problem of data dissemination on adaptive overlays has been treated in [16]. Our main contribution is the achievement of efficient topology-aware multicast trees with no or very little active measurement using distributed algorithms, while others aiming at similar goals require extensive probing [17], or rely on a much wider knowledge of the peer population [17] [18] than ours.

In the next section we present the main aspects of TOPLUS. In Section 3 we describe MULTI+. In Section 4 we comment some results on MULTI+ Multicast trees properties, before we conclude and sketch future work in Section 5.

2   TOPLUS Overview

TOPLUS [19] is based on the DHT paradigm, in which a resource is uniquely identified by a key, and each key is associated with a single peer in the network. Keys and peers share a numeric identifier space, and the peer with the identifier closest to a key is responsible for that key. The principal goal of TOPLUS is simple: each routing step takes the message closer to the destination.

Let I be the set of all 32-bit IP addresses. Let $\mathcal{G}$ be a collection of sets such that $G \subseteq I$ for each $G \in \mathcal{G}$. Thus, each set $G \in \mathcal{G}$ is a set of IP addresses. We refer to each such set G as a *group*. Any group $G \in \mathcal{G}$ that does not contain another group in $\mathcal{G}$ is said to be an *inner group*. We say that the collection $\mathcal{G}$ is a *proper nesting* if it satisfies all the following properties:

1. $I \in \mathcal{G}$.
2. For any pair of groups in $\mathcal{G}$, the two groups are either disjoint, or one group is a proper subset of the other.
3. Each $G \in \mathcal{G}$ consists of a set of contiguous IP addresses that can be represented by an IP prefix of the form $w.x.y.z/n$ (for example, 123.13.78.0/23).

The collection of sets $\mathcal{G}$ can be created by collecting the IP prefix networks from BGP tables and/or other sources [20]. In this case, many of the sets $\mathcal{G}$ would correspond to ASes, other sets would be subnets in ASes, and yet other sets would be aggregations of ASes. This approach of defining $\mathcal{G}$ from BGP tables require that a proper nesting is created. Note that the groups differ in size, and in number of subgroups (the fanout). If $\mathcal{G}$ is a proper nesting, then the relation $G \subset G'$ defines a partial ordering over the sets in $\mathcal{G}$, generating a partial-order tree with multiple tiers. The set I is at tier-0, the highest tier. A group G belongs to tier 1 if there does not exist a G' (other than I) such that $G \subset G'$. We define the remaining tiers recursively in the same manner (see Figure 1).

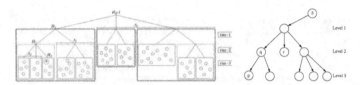

Fig. 1. A sample TOPLUS hierarchy (inner groups are represented by plain boxes) **Fig. 2.** A simple multicast tree.

Peer State. Let L denote the number of tiers in the tree, let U be the set of all current active peers and consider a peer $p \in U$. Peer p is contained in a collection of telescoping sets in $\mathcal{G}$; denote these sets by $H_N(p), H_{N-1}(p), \cdots, H_0(p) = I$, where $H_N(p) \subset H_{N-1}(p) \subset \cdots \subset H_0(p)$ and $N \leq L$ is the tier depth of p's inner group. Except for $H_0(p)$, each of these telescoping sets has one or more siblings in the partial-order tree (see Figure 1). Let $\mathcal{S}_i(p)$ be the set of siblings groups of $H_i(p)$ at tier i. Finally, let $\mathcal{S}(p)$ be the union of the sibling sets $\mathcal{S}_1(p), \cdots, \mathcal{S}_N(p)$.

Peer p should know the IP address of at least one peer in each group $G \in \mathcal{S}(p)$, as well as the IP addresses of all the other peers in p's inner group. We refer to the collection of these two sets of IP addresses as peer p's *routing table*, which constitutes peer p's state. The total number of IP addresses in the peer's routing table in tier L is $|H_L(p)| + |\mathcal{S}(p)|$. In [19] we describe how a new peer can join an existing TOPLUS network.

XOR Metric. Each key k' is required to be an element of I', where I' is the set of all s-bit binary strings ($s \geq 32$ is fixed). A key can be drawn uniformly randomly from I', or it can be biased as we shall describe later. For a given key $k' \in I'$, let k be the 32-bit suffix of k' (thus $k \in I$ and $k = k_{31}k_{30} \ldots k_1 k_0$). Throughout the discussion below, we will refer to k rather than to the original k'.

The XOR metric defines the distance between two IDs j and k as $d(j,k) = \sum_{\nu=0}^{31} |j_\nu - k_\nu| \cdot 2^\nu$. The metric $d(j,k)$ has the following properties, for IDs i,j and k:

– If $d(i,k) = d(j,k)$ for any k, then $i = j$.
– Let $p(j,k)$ be the number of bits in the common prefix of j and k. If $p(j,k) = m$,
  $d(j,k) \leq 2^{32-m} - 1$.

– If $d(i, k) \leq d(j, k)$, then $p(i, k) \geq p(j, k)$.

$d(j, k)$ is a refinement of longest-prefix matching. If j is the unique longest-prefix match with k, then j is the closest to k in terms of the metric. Further, if two peers share the longest matching prefix, the metric will break the tie. The peer p^* that minimizes $d(k, n)$, $p \in U$ is "responsible" for key k.

3   MULTI+: Multicast on TOPLUS

A Multicast Tree. First we assume that all peers are connected through links providing enough bandwidth. A simple multicast tree is shown in Figure 2. Let S be the source of the multicast group m. Peer p is receiving the flow from peer q. We say that q is the parent of p in the multicast tree. Conversely, we say that p is a child of q. Peer p is in level-3 of the multicast tree and q in level-2. It is important to note that, in principle, the *level* where a peer is in the multicast tree has nothing to do with the *tier* the peer belongs to in the TOPLUS tree.

In the kind of multicast trees we aim at building, each peer should be close to its parent in terms of network delay, while trying to join the multicast tree as high (close to the source) as possible. Each peer attempts at join time to minimize the number of hops from the source, and the length of the last hop. In the example of Figure 2, if p is a child of q and not of r, that is because p is closer to q than to r. By trying to minimize the network delay for data transmission between peers, we also avoid rearranging peers inside the multicast tree, except when a peer fails or disconnects.

Building Multicast Trees. We use the TOPLUS network and look-up algorithm in order to build the multicast trees. Consider a multicast IP address m, and the corresponding key that, abusing the notation, we also denote m. Each tier-i group G_i is defined by an IP network prefix a_i/b where a_i is an IP address and b is the length of the prefix in bits. Let m_i be the key resulting from the substitution of the first b bits of m by those of a_i. The inner group that contains the peer responsible for m_i (obtained with a TOPLUS look-up) is the *responsible inner group*, or RIG, for m in G_i (note that this RIG is contained in G_i.) Hereafter, we assume a single m, and for that m and a given peer p we denote the RIG in $H_i(p) \in$ tier-i simply as RIG-i of p. This RIG is a rendezvous point for all peers in $H_i(p)$. The deeper that a tier-i of a RIG-i is in the TOPLUS tree, the narrower the scope of the RIG as a rendezvous point (fewer peers can potentially use it).

In the simple 3-tier example of Figure 3, we have labeled the RIGs for a given multicast group (peers in grey are members of the multicast group), where all inner groups are at tier-3. The RIG-i of a peer can be found following the arrows. The arrows represent the process of asking the RIGs for a parent in the multicast tree. For example, p and q share the same RIG-1 because they are in the same tier-1 group. t's inner group is its RIG-1, but t would first contact a peer x (white) in its RIG-2 to ask for a parent. Note that this last peer is not in the multicast tree (Figure 4).

Assume a peer p in tier-$(i + 1)$ (i.e., a peer whose inner group is at tier-$(i + 1)$ of the TOPLUS tree) wants to join a multicast tree with multicast IP address m, which we call group m.

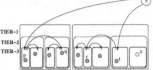

Fig. 3. The RIGs in a sample TOPLUS network.   **Fig. 4.** Sample multicast tree.

1. The peer p broadcasts a query to join group m inside its inner group. If there is a peer p' already part of group m, p connects to p' to receive the data.
2. If there is not such peer p', p must look for its RIG-i. A look-up of m_i inside p's tier-i group (thus among p's sibling groups at tier-$(i + 1)$) locates the RIG-i responsible for m. p contacts any peer p_i in RIG-i, and asks for a peer in multicast group m. If peer p_i knows about a peer p'' that is part of m, it sends the IP address of p'' to p, and p connects to p''. Note that p'' is not necessarily a member of the RIG-i inner group. In any case p_i adds p to the peers listening to m, and shares this information with all peers in RIG-i. If p'' does not exist, p proceeds similarly for RIG-$(i - 1)$: p looks up m_{i-1} inside p's tier-$i - 1$ group (i.e., among p's sibling groups at tier i). This process is repeated until a peer receiving m is found, or RIG-1 is reached. In the latter case, if there is still no peer listening to m, peer p must connect directly to the source of the multicast group. One can see that the search for a peer to connect to is done bottom up.

Property 1. *When a peer p in tier $i + 1$ joins the multicast tree, by construction, from all the groups $H_{i+1}(p), H_i(p), \cdots, H_1(p)$ that contain p, p connects to a peer $q \in H_k$ where $k = \max\{l = 1, \ldots, i + 1\} : \exists r \in H_l$ and r is a peer already connected to the multicast tree. That is, p connects to a peer in the deepest tier group which contains both p and a peer already connected to the multicast tree.*

This assures that a new peer connects to the closest available peer in the network. Notice that even in the case of failure of a peer in a RIG-i, the information is replicated in all other peers in the RIG-i. If a whole RIG-i group fails, although MULTI+ is undeniably affected, the look up process can continue in RIG-$(i-1)$. We believe this property makes MULTI+ a resilient system.

Property 2. *Using multicast over TOPLUS, the total number of flows in and out of a group defined by an IP network prefix is bounded by a constant.*

Due to lack of space, we do not further develop this important aspect of MULTI+. We refer the interested reader to the Technical Report [21]. However, in the experiments below we will notice the tight number of flows per network prefix.

Membership Management. Each peer p knows its parent q in the multicast tree, because there is a direct connection between them. Because p knows the RIG where it got its parent's address, if p's parent q in level i of the multicast tree fails or disconnects, p directly goes to the same RIG and asks for a new parent. If there is none, p becomes the new tree node at level i, replacing q. Then p must find a parent in level $i - 1$ of the multicast tree, through a join process starting at said RIG. If p had any siblings under its former parent, those siblings will find p as the new parent when they proceed like p. If more than one peer concurrently tries to become the new node at level i, peers in the RIG must consensually decide on one of them. It is not critical if a set of peers sharing a parent q are divided in two subsets with different parents upon q's depart.

Join and leave is a frequent process in a P2P network, but we expect the churn to be rather low due to the fact that in a multicast tree, all peers seek the same content concurrently, throughout the duration of the session.

Parent Selection Algorithms. From the ideas exposed before, we retain two main parent selection algorithms for testing the construction of multicast trees.

- FIFO, where a peer joins the multicast tree at the first parent found with a free connection. When a peer gets to a RIG to find a parent, the RIG answers with a list of already connected peers. This list is ordered by arrival time to the RIG. Obviously, the first to arrive connects closer (in hops) to the source. The arriving peer tests each possible peer in the list starting with the first one until it finds one that accepts a connection.
- Proximity-aware, where, *when the first parent in the list has all connections occupied,* a peer connects to the closest parent in the list still allowing one extra connection.

Note that we do not always verify if we are connecting to the closest parent in the list. The idea behind this is that, while we implicitly trust MULTI+ to find a close parent, we prefer to connect to a peer higher in the multicast tree (fewer hops from the source) than to optimize the last hop delay. If MULTI+ works correctly, the difference between these two policies should not be excessive, because the topology-awareness is already embedded in the protocol through TOPLUS.

4   MULTI+ Performance

Obviously, the $O(n^2)$ cost of actively measuring the full inter-host distance matrix for n peers limits the size of the peer sets we can use [21]. P2P systems must be designed to be potentially very large, and experiments should reflect this property by using significant peer populations. Methods like [22] map hosts into a M-dimensional coordinate space. The main advantage is that given a list of n hosts, the coordinates for all of them can be actively measured in $O(Mn)$ time (the distances of the hosts to a set of M *landmark* hosts, with $M \ll n$).

TC Coordinates. CAIDA [23] offers to researchers a set of network distance measurements from so-called *Skitter* hosts to a large number of destinations. Skitter is a traffic measurement application developed by CAIDA. In a recent paper [22], the authors have used these and other data to obtain a multi-dimensional coordinate space representing the Internet. A host location is denoted by a point in the coordinate space, and the latency between two hosts can be calculated as the distance between their corresponding points. The authors of [22] have kindly provided us with the coordinates of $196,297$ IP addresses for our study. Hereafter we call this space the TC (from Tang and Crovella) coordinate space. We calculate distances using a Euclidean metric, defined $D(x_i, x_k) = \sqrt{\sum_{k=1,\ldots,M}(x_{ik} - x_{jk})^2}$, for any two hosts identified by their M-coordinate vectors x_i and x_j.

5000 Peers Multicast Tree. In this experiment we test the characteristics of Multicast trees built with MULTI+ using the TC coordinate space and a set of 5,000 peers. We use the coordinate space to measure the distance between every pair of hosts. In order to make the experiment as realistic as possible, we use a TOPLUS tree with routing tables of reduced size, obtained from the grouping of small and medium-sized tier-1 groups into virtual groups, and this process introduces a distortion in the topological fidelity of the resulting tree [19]. The 5,000 peers are organized into a TOPLUS tree with 59 tier-1 groups, 2,562 inner-groups, and up to 4 tiers. We evaluate the two different parent selection policies described before: FIFO and proximity-aware. We also compare these two approaches with random parent selection. In all cases we test MULTI+ when we do not set a limit on the maximum number of connections a peer can accept, and for a limited number of connections, from 2 to 8 per peer. In the test we measure the following parameters, presented here using their CDF (Cumulative Distribution Function):

- The percentage of the peers in the total system, when the full multicast tree is built, closer to one peer that this peer's parent. Those figures *exclude* the peers directly connected to the source (Figure 5).
- The level peers occupy in the multicast tree. The more levels in the multicast tree, the more delay we incur in along the transmission path and the more the transmission becomes subject to losses due to peer failure (Figure 6).
- The latency from the root of the multicast tree to each receiving peer (Figure 7).
- The number of multicast flows that go into and out of each TOPLUS group (network) (Figure 8).

From our experiments we obtain very satisfactory results. From Figures 5 to 8 we draw a number of conclusions:

- Individual peers do not need to support a large number of outgoing connections to benefit from MULTI+ properties: three connections are feasible for broadband users, and the marginal improvement of 8 connections is not very significant.
- The proximity-aware policy performs better than FIFO in terms of end-to-end latency (Figure 7) and connection to closest parent (Figure 5). However, with respect to the number of flows per group (Figure 8) and level distribution in the multicast tree (Figure 6), they are very similar. That is because both trees follow the TOPLUS structure, but the proximity-aware policy takes better decisions when the optimal parent peer has no available connections.
- In Figure 5(c) we can see that having no connection restrictions makes closeness to parent less optimal than having restrictions, for the proximity-aware policy. This is normal, since when we have available connections, a peer's main goal is to connect

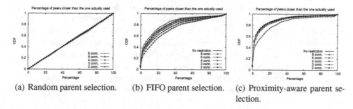

(a) Random parent selection.   (b) FIFO parent selection.   (c) Proximity-aware parent selection.

Fig. 5. Percentage of peers in the whole system closer than the one actually used (for those not connected to the source.)

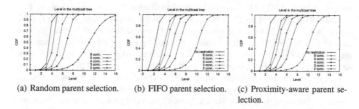

(a) Random parent selection.   (b) FIFO parent selection.   (c) Proximity-aware parent selection.

Fig. 6. Level of peers in the multicast tree.

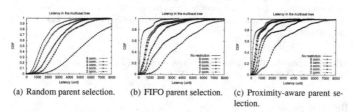

(a) Random parent selection.   (b) FIFO parent selection.   (c) Proximity-aware parent selection.

Fig. 7. Latency from root to leaf (in TC coordinate units) in the Multicast tree.

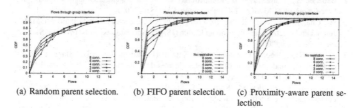

(a) Random parent selection.   (b) FIFO parent selection.   (c) Proximity-aware parent selection.

Fig. 8. Number of flows through group interface.

as high in the multicast tree as possible. See in Figure 6 how peers are organized in fewer levels, and in Figure 7 how the root-to-leaf latency is better for the unrestricted connection scheme. Still, we can assert that the multicast tree is following (when possible) a topology-aware structure, because most peers connect to nearby parents.
- The random parent selection policy organizes the tree in fewer levels than the other two policies (Figure 6(a)), because connections are not constrained to follow the TOPLUS structure. However those connections are not optimized, and the resulting end-to-end delay performance in any aspect is considerably poorer.

5   Conclusion and Future Work

We have presented MULTI+, a method to build application-level multicast trees on P2P systems. MULTI+ relies on TOPLUS in order to find a proper parent for a peer in the multicast tree. MULTI+ exhibits the advantage of being able to create topology-aware content distribution infrastructures without introducing extra traffic for active measurement. Admittedly, out-of-band information regarding the TOPLUS routing tables must be calculated offline (a simple process) and downloaded (like many P2P systems today require to download a list of peers for the join process). The proximity-aware scheme improves the end-to-end latency, and using host coordinates calculated offline and obtained at join time (as is done for TOPLUS) avoids the need for any active measurement. MULTI+ also decreases the number of redundant flows that must traverse a given network, even when only few connections per peer are possible, which allows for better bandwidth utilization. As future work, we plan to evaluate the impact of leaving and failing peers on the multicast tree performance, as well as comparing its properties with other systems.

Acknowledgments. This work could not have been done without the kind help of Prof. Mark Crovella, who provided us with the TC coordinate space data for our experiments.

References

1. Banerjee, S., Bhattacharjee, B., Kommareddy, C.: Scalable application layer multicast. In: Proceedings of ACM SIGCOMM. Pittsburgh, PA, USA (2002)
2. hua Chu, Y., Rao, S.G., Seshan, S., Zhang, H.: A case for end system multicast. IEEE Journal on Selected Areas in Communication (JSAC), Special Issue on Networking Support for Multicast **20** (2003)
3. Zhuang, S.Q., Zhao, B.Y., Joseph, A.D., Katz, R.H., Kubiatowicz, J.D.: Bayeux: An architecture for scalable and fault-tolerant wide area data dissemination. In: Proceedings of NOSSDAV'01. Port Jefferson, NY, USA (2001) 124–133
4. Ratnasamy, S., Handley, M., Karp, R.M., Shenker, S.: Application-level multicast using content-addressable networks. In: Proceedings of NGC, London, UK (2001) 14–29
5. Castro, M., Druschel, P., Kermarrec, A.M., Rowstron, A.: Scribe: a large-scale and decentralized application-level multicast infrastructure. IEEE Journal on Selected Areas in Communications **20** (2003) 1489–1499

6. Stoica, I., Morris, R., Karger, D., Kaashoek, M., Balakrishnan, H.: Chord: A scalable peer-to-peer lookup service for internet applications. In: Proceedings of ACM SIGCOMM, San Diego, CA, USA (2001)
7. Ratnasamy, S., Handley, M., Karp, R., Shenker, S.: A scalable content-addressable network. In: Proceedings of ACM SIGCOMM, San Diego, CA, USA (2001)
8. Rowstron, A., Druschel, P.: Pastry: Scalable, distributed object location and routing for large-scale peer-to-peer systems. In: IFIP/ACM International Conference on Distributed Systems Platforms (Middleware), Heidelberg, Germany (2001) 329–350
9. Harvey, N.J.A., Jones, M.B., Saroiu, S., Theimerm, M., Wolman, A.: Skipnet: A scalable overlay network with practical locality properties. In: Proceedings of the Fourth USENIX Symposium on Internet Technologies and Systems (USITS '03), Seattle, WA, USA (2003)
10. Freedman, M., Mazieres, D.: Sloppy hashing and self-organizing clusters. In: Proceedings of 2nd International Workshop on Peer-to-Peer Systems (IPTPS '03), Berkeley, CA, USA (2003)
11. Castro, M., Druschel, P., Hu, Y.C., Rowstron, A.: Topology-aware routing in structure peer-to-peer overlay network. In: International Workshop on Future Directions in Distributed Computing (FuDiCo), Bertinoro, Italy (2002)
12. Shenker, S., Ratnasamy, S., Handley, M., Karp, R.: Topologically-aware overlay construction and server selection. In: Proceedings of IEEE INFOCOM, New York City, NY (2002)
13. Castro, M., Jones, M.B., Kermarrec, A.M., Rowstron, A., Theimer, M., Wang, H., Wolman, A.: An evaluation of scalable application-level multicast built using peer-to-peer overlays. In: Proceedings of IEEE INFOCOM, San Francisco, USA (2003)
14. Castro, M., Druschel, P., Kermarrec, A.M., Nandi, A., Rowstron, A., Singh, A.: Splitstream: High-bandwidth multicast in a cooperative environment. In: Proceedings of SOSP'03, New York, USA (2003)
15. Byers, J.W., Considine, J., Mitzenmacher, M., Rost, S.: Informed content delivery across adaptive overlay networks. In: Proceedings of ACM SIGCOMM, Pittsburgh, PA, USA (2002) 47–60
16. Zhu, Y., Guo, J., Li, B.: oEvolve: Towards evolutionary overlay topologies for high bandwidth data dissemination. IEEE Journal on Selected Areas in Communications, Special Issue on Quality of Service Delivery in Variable Topology Networks (2004)
17. Rodriguez, A., Kostic, D., Vahdat, A.: Scalability in adaptive multi-metric overlays. In: Proceedings of IEEE ICDCS, Tokyo, Japan (2004) 112–121
18. Riabov, A., Liu, Z., Zhang, L.: Overlay multicast trees of minimal delay. In: Proceedings of IEEE ICDCS, Tokyo, Japan (2004) 654–664
19. Garcés-Erice, L., Ross, K.W., Biersack, E.W., Felber, P.A., Urvoy-Keller, G.: Topology-centric look-up service. In: Proceedings of COST264/ACM Fifth International Workshop on Networked Group Communications (NGC), Munich, Germany (2003) 58–69
20. Krisnamurthy, B., Wang, J.: On network-aware clustering of web sites. In: Proceedings of ACM SIGCOMM, Stockholm, Sweden (2000)
21. Garcés-Erice, L., Biersack, E.W., Felber, P.A.: Multi+: Building topology-aware overlay multicast trees. Technical Report RR-04-107, Institut Eurecom, Sophia-Antipolis, France (2004)
22. Tang, L., Crovella, M.: Virtual landmarks for the internet. In: Proceedings of the ACM SIGCOMM Internet Measurement Conference (IMC-03), Miami Beach, Florida, USA (2003)
23. CAIDA (http://www.caida.org/)

Predicting the Perceptual Service Quality
Using a Trace of VoIP Packets*

Christian Hoene, Sven Wiethölter, and Adam Wolisz

Telecommunication Networks Group, Technical University Berlin
Einsteinufer 25, 10587 Berlin, Germany
hoene@ieee.org

Abstract. We present an instrumental approach on how to assess the
perceptual quality of voice transmissions in IP-based communication networks. Our approach is end-to-end and uses combinations of common
codecs, loss concealment algorithms, playout schedulers, and ITU's quality assessment algorithms E-Model and PESQ. It is the first method that
takes the impact of playout rescheduling and *non-random packet loss
distributions* into account. Non-random packet losses occur if a rate-
distortion optimized multimedia streaming algorithm forwards packets
dependent on the packets' importance.
Our approach is implemented in open-source software. We have conducted formal listening-only tests to verify the accuracy of our quality
model. In the majority of cases, the human test results show a high
correlation with the calculated predictions.

Keywords: VoIP, quality assessment, playout scheduling, rate-
distortion optimized streaming.

1 Introduction

Instrumental perceptual assessment methods predict the behavior of humans
rating the quality of multimedia streams. The ITU has standardized a psycho
acoustic quality model called PESQ, which predicts the human rating of speech
quality and calculates a mean opinion score (MOS) value [1]. Another model,
the E-Model [2], evaluates the configurations of telephone systems. Among other
factors it takes coding mode, packet loss rate, and absolute transmission delay
into account to give an overall rating of the quality of telephone calls. Both
models consider most sources of impairment which could occur in a telephone
system. For example, they can predict the impact of the mean packet loss rate on
speech quality. However, they do not consider packet losses if the loss depends on
the packets' content or importance. Furthermore, they cannot be directly applied
to traces of VoIP packets, which are produced by experimental measurements
or network simulations.

---

* This work has been supported by Deutsche Forschungsgemeinschaft (DFG) via the
AKOM program.

J. Solé-Pareta et al. (Eds.): QofIS 2004, LNCS 3266, pp. 21–30, 2004.
© Springer-Verlag Berlin Heidelberg 2004

To overcome these deficiencies we have developed a systematic approach that combines ITU's E-model, ITU's PESQ algorithm, and various implementations of codecs and playout schedulers. Our software encodes a speech sample, analyzes a given trace of VoIP packets, simulates multiple playout schedulers, and finally assesses the quality of telephone services (coding distortion, packet loss, transmission delay and playout rescheduling). Thus, it can determine the final packet loss rate, speech quality, mean transmission delay and conversational call quality. In this regard, we have achieved the following contributions, which this paper subsumes and describes as they are presented in relation to each other.

- We developed a formula in [3] on how to include PESQ into the E-Model. The ITU approved this formula as a standard extension.
- We conducted formal listening-only tests to verify the prediction performance of PESQ for impairments due to non-random packet losses [4] and playout rescheduling, which are caused by rate-distortion optimized streaming and adaptive playout scheduling respectively. The overall correlations are R=0.94 and R=0.87 respectively.
- Finally we implemented the most common playout schedulers and provide them to the research community as open-source software [5]. Because perceptual speech quality assessment is computational complex, we provide a tool which runs the calculations in parallel.

Our approach outperforms previous algorithms because it does not only consider the impact of playout rescheduling but also takes transmission delay, speech quality and non-random packet loss distribution into account. Altogether, we are able to predict the quality of VoIP transmissions at a high precision that has not been reached before.

The paper is structured as follows: In section 2 we present the technical background and discuss related work. How to combine PESQ, E-Model and playout schedulers is explained in section 3. The next section contains the results of listening-only tests. Finally, in section 5, we draw conclusions.

2   Background

2.1   Internet Telephony

The principle components of a VoIP system, which covers the end-to-end transmission of voice, are displayed in Fig. 1. First, at the source the analogue processing, digitalization, encoding, packetization, and protocol processing (RTP, UDP, and IP) are conducted. Then, the resulting packets are transmitted through the network, consisting of an Internet backbone and access networks. At the receiver, protocols process the packets and deliver them to the playout scheduler/buffer. In the next step, the multimedia frames are decoded and played out. Because telephony consists of bidirectional transmission a similar transmission is presented in the reverse direction. In the following paragraphs we will discuss some components in more detail to show how they cause the service quality to degrade.

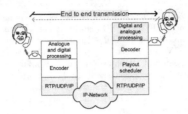

Fig. 1. VoIP transmission of a telephone call

Network: On the Internet and in the access networks packets can get lost because of congestion or (wireless) transmission errors. The packet loss process can be controlled to optimize the perceived service quality:

Chou et al. [6] suggest to forward multimedia packets according to their estimated distortion and error propagation. He proposed a rate-distortion optimized multimedia streaming framework for packetized and lossy networks.

De Martin [7] proposed an approach called Source-Driven Packet Marking, which controls the priority marking of speech packets in a DiffServ network. If packets are assumed to be perceptually critical, they are transmitted at a premium traffic class.

Sanneck used a modified Random-Early-Dropping (RED) at packet forwarding nodes [8]. If a node is congested, the probability of packet dropping should depend on the packet markings. Additionally, Sanneck proposes to mark G.729 coded voice packets according to their estimated importance.

All three algorithms handle packets in a content-sensitive manner. Therefore dropping of packets might depend on their marking and content. Thus, it is inadequately to measure only the mean packet loss rate for predicting the speech quality.

Playout scheduler: At the receiver, a playout buffer stores packets so that they can be played out in a time-regular manner, concealing variations in network delay (jitter). As the playout buffer contributes to the end-to-end delay it should not store packets longer than necessary. Instead, the playout buffer should drop packets that arrive too late to be played out at the scheduled time.

The playout scheduling can be static: If packets exceed a given transmission time they will be discarded (we will refer to this scheme as *fixed playout buffer*). Alternatively *adaptive playout buffers* re-define the playout time in accordance to the delay process of the network [9,10]. We refer to this kind of adaptation as *rescheduling*. The playout schedule can be adjusted easily during silence because then it is not notable. Adjustments during voice activity require more sophisticated concealment algorithms [11].

2.2 Quality Assessment

The perceived quality of a service can be measured with subjective tests. Humans evaluate the quality of service according to a standardized quality assessment process [12]. Often the quality is described by a *mean opinion score (MOS)* value, which scales from 1 (bad) to 5 (excellent). *Listening-only tests* are time consuming. Especially if many tests have to be made, the effort of subjective evaluation is prohibitive. Fortunately, in the last years, considerable effort has been made to develop instrumental measurement tools, which predict the human rating behavior. We will explain shortly the approaches used in this paper.

The *perceptual assessment of speech quality (PESQ)* algorithm predicts human rating behavior for narrow band speech transmission [1]. It compares an original speech fragment with its transmitted and thus degraded version to determine an estimated MOS value. For multiple known sources of impairment (typical for analogue, digital and packetized voice transmission systems) it shows a high correlation (about 0.94) with human ratings.

The quality of a telephone call cannot be judged by the speech quality alone. The ITU *E-Model* [2] additionally considers end-to-end delay, echoes, side-tones, loudness and other factors to calculate the so called *R-factor*. A higher R-factor corresponds to a better telephone quality, being 0 the worst value, 70 the minimal quality of telephone calls ("toll quality"), and 100 the best value.

2.3 Related Work

Markopoulou et al. [13] measured the performance of a couple of Internet backbone links and analyzed them with ITU's E-Model. Their findings include not only that the quality of VoIP depends largely on the provider's link quality but also on the playout buffer scheme.

Hammer et al. [14] suggest to use PESQ to assess the speech quality of a VoIP packet trace. He proposes to split the trace into overlapping subparts. The benefit of this approach is that different coding schemes and also packet marking algorithms can be judged. Also, FEC or different playout schedulers can be supported in principle.

An approach that also considers interactivity is presented by Sun and Ifeachor [15]. The authors suggest to combine the E-Model and PESQ and describe a set of equations, which they derived by linear approximations to the rating behavior of PESQ, E-Model and the correlation between packet loss rate and speech quality.

3   Combining E-Model, PESQ, and Playout Schedulers

Considering the characteristics of VoIP packet transmissions on the one side and the capability of perceptual models on the other side, we identify the following aspects as incomplete:

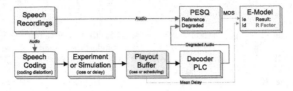

Fig. 2. Speech and delay assessment

- Perceptual quality assessment has to take into account the entire processing chain from source to sink, including encoding, routing across the Internet, de-jittering, decoding and playing at the receiving side, because only this reflects human-to-human conversation. Thus, when studying a transmission of VoIP packets the entire transmission system has to be considered.
- The end-to-end quality depends largely on the playout buffer scheme [13]. However, until now an "ideal" playout scheduler has not been identified and any implementer of a VoIP phone is free to choose any scheme. Thus, to predict the impact of playout scheduling one has to consider all, and if not possible, the most common playout schedulers.
- Rescheduling of adaptive playout schedulers harms the speech quality because of temporal discontinuities. The E-Model does not take the dynamics of a transmission into account because it relies on static transmission parameters. PESQ instead considers playout adaptation but does not include the absolute delay into its rating algorithm. PESQ has been designed to judge the impact of playout scheduling but has not been validated yet for this purpose [16].
- The E-Model does not consider non-random packet losses and PESQ has not been verified for this kind of distortion and the prediction accuracy is unknown.

To overcome these shortcomings we combine the E-Model, PESQ and playout schedulers as shown in Fig. 2: First, a set of the most common playout scheduler schemes (including fixed-deadline and adaptive algorithms [9,10] of Van Jacobsen, Mills, Schulzrinne, Ramjee, and Moon) calculates the packets' playout times and the mean transmission delay. One should note that only speech frames during voice activity are considered because during silence a human cannot identify the transmission delay.[1]

Next, PESQ calculates the speech quality that depends on coding distortion, non-random packet loss and playout rescheduling. Because PESQ has not been verified for non-random packet loss and playout rescheduling, we conduct formal listening tests to verify its accuracy (see section 4). Last, both the speech quality and the mean transmission delay are fed into the E-Model. We assume the

---

[1] Indeed, some playout schedulers change the playout time at the start of a talk spurt. Others change it at the beginning of silence periods. Both have to be considered as equal with respect to the transmission delay.

acoustic processing as optimal [3]. Therefore we can simplify the E-Model to a model with only few parameters. The computation of R_{factor} is then given by:

$$R_{factor} = \mathrm{MOS_2R} \left(\mathrm{MOS_{PESQ}}\right) - I_{dd}\left(t\right) \tag{1}$$

Reference [3] describes the function $\mathrm{MOS_2R}$ and the conditions under which (1) can be applied. For a definition of the function I_{dd} we like to refer to [2].

Software Package: We have implemented the approach discussed above and provide it as open-source to the research community. The software covers the digital processing chain of VoIP. To be fully operational, the PESQ algorithm and a G.729 codec have to be bought from its rights owners. Alternatively they can be downloaded at no costs from ITU's web page for trials only. Further information can be found at our web page and in the manual [5].

We try to verify the correctness of our software by several means. Publishing of this software together with its source code ensures that more users are going to use it and to study its code. Thus, the pace of finding potential errors will be increased. Last not least, we have tested our tool-set on various projects which include the assessment of voice over WLAN, the impact of handover and wireless link scheduling. Overall, we are confident of the correctness of our implementation.

4   Listening-Only Tests

PESQ has not been verified for all causes of impairment. In [4] we have conducted formal listening-only test to determine PESQ's prediction performance in cases of single or non-random packet losses. In this section we verify whether PESQ can measure the impairment of playout rescheduling. This verification is important because PESQ was not designed for this kind of impairments and operates outside the scope of its operational specification [16].

To verify PESQ, we construct artificially degraded samples and conduct both listening-only tests and instrumental predictions. If both PESQ and human tests yield similar results for the samples, PESQ is verified. Usually, the results of speech quality tests are compared via correlation. Thus, the amount of correlation (R) between subjective and objective speech quality results is our measure of similarity. R=1 means that the results are perfectly related. If no correlation is present, R is equal to zero. To compare absolute subjective and instrumental MOS values, we apply linear regression to one set of values which is a usual practice. The correlation R does not change after linear regression.

Sample Design: Analysis of Internet traces has shown that sometimes packet delays show a sharp, spike-like increase [9,13] which cannot be predicted in advance. Delay spikes are a short increase of the packet transmission times which usually occur after congestion or on a wireless link after fading. Soon after the spike the following packets arrive shortly one after the other until the transmission delay has returned to its normal value (Fig. 3). We like to consider the question

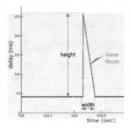

Fig. 3. Delay spike

whether to adjust the playout of speech frames to delay spikes by concentrating on the non-trivial case of delay spikes during voice activity.

For constructing the samples we have used the software package described in this paper. It generates artificial packet traces that contain delay spikes. One can control the frequency, the height and weight of delay spike. Further, three different playout strategies are analyzed: First, we drop every packet that is affected by the spike and thus arrives too late. Second, in case of a delay spike, the playout is re-scheduled so that no packet will be dropped. As a consequence, the playout delay will be increased. The last strategy is similar to the second, but after any spikes, the playout delay is adjusted during silence periods until the playout delay returns to normal.

We construct 220 samples (length approx. 5-10s), containing samples encoded with G.711, G.729 and containing one delay spike with a height of 50 to 300ms and a width of 55 to 330ms.

Formal Listening Tests: The listening-only tests followed closely the ITU recommendations [12], Appendix B, that describes methods for subjective assessment of quality. The tests took place in a professional sound studio ($46\,m^2$, low environmental noise, etc.). Nine persons judged the quality of 164 samples. The samples' language is German, which all listeners understand.

We do not follow the ITU's recommendations when scientific results suggest changes that improve the rating performance. For example, we have used high quality studio headphones instead of an Intermediate Reference System, because headphones have a better sound quality. Further, multiple persons were in the room at the same time to reduce the duration of the experiment.

Last but not least we do not apply the "Absolute Category Rating" because a discrete MOS makes it difficult to compare two only slightly different samples. The impact of a single frame loss is indeed very small. We allow intermediate values and use a linear MOS scale. PESQ calculates a MOS value with a resolution of up to 10^{-6} at the MOS scale.

Results: Ten persons gave a total of 2210 judgments. We could use only 2033 judgments because some test persons failed to get track with the sample number.

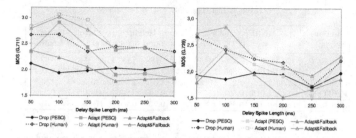

Fig. 4. Playout strategy: delay spike height vs. speech quality.

The rating performance during the second half of the test was significantly worse than during the first half. We also compared a group of native speakers and a group of foreign students. Both have shown a similar rating performance. This leads us to the conclusion that being concentrated is more important than being a native speaker.

Figure 4 displays the speech quality versus the spike height. We show the rating results of humans and PESQ for different adaptation policies. The black lines (drop) refer to the dropping of any late packet during the delay spike. The blue lines (adapt) display the results when delaying the playout after a delay spike. Last, the red lines (adapt&fallback) include the effect of falling back to the original playout time as soon as possible. The later rescheduling occurs only during periods of silenced speech.

Analysis: In the experiments the sample content is varied for each different delay spike height. Because the sample content has a large influence on the speech quality ratings, one cannot compare absolute MOS values on the horizontal axis (that displays the delay spike height) in Figure 4. However, the playout strategies can be ranked against each other if the delay spike height remains the same.

If the delay-spike's height is 200 ms or larger, dropping packets is more beneficial than delaying the playout. Further, the "fallback" adjustments during silence degrade the speech quality. The "adapt" algorithm performs always better.

Table 1 displays the prediction performance of PESQ. The overall correlation is $R = 0.866$. If one considers only samples that contain modulated noise (MNRU), the correlation is nearly perfect ($R = 0.978$). Also, we identify a coherency between the MOS variance and the correlation. For a given sample set, the more the samples differ the higher is the correlation. Considering the "drop" strategy for example, both the PESQ variance and the correlation are low. We assume that humans cannot distinguish degraded samples which are only slightly different.

Comparing the absolute MOS values in Table 1, one can see that there exists a constant offset between instrumental and subjective MOS values. As we can

Table 1. Listening-only test: delay spikes

Selection criteria:	all	MNRU yes	Coding G.711	G.729	Spike Height 100ms	200ms	300ms	Playout strategy drop	adapt	&fallback
Samples	113	13	33	63	15	18	18	32	32	32
MOS	2.518	3.013	2.565	2.284	2.844	2.228	2.280	2.220	2.453	2.469
PESQ MOS	2.280	2.823	2.277	2.028	2.680	1.873	1.998	2.039	2.243	2.058
PESQ var.	0.723	1.015	0.564	0.373	0.882	0.141	0.246	0.088	0.717	0.541
Correlation	0.866	0.978	0.856	0.668	0.906	0.737	0.768	0.476	0.838	0.799

not understand the reasons for this offset, we assume that it can be explained due to the social behavior and emotions of our listening personal. Their ratings are severer or more indulgent as compared to the ratings used during the development of PESQ.

5   Conclusions

In this paper we have presented an approach on how to assess the quality of VoIP transmissions. We identify important sources of quality degradation that can occur in a VoIP system; especially the impact of playout rescheduling and non-random packet losses has not been considered in previous approaches.

We combine PESQ, the E-model, different coding schemes and playout schedulers to analyze VoIP packet traces. The ITU approved the mathematical combination of PESQ and E-Model as a standard extension.

PESQ is verified with formal listening-only tests to identify its prediction accuracy[2]. The listening-only tests lead to manifold results. They show that PESQ indeed predicts in general the speech quality well. However, we identified that in the same cases, PESQ has to be improved. These improvements are beyond the scope of this paper. To enable other researchers the verification as well as the tuning of their algorithms, the complete experimental data including all samples and ratings are available on request.

Beyond the scope of this paper are also various performance evaluations that our approach enables. For example, the assessment of playout schedulers can be used to identify the ideal one. Also, Internet backbone traces can be assessed and novel VoIP over WLAN systems can be developed. Especially, if the importance of speech frames is utilized [4] and non-random packet losses are enforced, we will show that the performance of VoIP over wireless can be enhanced significantly.

References

1. ITU-T Recommendation P.862: Perceptual Evaluation of speech quality (PESQ), an Objective Method for End-to-end Speech Quality Assessment of Narrowband Telephone Networks and Speech Codecs (2001)

---

[2] We like to thank our students L. Abdelkarim and T. Dulamsuren-Lalla

2. ITU-T Recommendation G.107: The E-model, a Computational Model for Use in Transmission Planning (2000)
3. Hoene, C., Karl, H., Wolisz, A.: A perceptual quality model for adaptive VoIP applications. In: SPECTS, San Jose, California, USA (2004)
4. Hoene, C., Dulamsuren-Lalla, E.: Predicting performance of PESQ in case of single frame losses. In: Measurement of Speech and Audio Quality in Networks Workshop (MESAQIN), Prague, CZ (2004)
5. Hoene, C.: Simulating playout schedulers for VoIP - software package. URL:http://www.tkn.tu-berlin.de/research/qofis/ (2004)
6. Chou, P., Mohr, A., Wang, A., Mehrotra, S.: Error control for receiver-driven layered multicast of audio and video. IEEE Transactions on Multimedia 3 (2001) 108–122
7. Martin, J.D.: Source-driven packet marking for speech transmission over differentiated-services networks. In: IEEE ICASSP. Volume 2. (2001) 753–756
8. Sanneck, H., Tuong, N., Le, L., Wolisz, A., Carle, G.: Intra-flow loss recovery and control for VoIP. In: ACM Multimedia. (2001) 441–454
9. Ramjee, R., Kurose, J.F., Towsley, D.F., Schulzrinne, H.: Adaptive playout mechanisms for packetized audio applications in wide-area networks. In: IEEE Infocom, Toronto, Canada (1994) 680–688
10. Moon, S.B., Kurose, J., Towsley, D.: Packet audio playout delay adjustments: performance bounds and algorithms. ACM/Springer Multimedia Systems 27 (1998) 17–28
11. Liang, Y.J., Färber, N., Girod, B.: Adaptive playout scheduling and loss concealment for voice communication over IP networks. IEEE Transactions on Multimedia 5 (2003) 532–543
12. ITU-T Recommendation P.800: Methods for subjective determination of transmission quality (1996)
13. Markopoulou, A., Tobagi, F., Karam, M.: Assessing the quality of voice communications over internet backbones. IEEE/ACM Transactions on Networking 11 (2003) 747–760
14. Hammer, F., Reichl, P., Ziegler, T.: Where packet traces meet speech samples: An instrumental approach to perceptual QoS evaluation of VoIP. In: IEEE IWQoS, Montreal, Canada (2004)
15. Sun, L., Ifeachor, E.: New models for perceived voice quality prediction and their applications in playout buffer optimization for VoIP networks. In: IEEE ICC, Paris, France (2004)
16. Rix, A.W., Hollier, M.P., Hekstra, A.P., Beerends, J.G.: Perceptual evaluation of speech quality (PESQ), the new ITU standard for end-to-end speech quality assessment, part I - time alignment. Journal of the Audio Engineering Society 50 (2002) 755

Evaluating the Utility of Media–Dependent FEC in VoIP Flows

Gerardo Rubino and Martín Varela

Irisa - INRIA/Rennes
Campus universitaire de Beaulieu
35042 Rennes CEDEX, France
{rubino|mvarela}@irisa.fr

Abstract. In this paper, we present an analysis of the impact of using media–dependent Forward Error Correction (FEC) in VoIP flows over the Internet. This error correction mechanism consists of piggy-backing a compressed copy of the contents of packet n in packet $n + i$ (i being variable), so as to mitigate the effect of network losses on the quality of the conversation. To evaluate the impact of this technique on the perceived quality, we propose a simple network model, and study different scenarios to see how the increase in load produced by FEC affects the network state. We then use a *pseudo–subjective* quality evaluation tool that we have recently developed in order to assess the effects of FEC and the affected network conditions on the quality as perceived by the end–user.

1 Introduction

In recent years, the growth of the Internet has spawned a whole new generation of networked applications, such as VoIP, videoconferencing, video on demand, music streaming, etc. which have very specific, and stringent, requirements in terms of network QoS. In this paper we will focus on VoIP technology, which has some particularities with respect to other real–time applications, and it is one of the most widely deployed to date. The current Internet infrastructure was not designed with these kinds of applications in mind, so multimedia applications' quality is very dependent on the capacity, load and topology of the networks involved, as QoS provisioning mechanisms are not widely deployed. Therefore, it becomes necessary to develop mechanisms which allow to overcome the technical deficiencies presented by current networks when dealing with real–time applications.

Voice–over–IP applications tend to be sensitive to packet losses and end–to–end delay and jitter. In this paper we will concentrate on the effect of FEC on packet loss, and the effect of both on the perceived quality. While it has been shown [1] that delay and jitter have a significant impact on the perceived quality, we will focus on one–way flows, whose quality is largely dominated (at the network level) by the packet loss process found in the network. The effects of FEC on interactive (two–way) VoIP applications is the subject of future studies.

J. Solé-Pareta et al. (Eds.): QofIS 2004, LNCS 3266, pp. 31–43, 2004.
© Springer-Verlag Berlin Heidelberg 2004

In order to assess the variations in perceived quality due to the use of FEC, we will use a technique we have recently developed [2,3]. The idea is to train an appropriate tool (a Random Neural Network, or RNN) to behave like a "typical" human evaluating the streams. This is not done by using biological models of perception organs but by identifying an appropriate set of input variables related to the source and to the network, which affect the quality, and to teach the RNN the relationships between these variables and the perceived quality. One of the main characteristics of this approach is that the result is extremely accurate (as it matches very well the result obtained by asking a team of humans to evaluate the streams). In [4], we applied this method to analyze the behavior of audio communications on IP networks, with very good results after comparison with real human evaluations.

The rest of the paper is organized as follows. Section 2 presents the tool we used to assess the perceived quality of the flows. Section 3 presents the network model we used for our analysis, and the effects of adding FEC to the audio traffic. In Section 4, we present our analysis of the effects of FEC on the quality of the flows. Finally, Section 5 presents our conclusions.

2   Assessing the Perceived Quality

Correctly assessing the perceived quality of a speech stream is not an easy task. As quality is, in this context, a very subjective concept, the best way to evaluate it is to have real people do the assessment. There exist standard methods for conducting *subjective* quality evaluations, such as the ITU-P.800 [5] recommendation for telephony. The main problem with subjective evaluations is that they are very expensive (in terms of both time and manpower) to carry out, which makes them hard to repeat often. And, of course, they cannot be a part of an automatic process.

Given that subjective assessment is expensive and impractical, a significant research effort has been done in order to obtain similar evaluations by *objective* methods, i.e., algorithms and formulas that measure, in a certain way, the quality of a stream. The most commonly used objective measures for speech/audio are Signal-to-Noise Ratio (SNR), Segmental SNR (SNRseg), Perceptual Speech Quality Measure (PSQM) [6], Measuring Normalizing Blocks (MNB) [7], ITU E–model [8], Enhanced Modified Bark Spectral Distortion (EMBSD) [9], Perceptual Analysis Measurement System (PAMS) [10] and PSQM+ [11]. These methods have three main drawbacks: (i) they generally don't correlate well with human perception [12,8]; (ii) virtually all of them (one exception is the E-model) are *comparing techniques* between the original and the received stream (so they need the former to perform the evaluation, which precludes their use in a live, real-time networking context), and (iii) they generally don't take into account network parameters. Points (ii) and (iii) are due to the fact that they have been mainly designed for analyzing the effect of coding on the streams' quality.

The method used here [2,3] is a hybrid between subjective and objective evaluation. The idea is to have several distorted samples evaluated subjectively, and then use the results of this evaluation to teach a RNN the relation between the parameters that cause the distortion and the perceived quality. In order for it to work, we need to consider a set of P parameters (selected *a priori*) which may have an effect on the perceived quality. For example, we can select the codec used, the packet loss rate of the network, the end–to–end delay and/or jitter, etc. Let this set be $\mathcal{P} = \{\pi_1, \dots, \pi_P\}$. Once these *quality–affecting* parameters are defined, it is necessary to choose a set of representative values for each π_i, with minimal value $\pi_{\min}$ and maximal value $\pi_{\max}$, according to the conditions under which we expect the system to work. Let $\{p_{i1}, \cdots, p_{iH_i}\}$ be this set of values, with $\pi_{\min} = p_{i1}$ and $\pi_{\max} = p_{iH_i}$. The number of values to choose for each parameter depends on the size of the chosen interval, and on the desired precision. For example, if we consider the packet loss rate as one of the parameters, and if we expect its values to range mainly from 0% to 5%, we could use 0, 1, 2, 5 and perhaps also 10% as the selected values. In this context, we call *configuration* a set with the form $\gamma = \{v_1, \dots, v_P\}$, where v_i is one of the chosen values for p_i.

The total number of possible configurations (that is, the number $\prod_{i=1}^{P} H_i$) is usually very large. For this reason, the next step is to select a subset of the possible configurations to be subjectively evaluated. This selection may be done randomly, but it is important to cover the points near the boundaries of the configuration space. It is also advisable not to use a uniform distribution, but to sample more points in the regions near the configurations which are most likely to happen during normal use. Once the configurations have been chosen, we need to generate a set of "distorted samples", that is, samples resulting from the transmission of the original media over the network under the different configurations. For this, we use a testbed, or network simulator.

Formally, we must select a set of M media samples (σ_m), $m = 1, \cdots, M$, for instance, M short pieces of audio (subjective testing standards advise to use sequences having an average 10 sec length -following [5], for instance). We also need a set of S configurations denoted by $\{\gamma_1, \cdots, \gamma_S\}$ where $\gamma_s = (v_{s1}, \cdots, v_{sP})$, v_{sp} being the value of parameter π_p in configuration γ_s. From each sample σ_i, we build a set $\{\sigma_{i1}, \cdots, \sigma_{iS}\}$ of samples that have encountered varied conditions when transmitted over the network: sequence σ_{is} is the sequence that arrived at the receiver when the sender sent σ_i through the source-network system where the P chosen parameters had the values of configuration γ_s.

Once the distorted samples are generated, a subjective test [5] is carried out on each received piece σ_{is}. After a statistical screening of the answers (to eliminate "bad" observers), the sequence σ_{is} receives the value μ_{is} (often, this is a *Mean Opinion Score*, or MOS), the average of the values given to it by the set of observers. The idea is then to associate each configuration γ_s with the value $\mu_s = (1/M) \sum_{m=1}^{M} \mu_{ms}$.

At this step we have a set of S configurations $\gamma_1, \dots, \gamma_S$, and we associate μ_s with configuration γ_s. We randomly choose S_1 configurations among the S available. These, together with their values, constitute the "Training Database". The remaining $S_2 = S - S_1$ configurations and their respective values constitute the "Validation Database", reserved for further (and critical) use in the last step of the process.

The next step is to train a specific statistical learning tool (a RNN) to learn the mapping between configurations and values as defined by the Training Database. Assume that the selected parameters have values scaled into [0,1] and the same with quality. Once the tool has "captured" the mapping, that is, once the tool is trained, we have a function $f()$ from $[0,1]^P$ into $[0,1]$ mapping now any possible value of the (scaled) parameters into the (also scaled) quality metric. The last step is the validation phase: we compare the value given by $f()$ at the point corresponding to each configuration γ_s in the Validation Database to μ_s; if they are close enough for all of them, the RNN is validated (in Neural Network Theory, we say that the tool *generalizes well*). In fact, the results produced by the RNN are generally closer to the MOS than that of the human subjects (that is, the error is less than the average deviation between human evaluations). As the RNN generalizes well, it suffices to train it with a small (but well chosen) part of the configuration space, and it will be able to produce good assessments for any configuration in that space. The choice of the RNN as an approximator is not arbitrary. We have experimented with other tools, namely Artificial Neural Networks, and Bayesian classifiers, and found that RNN are more performant in the context considered. ANN exhibited some performance problems due to over-training, which we did not find when using RNN. As for the Bayesian classifier, we found that while it worked, it did so quite roughly, with much less precision than RNN. Besides, it is only able to provide discrete quality scores, while the NN approach allows for a finer view of the quality function.

For this study, we will use a RNN trained with results from a subjective tests campaign carried out with 17 subjects. The subjects were presented 115 sets of speech samples that had been generated using the Robust Audio Tool (RAT [13]), corresponding to different network and coding configurations. A MOS test was performed and the results screened as per [5]. About 90 of the results obtained were used to train the RNN, and the remaining ones were used to validate it. The parameters considered for our experiment are listed on table 1, and are described below.

Codec – the primary codec (16 bit linear PCM, and GSM),
FEC – the secondary codec (GSM), if any,

Table 1. Network and encoding parameters and values used

Parameter	Values
Loss rate	0% ... 15%
Mean loss burst size	1 ... 2.5
Codec	PCM Linear 16 bits, GSM
FEC	ON(GSM)/OFF
FEC offset	1 ... 3
Packetization interval	20, 40, and 80ms

FEC offset – the offset, in packets, of the redundant data (we used offsets of 1 and 3 packets for the Forward Error Correction (FEC) scheme presented in [14,15].

Packetization Interval (PI) – the length (in milliseconds) of audio contained in each packet (we considered packets containing 20, 40 and 80ms).

Packet loss rate – the percentage of lost packets (2, 7, 10 and 15%).

Mean loss burst size – the average size of packet loss bursts (1, 1.7 and 2.5 packets); we consider this parameter to have a finer view on the packet loss process than the one reduced to the packet loss rate only.

So, in order to obtain an estimation of the perceived quality, all that is needed is to feed the trained RNN with values for those parameters, and it will output a MOS estimation, very close to the actual MOS.

3   The Network Model

The tool described in the last section gives us a way to explore the perceived quality as a function of the 6 selected parameters. For instance, this allows to plot MOS against, say, packet loss rate in different cases (parameterized by the 5 remaining parameters), etc. For performance evaluation purposes, we want to know which is the impact on quality of typical *traffic* parameters (such as throughputs) and the parameters related to the *dimensions* (such as windows, buffer sizes, etc.). This gap is here bridged by adding a network model.

In this paper we will consider a very simple network model, much like the one presented in [16,17]. It consists of an $M/M/1/H$ queue which represents the bottleneck router in the network path considered. In spite of its simplicity, this model will allow us to capture the way FEC affects *perceived* quality. Moreover, it appears to be quite robust (see the comments in 3.3). We will concern ourselves with two classes of packets, namely audio packets and background traffic. Audio packets can have FEC or not, but we will consider that if FEC is on, then all flows are using it. Our router will have a drop–tail buffer policy, which is common in the current Internet.

3.1   Transmission Without FEC

First, consider the case of audio without FEC. We will take the standard environment in the $M/M/1/H$ case: Poisson arrivals, exponential services, and the usual independence assumptions. The arrival rates of class i units is λ_i pps (packets per second) and the link has a transmission capacity of c bps. The average packet length for class i packets is B_i bits. In order to be able to use analytical expressions, we consider in the model the global average length of the packets sharing the link, B, given by $B = \alpha_1 B_1 + \alpha_2 B_2$, where $\alpha_i = \lambda_i/\lambda$, with $\lambda = \lambda_1 + \lambda_2$. The service rate of the link *in pps* is then $\mu = c/B$.

Let us assume that the buffer associated with the link has capacity equal to N bits. Then, in packets, its capacity will be taken equal to $H = N/B$. To simplify the analysis, we will use the expressions for the performance metrics of the $M/M/1/H$ models even if H is not an integer; this does not significantly change the results and the exposition is considerably more clear.

If we denote $\varrho = \lambda/\mu$, then the packet loss probability p is

$$p = \frac{1 - \varrho}{1 - \varrho^{H+1}} \varrho^H$$

(assuming $\varrho \neq 1$). We also need to compute the mean size of loss bursts for audio packets, since the correlation between losses influences the speech quality. Here, we must discuss different possible definitions for the concept of loss burst, because of the multi-class context. To this end, let us adopt the following code for the examples we will use in the discussion: a chunk of the arrival stream at the link will be denoted by "$\dots, x, y, z, \dots$" where symbols $x, y, z, \dots$ are equal to i if a class-i packet arrives and is accepted (the queue was not full) and to $\bar{i}$ if the arrival is a class-i one and it is rejected (because there was no room for it).

Assume a packet of class 1 arrives at a full queue and is lost, and assume that the previous class 1 packet was not lost. Then, a burst of class 1 losses starts. Strictly speaking, the burst is composed of a set of consecutive audio packets all lost, whatever happens to the class-2 packets arriving between them. For instance, in the path "$\dots, 1, 2, \bar{1}, \bar{1}, 2, 2, \bar{1}, \bar{2}, 2, \bar{2}, 1, \dots$" there is a class 1 burst with loss burst size LBS $= 3$. If we use this definition of burst, when audio packets are a small fraction of the total offered traffic, this definition can exaggerate the effective impact of correlation. Even in the case of many audio packets in the global arrival process, allowing the class-2 packets to merge inside audio loss bursts can be excessive. On the other extreme, we can define an audio loss burst as a consecutive set of class-1 arrivals finding the buffer full, without class-2 packets between them. In the path shown before, if we consider this burst definition, there is a first burst of class-2 losses with size 2, then a second one composed of only one packet. Consider now the piece of path "$\dots, 1, 2, 2, \bar{1}, \bar{1}, \bar{2}, \bar{2}, \bar{1}, 2, 2, 2, \bar{1}, \bar{1}, 1, \dots$". An intermediate definition consists of considering that we accept class-2 packets inside the same audio burst only if they are also lost (because this corresponds basically to the same congestion period). In the last example, this means that we have a loss burst with size 4. We will keep this last definition for our analysis.

Let us denote by LBS the loss burst size (recall that we focus only on class 1 units). The probability that a burst has size strictly greater than n is the probability that, after a class 1 loss, the following n class-1 arrivals are losses, accepting between them class-2 losses in any number. This means

$$\Pr(\text{LBS} > n) = p^n,$$

where

$$p = \sum_{k \geq 0} \left(\frac{\lambda_2}{\lambda_1 + \lambda_2 + \mu} \right)^k \frac{\lambda_1}{\lambda_1 + \lambda_2 + \mu} = \frac{\lambda_1}{\lambda_1 + \mu}.$$

The last relationship comes from the fact that we allow any number of class 2 units to arrive as far as their are lost, between two class 1 losses (that is, while the burst is not finished, no departure from the queue is allowed). Using the value of p, we have

$$E(LBS) = \sum_{n \geq 0} \Pr(LBS > n) = \sum_{n \geq 0} \left(\frac{\lambda_1}{\lambda_1 + \mu} \right)^n = 1 + \frac{\lambda_1}{\mu}.$$

3.2   Transmission With FEC

If FEC is used, each audio packet has supplementary data, and we denote this overhead by r. If B' is the mean audio packet length when FEC is used, then $B_1' = B_1(1 + r)$. The rest of the parameters are computed as before. We have $B' = \alpha_1 B_1' + \alpha_2 B_2$, $\mu' = c/B'$, $H' = N/B'$ and $\varrho' = \lambda/\mu'$. This leads to the corresponding expression for the loss probability $p' = \varrho'^{H^\bullet}(1 - \varrho')/(1 - \varrho'^{H^\bullet+1})$ and $E(LBS') = 1 + \lambda_1/\mu'$.

3.3   About the Model Robustness

The simplicity of the classical $M/M/1/H$ single class model hides its capacity a capture the dynamics of the system. We also explored the more directly multi-class $M/M/1/H$ FIFO queue where the service rate is μ_i for class-i packets. This model can be easily numerically analyzed by writing and solving the associated equilibrium equations.

Let us denote

$q_i = \Pr(\text{in steady state, the queue is saturated and a class-}i \text{ packet is being}$
$\quad \text{transmitted}).$

Then, using the previously definition of class-k loss bursts, their average length $E(LBS_k)$ is derived exactly as before. First, conditioned on the class of the packet being transmitted,

$$E(LBS_k \mid \text{a class-}i \text{ packet is being transmitted}) = 1 + \lambda_k/\mu_i.$$

Then, given the low loss rate considered, we average the average loss burst length for class-k as

$$E(LBS_k) = \frac{\sum_i q_i(1 + \lambda_k/\mu_i)}{\sum_j q_j}.$$

We explored the perceived quality as described before within this multi-class model and the numerical results we obtained were very similar to those presented in next section.

4   The Impact of FEC on VoIP Flows

The main idea of using FEC is that as the real–time requirements of VoIP make it impossible to retransmit lost packets, it is possible to mitigate the effect of the losses by transmitting the information contained in each packet more than once. To this end, we send one (maybe more) copy of packet n's contents in packet $n + i$ ($n + i + 1$, and so on if several copies are sent), i being variable. The extra copies are normally compressed at a lower bit rate, so as to minimize the increase in bandwidth usage. When a packet is lost, if the packet containing the redundant copy of the lost one arrives in a timely fashion, the receiver can use said copy to recover from the loss, in a way that minimizes the perceived quality degradation. The variability of i is due to the bursty nature of packet losses on IP networks [18,19], and it allows to improve the performance of the FEC scheme described above by avoiding the loss of all copies of a given voice packet in the same loss burst. However, i should remain as close to 1 as possible, in order to minimize the increase of the end–to–end delay, which is known to degrade the quality of a two–way VoIP session.

In our study, we focus on the perceived audio quality, and so it is sufficient to consider only one–way flows. Therefore, we didn't consider the effects of FEC on the end–to–end delay.

In order to assess the impact on FEC at the network loss level, we must consider two factors:

- the amount of redundancy introduced by the use of FEC (and therefore the increase in packet size for the flows using FEC),
- the proportion of flows using FEC, which allows us to derive the new network load.

While the increase in size is likely to be small, more or less equal for different applications, and certainly bounded by a factor of 1, the amount of flows using FEC is very hard to determine. Estimations of the number of VoIP flows on the Internet are not readily available in the literature, but there are some estimations [20] that say that UDP traffic only accounts for about 15 to 20% of the total network traffic. Even then, VoIP is probably still a modest fraction of that UDP traffic (some studies [21] suggest that about 6% of traffic corresponds to streaming applications). However, being that VoIP applications have a growing user base, it seems reasonable that in some time they may account for a higher fraction of the total Internet traffic.

We studied different scenarios, with increasing volumes of VoIP traffic in order to assess the impact of FEC in the quality perceived by the end–user. For simplicity's sake, we assumed that if FEC is in use, then all VoIP flows are using it.

For our example, we'll consider an T3–type line which runs at 45Mbps (actual speeds are not really relevant, since we will concern ourselves with the load of the network – numbers are used to better set the example). In practice, buffer space in core routers is not limited by physical memory, but rather by the administrators, who may wish to minimize delays (while this implies that the loss rate is higher than it could be, it is important to make TCP flows behave properly). This time is normally limited to a few milliseconds [22]. We will use a value of 200µs (which requires a buffer space of $N = 45\text{Mbps} * 0.2\text{s} = 9\text{Mbits}$, or about 1965 packets of 600B), which is on par with current practices.

Table 2. Packet size distribution

Size (Bytes)	Probability
40	50
512	0.25
1500	0.24
9180	0.01

Some studies [23] show that an important percentage (about 60%) of packets have a size of about 44B, and that about 50% of the volume of bytes transferred is transferred on 1500B or higher packet sizes. Table 2 shows the distribution we chose for our background traffic packet sizes, which yields an average packet size B_2 of about 600B.

We considered that network load varies between 0.5 and 1.15. Smaller values have no practical interest, since in the model we used, they result in negligible loss rates, and higher load values result in loss rates above 15%, which typically yield unacceptable quality levels.

As for the fraction of VoIP traffic, we studied different values, between 5% and 50% of the total packet count. Granted, the higher values are currently unrealistic, but they may make sense in a near future if more telephony providers move their services toward IP platforms, and other VoIP applications gain more acceptance.

We chose to use PCM audio with GSM encoding for FEC, and an offset of one packet for the redundant data. This is a bit pessimistic, since sample–based codecs are not very efficient for network transmission, but it allows us to get a worst–case scenario of sorts (since increasing the fraction of audio traffic does indeed increase network load). In this case, the increase in packet size is of about 10% when using FEC, which gives payloads of 353B, against 320B of PCM–only traffic for 20ms packets. We also tried GSM/GSM, which results in a much smaller packet size (66B and 33B with and without FEC respectively), but found that the results are qualitatively similar to those of PCM, and so we will discuss only the PCM ones.

4.1   Assessing the Impact of FEC on the Perceived VoIP Quality

We present our results as a series of curves, plotting the estimated MOS values against network load, and for different values of the proportion of voice packets.

As can be seen in the curves in Figure 1, using FEC is beneficial for the perceived quality in all the conditions considered. It can be seen that when the proportion of voice traffic becomes important (> 30%) the performance of the FEC protection decreases. We believe that this is related to the fact that a higher proportion of voice packets implies a higher mean loss burst size (which is still smaller than 2 in our model for the conditions considered), and being that we chose an offset of 1 packet for the FEC, it is logical to see a slight

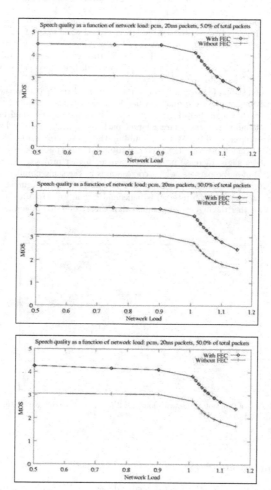

Fig. 1. MOS as a function of network load for different fractions of voice traffic, with and without FEC (20ms packets)

decrease in performance.We did not find, however, a negative impact of FEC on the perceived quality, as predicted in [16,17]. Even when most of the packets carry voice with FEC, we found that it is better to use FEC than not to use it. It might be that for extremely high values of the network load, this will not hold, but in that case quality will be be already well below acceptable values, FEC or no FEC, so it doesn't really matter.

The second strong fact coming from these numerical values is that the qualitative behaviour of perceived quality when load increases is basically the same in all cases, with a significant degradation when load approaches 1 and beyond.

We also found that using a larger larger packetization interval can help improve the perceived quality. We tried doubling the packet size, obtaining 40ms packets (20ms and 40ms packets are commonly used in telephony applications), and obtained a slight increase in quality even for higher proportions of audio packets (25% of audio packets, which corresponds to the same number of flows of a 50%–20ms audio packet proportion). This can be seen in Figure 2. While increasing the packetization interval is beneficial in one–way streams, it should be studied whether the higher delay values that this creates do not counter these benefits.

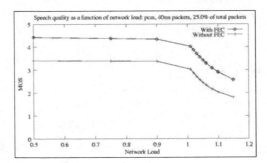

Fig. 2. MOS as a function of network load for 25% of voice traffic, with and without FEC (40ms packets)

5   Conclusions

In this paper we analyze the effect of media–dependent FEC on one–way speech streams. To this end, we studied the effects of the increase in network load generated by adding FEC to voice streams, using a simple queueing model to represent the bottleneck router, in a similar fashion as in[16,17].

In order to estimate the voice quality perceived by the end–user, we used a method we have recently proposed, based on Random Neural Networks trained with data obtained from subjective quality assessments. We fed the RNN with coding parameters and with packet loss parameters derived from the network model, and obtained MOS estimates for each configuration.

We considered a range of network loads that yielded reasonable loss rates, and several values for the fraction of packets corresponding to speech flows. The results obtained indicate that, as expected, the use of FEC is always beneficial to the perceived quality, provided that the network parameters stay within reasonable ranges. Our approach allows to provide actual quantitative estimates of this gain, as a function of different parameters. In the paper we focused on the network load, but similar analysis can be done to explore many other interesting aspects, such as delay/jitter for interactive flows.

One of the strong points of this approach is the coupling of an accurate technique to assess perceived quality (avoiding the use of abstract "utility functions") with a model of the network allowing to obtain information about the loss process. If for some reason the model used yesterday is not considered appropriate today, one can move to some other (possibly more detailed) representation of the bottleneck (or perhaps to a tandem queueing network corresponding to the whole path followed by the packets), and use the same approach to finally map traffic parameters to final, end–to–end quality *as perceived by the final user*.

References

1. Claypool, M., Tanner, J.: The effects of jitter on the perceptual quality of video. In: Proceedings of ACM Multimedia Conference. (1999)
2. Mohamed, S., Cervantes, F., Afifi, H.: Integrating networks measurements and speech quality subjective scores for control purposes. In: Proceedings of IEEE INFOCOM'01, Anchorage, AK, USA (2001) 641–649
3. Mohamed, S., Rubino, G.: A study of real–time packet video quality using random neural networks. IEEE Transactions On Circuits and Systems for Video Technology **12** (2002) 1071 –1083
4. Mohamed, S., Rubino, G., Varela, M.: Performance evaluation of real-time speech through a packet network: a random neural networks-based approach. Performance Evaluation **57** (2004) 141–162
5. ITU-T Recommendation P.800: (Methods for subjective determination of transmission quality)
6. Beerends, J., Stemerdink, J.: A perceptual speech quality measure based on a psychoacoustic sound representation. Journal of Audio Eng. Soc. **42** (1994) 115–123
7. Voran, S.: Estimation of perceived speech quality using measuring normalizing blocks. In: IEEE Workshop on Speech Coding For Telecommunications Proceeding, Pocono Manor, PA, USA (1997) 83–84
8. ITU-T Recommendation G.107: (The E-model, a computational model for use in transmission planning)
9. Yang, W.: Enhanced Modified Bark Spectral Distortion (EMBSD): an Objective Speech Quality Measrure Based on Audible Distortion and Cognition Model. PhD thesis, Temple University Graduate Board (1999)

10. Rix, A.: Advances in objective quality assessment of speech over analogue and packet-based networks. In: the IEEE Data Compression Colloquium, London, UK (1999)
11. Beerends, J.: Improvement of the p.861 perceptual speech quality measure. ITU-T SG12 COM-34E (1997)
12. Hall, T.A.: Objective speech quality measures for Internet telephony. In: Voice over IP (VoIP) Technology, Proceedings of SPIE. Volume 4522., Denver, CO, USA (2001) 128–136
13. London, U.C.: Robust Audio Tool website. (http://www-mice.cs.ucl.ac.uk/multimedia/software/rat/index.html)
14. IETF Network Working Group: RTP payload for redundant audio data (RFC 2198) (1997)
15. Bolot, J.C., Fosse-Parisis, S., Towsley, D.: Adaptive FEC–based error control for Internet telephony. In: Proceedings of INFOCOM '99, New York, NY, USA (1999) 1453–1460
16. Altman, E., Barakat, C., R., V.M.R.: On the utility of FEC mechanisms for audio applications. Lecture Notes in Computer Science **2156** (2001)
17. Altman, E., Barakat, C., R., V.M.R.: Queueing analysis of simple FEC schemes for IP telephony. In: Proceedings of INFOCOM '01. (2001) 796–804
18. Yajnik, M., Moon, S., Kurose, J., Towsley, D.: Measurement and modeling of the temporal dependence in packet loss. In: Proccedings of IEEE INFOCOM '99. (1999) 345–352
19. Hands, D., Wilkins, M.: A study of the impact of network loss and burst size on video streaming quality and acceptability. In: Interactive Distributed Multimedia Systems and Telecommunication Services Workshop. (1999)
20. Fomenkov, M., Keys, K., Moore, D., Claffy, K.: Longitudinal study of Internet traffic in 1998-2003. http://www.caida.org/outreach/papers/2003/nlanr/nlanr_overview.pdf (2003) CAIDA, San Diego Super Computing Center, University of California San Diego.
21. Fraleigh, C., Moon, S., Lyles, B., Cotton, C., Kahn, M., Moll, D., Rockell, R., Seely, T., Diot, C.: Packet-level traffic measurements from the sprint ip backbone. IEEE Network **17** (2003) 6–17
22. Various Authors: Discussion on the e2e mailing list: "Queue size for routers?". (http://www.postel.org/pipermail/end2end-interest/2003-January/002643.html)
23. Claffy, K., Miller, G., Thompson, K.: The nature of the beast: Recent traffic measurements from an Internet backbone. In: INET '98, Geneva, Switzerland (1998)

Architecture of End-to-End QoS for VoIP Call Processing in the MPLS Network*

Chinchol Kim[1], Sangcheol Shin[1], Sangyong Ha[1],
Kyongro Yoon[2], and Sunyoung Han[2]

[1]National Computerization Agency,
NCA Bldg, 77, Mugyo-Dong, Jung-Gu, Seoul, 110-775, Korea
{cckim, ssc, yong}@nca.or.kr
[2]Department of Computer Science and Engineering, Konkuk University,
1, Hwayangdong, Kwangin-gu, Seoul, 143-701, Korea
yoonk@konkuk.ac.kr, syhan@cclab.konkuk.ac.kr

Abstract. This paper proposes the architecture of end-to-end QoS for VoIP call processing in the MPLS (Multiprotocol Label Switching)-based NGN (Next Generation Network). End-to-end QoS for VoIP call processing in the MPLS network is guaranteed through the router and VoIP server. However, QoS is not presently supported in the VoIP server. In order to resolve this problem, this paper proposes QoS resource management and differentiated call processing technology by extending VoIP signaling protocol SIP (Session Initiation Protocol). QoS resource management coordinates service priority in call processing, while differentiated call processing technology processes calls, applying the service priority negotiated in the SIP server. Service priority is set up through the flow label field of the IPv6 header, considering future MPLS label mapping. A performance analysis shows that there is a considerable difference in end-to-end call setup delay depending on service priority, in setting up SIP calls in the MPLS network. We will be able to ensure excellent performance by applying this result to call setups that requires real-time or emergency response in the future NGN environment.

1 Introduction

With the fast development of network technology, the current network architecture is evolving into NGN (Next Generation Network) that allows transfer of broadband multimedia to meet the various needs of users. VoIP, which is one of the core technology of NGN, requires QoS (Quality of Service) support, such as MPLS (Multiprotocol Label Switching) [1] or DiffServ (Differentiated Service) [2], to ensure voice service quality comparable to the quality provided by the PSTN (Public Switched Telephone Network).

---

* This work is supported by the Korean Science and Engineering Foundation under grant number R01-2001-000-00349-0(2003)

J. Solé-Pareta et al. (Eds.): QofIS 2004, LNCS 3266, pp. 44–53, 2004.
© Springer-Verlag Berlin Heidelberg 2004

VoIP is a technology that accommodates the voice service of PSTN in IP network. Compared with PSTN, it has an advantage of providing long distance or overseas call services at relatively low costs [3]. VoIP's core protocols are H.323 and SIP (Session Initiation Protocol) signaling protocols [4][5]. At present, the text-based SIP has been adopted as the standard for NGN. It is also expected to be integrated or to provide compatibility with the current services such as web, instance messaging, SNMP (Simple Network Management Protocol) and SMTP (Simple Mail Transport Protocol) services [6][7]. Therefore, SIP must provide a service quality better than the quality provided by the PSTN for call setup in NGN, and offer priority-based call processing, depending on the traffic properties of application services.

The VoIP service in NGN must guarantee call quality for voice data transfer and call setup quality for call setup [8], in which the latter must precede the former. Call setup is completed before voice data are transferred, and even one bit error or long delay results in the failure of call setup, while a partial loss of voice data does not cause any serious problem. However, if an error occurs, even in one bit or a long delay, the call setup fails and fatal problems may arise.

MPLS-based NGN guarantees end-to-end QoS, since voice data transferring only goes through the router in MPLS network. On the other hand, call setup does not guarantee end-to-end QoS, since it goes through MPLS router and multiple SIP servers. That is, the routers in the MPLS network guarantee QoS. However, as a SIP server that processes call authentication, address transformation, and call routing that does not provide QoS for call processing, it does not guarantee end-to-end QoS.

In order to resolve this problem, this paper proposes the architecture of end-to-end QoS for VoIP call processing in an MPLS-based NGN, which supports QoS resource management and differentiated call processing technology by extending VoIP signaling protocol SIP. QoS resource management coordinates service priority in call processing, while differentiated call processing technology processes calls, applying the service priority negotiated in the SIP server. Service priority is set up through the flow label field of IPv6 header [9]. A performance analysis shows that there is a considerable difference in end-to-end call setup delay depending on service priority, in setting up SIP calls in the MPLS network.

This paper is composed of six chapters. The second chapter examines relevant research, and the third chapter explores system design. The fourth chapter explains implementation, and the fifth chapter explores performance experiments, and finally, the sixth chapter provides a conclusion and comments on future research tasks.

2   QoS Architecture for VoIP in the NGN [8]

The VoIP service in NGN must guarantee end-to-end QoS. VoIP end-to-end QoS is classified into call quality and call setup quality. Call quality refers to voice data security, speech coding distortion, terminal noise and overall delay including delays caused by packetization, buffering, and codecs. The call quality can be guaranteed by QoS technology, including MPSL, DiffServ, and IntServ (Integrated Service). On the other hand, call setup quality refers to guaranteeing call setup, which is classified into call setup delay in the network and call setup delay in VoIP server (SIP server). Call setup delay in the network can be guaranteed through QoS technology, such as MPLS and DiffServ. However, there is no standard technology that ensures call setup delay in the VoIP server. Therefore, we need various forms of QoS mechanisms in the VoIP server to guarantee end-to-end QoS for call setup quality.

3   Architecture of End-to-End QoS for SIP Call Signaling

This paper is based on the MPLS-based NGN environment. Fig. 1 shows the architecture of end-to-end QoS for VoIP call signaling, proposed in this paper. VoIP calls are transferred through the predetermined LSP via MPLS network routers and multiple SIP servers. This paper proposes QoS resource management and differentiated call processing, by extending SIP protocol, to support QoS in the VoIP server.

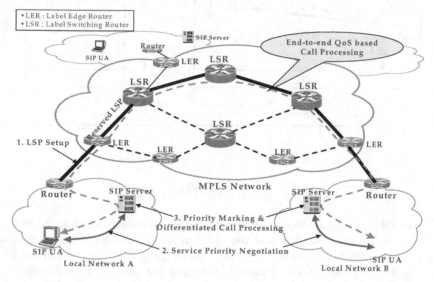

Fig. 1. Architecture of End-to-end QoS for Call Processing in the MPLS Network

The call processing procedure based on end-to-end QoS is as follows:

1. The network manager sets up the LSP (Label Switched Path) to transfer voice and signaling data through LDP/CR-LDP protocol.
2. SIP UA negotiates with the SIP server for the desired service priority, through QoS resource management, prior to session set up. In this paper, service priority is classified into premium level, assured level, and normal level.
3. SIP UA transfers session setup messages to the SIP server. Then the SIP server over SIP message pass processes messages according to the service priority requested by the user, through differentiated call processing technology.

As mentioned above, end-to-end QoS for SIP call setup is guaranteed through QoS in the MPLS network and differentiated call processing in the SIP server.

3.1 System Structure

Fig. 2 shows SIP system structures with extended QoS resource management and differentiated call processing technology, as proposed in this paper. This system consists of SIP6 Server and User Agent.

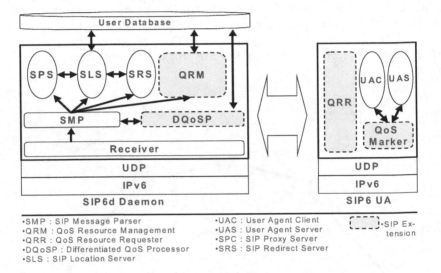

•SMP : SIP Message Parser
•QRM : QoS Resource Management
•QRR : QoS Resource Requester
•DQoSP : Differentiated QoS Processor
•SLS : SIP Location Server

•UAC : User Agent Client
•UAS : User Agent Server
•SPC : SIP Proxy Server
•SRS : SIP Redirect Server

•SIP Ex-
tension

Fig. 2. System structure of SIP Extension

SIP6d daemon includes the proxy server for call forwarding service, the redirect server for call redirect service, and the location server for user location and user information registration services. Receiver receives SIP6 message through the IPv6 UDP Socket interface. Those received messages are divided into Request and Response message types and are sent to the SIP message parser. The SIP message parser parses SIP messages, and calls SIP server or QoS resource management, depending on the method and header type. Differentiated QoS processor applies differentiated call processing technology (see Sect. 3.3) to all incoming messages. The differentiated call processing technology supports differentiated call processing, based on the service priority negotiated in SIP server. The QoS resource management (see Sect 3.2) coordinates service priority between SIP UA and SIP server depending on the extended method and header in this paper. User database manage various user information, including service priority, user-ID, and user location.

The SIP6 user agent is composed of the UA server/client sending and receiving calls in order to establish the sessions, QoS resource requester for priority negotiation, and QoS marker that sets up the negotiated service priority for SIP messages.

3.2 SIP Message Extension and Flow

This paper gives extended definitions of SIP method and header, which follow RFC 2543 message type and processing procedures [10][11], supporting QoS resource management and differentiated call processing technology. Table 1 shows the extended forms of SIP method and header.

Table 1. SIP Method Extension

Method Type	Description
QOSREQUEST	QOSREQUEST method is used by SIP UA to negotiate service priority with the SIP server. It sets up the service priority requested by the user through the Qosinfo header field.
QOSWITHDRAW	QOSWITHDRAW method is used by SIP UA to nullify negotiated service priority. It specifies the service priority to be nullified through Qosinfo header field.

Table 2 shows the extended SIP header field in this paper, and each header field is included in the method defined in Table 1 and the SIP response message.

Table 2. SIP Header and Header Option Extension

Header Type	Description
Qosinfo	**Syntax Formalism: Qosinfo:"desired/release"="ServiceLevel"** "Desired" header option is used to set up the service priority requested by the user in QOSREQUEST method. It is also used to set up the negotiated service priority in 200 OK response messages. "Release" header option is used to release the service priority negotiated by the user in QOSWITHDRAW method and 200 OK response messages.
Qosmark	**Syntax formalism: Qosmark : "ok / no"** This is included in request/response message transferred to SIP message pass. If the header option in Qosmark header field is "Ok," the SIP server applies differentiated call processing technology. If it is "No," the SIP server doesn't.

Fig. 3 shows an SIP message processing procedure extended in this paper. SIP UA negotiates service priority with the SIP server through QOSREQUEST message prior to session set up. The SIP server has the best available service priority. Using this information, it authenticates and assigns the service priority requested by the user. After negotiating service priority, SIP UA sets up the session beginning from INVITE message by setting up the service priority in the flow label field of the IPv6 header. In this process, differentiated call processing technology is applied. After all voice calls are finished and the session ends, SIP UA can nullify the service priority through QOSWITHDRAW message.

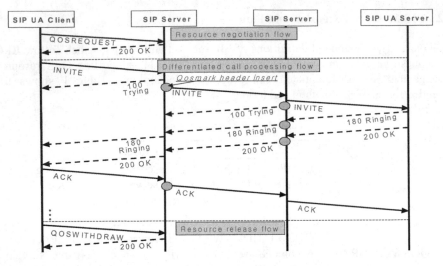

Fig. 3. SIP Extension Message Flow

3.3   Differentiated Call Processing Algorithm

The differentiated call processing technology proposed in this paper is implemented through priority scheduling technology in application level. Fig 4 shows the differentiated call processing algorithm.

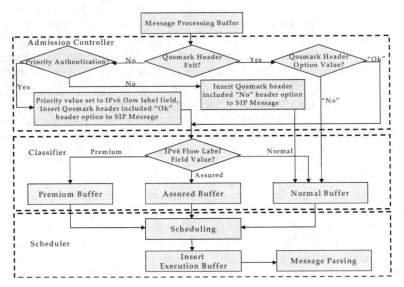

Fig. 4. Differentiated Call Processing Algorithm

The received SIP messages are stored in message processing buffer for processing. The Admission controller determines whether to classify them through Qosmark header field. If there is no Qosmark header option, it performs the priority authentication procedure negotiated with the user. Next, it inserts the Qosmark header field in the SIP messages. The Classifier classifies SIP messages processed by the admission controller into three buffers, according to the value set up in the flow label field of the IPv6 header. The scheduler accesses these three buffers and arranges SIP messages in an excution buffer in the order of priority. The SIP messages arranged in execution buffer are consecutively read by the SIP message parser to begin SIP transaction.

4   Implementation

We referred to the SIP source code from Columbia University. SIP6d is implemented using C language in Linux kernel 2.2.x system environment that supports IPv6. The GUI environment of SIP UA is implemented using Tcl/Tk language, and the controller is implemented using C++. MySQL server is used to manage user information, and major modules are implemented using POSIX thread technology, considering the simultaneous processing rate of messages.

5   Performance Analysis

This paper conducted an end-to-end call setup delay experiment for SIP call processing, by establishing an MPLS test network.

Fig. 5 shows the experiment environment. It is composed of two Linux servers with SIP6d, two PCs with test programs, and two Linux servers, which are used as routers with MPLS modules based on the software provided by Sourceforge.net.

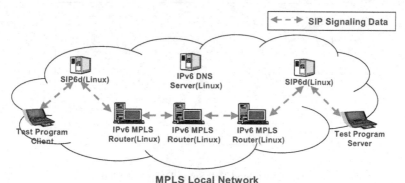

Fig. 5. MPLS based local test network

The experiment has adopted the following procedure. First, the test client program generate equal number of three different INVITE messages and simultaneously transfers to SIP6d in the number of 50, 100, 150, 200, 250, and 300 messages of each priority. The INVITE messages are sent to the test server program through two MPLS routers and two SIP6ds. Secondly, the two SIP6ds process the received messages, using differentiated call processing technology and transfer them to the test client server. Each message is moved along the LSP path predetermined in the three routers. Next, the test server program transfers 200OK response messages for INVITE messages. Then, the test client program receives 200OK response messages coming through two SIP6ds, and measures session setup time, i.e. the average end-to-end call setup delay time. For the comparison, INVITE messages without priority are generated and sent to the SIP6d without a differentiated call processing function in the number of 150, 300, 450, 600, 750, and 900.

Fig. 6 shows the results of an end-to-end call setup delay experiment in SIP6d that applied the differentiated call processing technology and an end-to-end call setup delay experiment in SIP6d that did not apply the technology over MPLS Network.

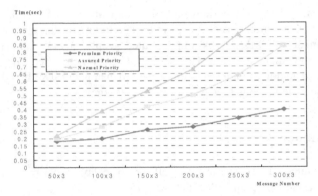

(A) End-to-end Call Setup Delay in SIP6d supporting Differentiated Call Processing

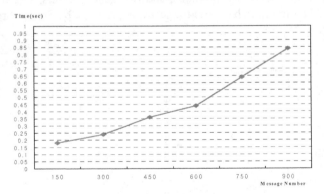

(B) End-to-end Call Setup Delay in SIP6d

Fig. 6. End-to-End Call Setup Delay in SIP6d over MPLS Network

Fig. 7 shows the results of an end-to-end call setup delay experiment in SIP6d over Non-MPLS Network. The INVITE messages are sent to the test server program through three IPv6 routers that use static routing and two SIP6ds.

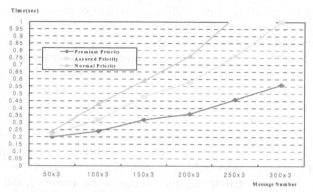

(A) End-to-end Call Setup Delay in SIP6d supporting Differentiated Call Processing

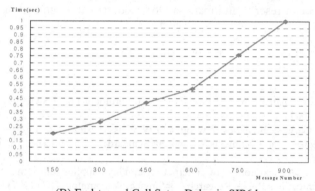

(B) End-to-end Call Setup Delay in SIP6d

Fig. 7. End-to-End Call Setup Delay in SIP6d over Non-MPLS Network

As one can see from the graph, the SIP6d that supports differentiated call processing technology shows a difference in call setup delay when processing messages. In particular, INVITE messages with premium priority have very short call setup delay. Therefore, we can see that INVITE messages with higher service priority have far shorter call setup delay than those with lower service priority. However, SIP6d that does not support differentiated call processing technology has no difference in call setup delay. Also, End-to-end call setup delay over MPLS Network show better performance than end-to-end call setup delay over Non-MPLS Network.

The results of the experiments prove that the SIP6d proposed in this paper can guarantee end-to-end QoS through a differentiated message processing policy when calls are set up in the MPLS network. They also prove that it can provide excellent performance in call setups requiring real time or service priority when providing voice services in the future NGN based on MPLS.

6   Conclusion

This paper proposes an architecture that can guarantee end-to-end QoS when VoIP calls are set up in the MPLS-based NGN, through QoS resource management and differentiated call processing technology, by extending SIP protocol.

The differentiated call processing technology proposed in this paper reserves resources by extending SIP, and minimizes end-to-end call setup delay for specific calls by using priority scheduling technology in the application level. It also has an advantage of setting up the service priority through the flow label field of IPv6 header, considering future MPLS label mapping. A performance analysis has showed that, unlike existing SIP servers, SIP6d provides a very fast processing rate for messages with high service priority, through differentiated QoS function in call processing. Also, in the MPLS experiment environment, it has been proven that calls with high service priority have very short end-to-end call setup delay. These results prove that we can provide excellent performance for call setups that require real-time or service priority when providing future voice service in the NGN based on MPLS.

As a follow-up research task, we perform research on SIP extension MPLS UNI protocol, which can reserve and manage QoS resources through SIP protocol in the MPLS network.

References

1. Rosen, E. Viswanathan, A. and R. Callon. : Multiprotocol Label Switching Architecture, RFC 3031 (2001)
2. S.Blake, M.Carlson, E.Davies, Z.wang, and W.Weiss. : An architecture for differentiated service, Request for Comments (Proposed Standard) 2475, Internet Engineering Task Force (1998)
3. D. Clark. : A taxonomy of internet telephony applications, in Proc. Of 25th Telecommunications Policy Research Conference, Washington, DC (1997)
4. Henning Schulzrinne and Jonathan Rosenberg. : Signaling for Internet Telephony (1998)
5. I. Dalgic and H. Fang. : Comparison of H.323 and SIP for IP Telephony Signaling, Photonics East, Proceeding of SPIE'99, Boston, Massachusetts (1999)
6. C. A. Polyzois, K. H. Purdy, P.-F. Yang, D. Shrader, H. Sinnreich, F. Mnard, and H. Schulzrinne. : From POTS to PANS – a commentary on the evolution to internet telephony, IEEE Network, Vol. 13, pp. 58-64 (1999)
7. International Softswitch Consortium. : ISC Reference Architecture V 1.2 (2002)
8. Chinchol Kim, Byunguk Choi. : Design and implementation of the VoIPv6 supporting the differentiated call processing, ICOIN2002 (2002)
9. R.Gilligan, S.Thomson, J.Bound, W.Stevens. : Basic Socket Interface Extensions for IPv6, RFC 2553 (1999)
10. Handley, H.Schulzrine, E.Schooler, and J.Rosenberg. : SIP: session initiation protocol, RFC 2543 (1999)
11. J. Rosenberg and H. Schulzrinne. : Guidelines for Authors of Extensions to the Session Initiation Protocol (SIP), Internet Draft (2002)
12. M. Handley and V. Jacobson. : SDP: Session Description Protocol, RFC2327 (1998)

Analysis of the Distribution of the Backoff Delay in 802.11 DCF: A Step Towards End-to-End Delay Guarantees in WLANs*

Albert Banchs

Universidad Carlos III de Madrid
Departamento de Ingeniería Telemática
banchs@it.uc3m.es

Abstract. In this paper we present an analytical method to study the distribution of the backoff delay in an 802.11 DCF WLAN under saturation conditions. We show that, with our method, the probability that the delay is below a given threshold can be computed accurately and efficiently. We also discuss how our analysis can be used to perform admission control on the number of accepted stations in the WLAN in order to provide delay assurances to real-time applications.

1 Introduction

As 802.11 WLANs see their capacity increased (from the traditional 2 Mbps channel capacity to 11 Mbps in 802.11b and 54 Mbps in 802.11a), these networks become better suited for the transport of real-time traffic. Since the performance of real-time applications is largely dependent on delay, there arises the need for an analysis of the delay in this type of networks.

To the date, the analysis of the delay in 802.11 WLAN has received some attention. The analyses of [1,2,3] are limited to the average delay, which is insufficient to assess the performance of real-time applications, as these applications require not only a low average delay but a low delay for all (or most of) their packets. The analyses of [4,5] overcome this limitation by introducing probability generating functions (pgf's), which allow the computation of the probability distribution function (pdf) of the delay. However, computing pdf values with this method is very costly computationally and hence the approaches of [4,5] are of little practical use to perform e.g. admission control functionality. This paper presents an original method to compute the delay distribution of 802.11 DCF that, in contrast to the previous analyses, is both accurate and efficient.

The analysis of the delay in this paper focuses on the backoff component of the delay under saturation conditions, hereafter referred to with *saturation delay*. By backoff delay we understand the time elapsed since a packet starts its backoff process until it is successfully transmitted[1]. This is one of the main components of the end-to-end delay. With saturation conditions we mean that all the stations in the WLAN always have packets to transmit. Note that assuming saturation conditions corresponds to the worst case and thus provides us with an upper bound on the backoff delay.

---

* This work has been performed in the context of the Daidalos European IP.
[1] In case the packet is discarded, we consider its backoff delay equal to ∞.

J. Solé-Pareta et al. (Eds.): QofIS 2004, LNCS 3266, pp. 54–63, 2004.
© Springer-Verlag Berlin Heidelberg 2004

The rest of the paper is structured as follows. In Section 2 we present a brief overview of the 802.11 DCF protocol. In Section 3 we propose a method to analyze the distribution of the saturation delay. In Sections 4 we evaluate the performance (namely, accuracy and computational efficiency) of the method proposed. The results obtained show that, with our method, the probability that the delay falls below a certain value can be computed accurately and efficiently. In Section 5 we discuss how our algorithm to compute the saturation delay distribution can be used to perform admission control in a WLAN with real-time traffic in order to provide this traffic type with end-to-end delay guarantees. Finally, in Section 6 we present our concluding remarks.

2   802.11 DCF

The DCF access method of the IEEE 802.11 standard [6] is based on the CSMA/CA protocol. A station with a new packet to transmit senses the channel and, if it remains free for a DIFS time, it transmits. If the channel is sensed busy, the station waits until the channel becomes idle for a DIFS time, after which it starts a backoff process. Specifically, it generates a random backoff time before transmitting.

The backoff time is chosen from a uniform distribution in the range $(0, CW - 1)$, where the CW value is called Contention Window, and depends on the number of transmissions failed for the packet. At the first transmission attempt, CW is set equal to a value CW_{min}, and it is doubled after each unsuccessful transmission, up to a maximum value CW_{max}.

The backoff time is decremented once every time interval T_e for which the channel is detected empty, "frozen" when a transmission is detected on the channel, and reactivated when the channel is sensed empty again for a DIFS time (if the transmission is detected as successful) or an EIFS time (if it is detected as unsuccessful). The station transmits when the backoff time reaches zero.

If the packet is correctly received, the receiving station sends an ACK frame after a SIFS time. If the ACK frame is not received within an ACK Timeout time, a collision is assumed to have occurred and the packet transmission is rescheduled according to the given backoff rules. If the number of retransmissions reaches a predefined Retry Limit, the packet is discarded. Upon completing the transmission (either with a success or with a discard), the transmitting station resets the CW to its initial value and starts a new backoff process; before this ends, a new packet cannot be transmitted.

The use of the Request to Send (RTS) / Clear to Send (CTS) mechanism is optional in 802.11. When this option is applied, upon the backoff counter reaching zero, the transmitting station sends an RTS frame to the receiving station, which responds with a CTS frame. The packet is then sent when the transmitting station receives the CTS.

3   Saturation Delay Analysis

In this section we propose an analytical model to compute the distribution of the saturation delay. We first analyze the simplified case in which all packets have the same fixed length and the RTS/CTS mechanism is not used, and then propose two extensions of the basic analysis to account for these cases.

3.1 Basic Analysis

Let us consider a WLAN with N stations operating under saturation conditions and sending packets of a fixed packet length l. Our objective is to compute the probability that, under these conditions, a packet transmission of a tagged station experiences a saturation delay smaller than a given value D. We denote this probability by $P(d < D)$.

Fig. 1 illustrates the different components of the saturation delay. Applying the theorem of the total probability, $P(d < D)$ can be decomposed as follows

$$P(d < D) = \sum_{i=0}^{R} P(d < D/i \text{ col}) P(i \text{ col}) \tag{1}$$

where $P(i \text{ col})$ represents the probability that a packet suffers i collisions before being successfully transmitted and R is the Retry Limit.

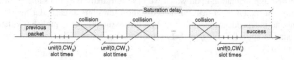

Fig. 1. Saturation delay.

Let us define a slot time as the time interval between two consecutive backoff time decrements of the tagged station. Note that, according to this definition, a slot time may be either empty or contain the transmission of one or more stations. Applying to the previous equation the theorem of the total probability on the total number of slot times the tagged station counts down before transmitting successfully, we have

$$P(d < D) = \sum_{i=0}^{R} \sum_{j=0}^{W_i} P(d < D/i \text{ col}, j \text{ slots}) P(j \text{ slots}/i \text{ col}) P(i \text{ col}) \tag{2}$$

where $W_i = \sum_{k=0}^{i} CW_k - 1$, with $CW_k = min(2^k CW_{min}, CW_{max})$, and $P(j \text{ slots}/i \text{ col})$ is the probability that the sum of the $i + 1$ backoff times of the packet equals j,

$$P(j \text{ slots}/i \text{ col}) = P\left(\sum_{k=0}^{i} unif(0, CW_k - 1) = j\right) \tag{3}$$

where $unif(0, C)$ represents a discrete random variable uniformly distributed on $\{0, 1, \dots, C\}$.

As the probability mass function (pmf) of a sum of discrete random variables is equal to the convolution of the individual pmf's, we can compute $P(j \text{ slots}/i \text{ col})$ as follows

$$P(j \text{ slots}/i \text{ col}) = (f_0 * f_1 * \dots * f_i)_j \tag{4}$$

being f_k the pmf of $unif(0, CW_k - 1)$. We compute the above convolution with Fast Fourier Transforms (FFT's), as FFT provides a very efficient means of computing convolutions.

Let τ be the probability that a station transmits in a slot time in a WLAN with N stations under saturation conditions. Following the analysis of [7], we compute τ by solving the non-linear equation resulting from the following two equations:

$$p = 1 - (1 - \tau)^{N-1} \tag{5}$$

and

$$\tau = \frac{2(1 - 2p)(1 - p^{R+1})}{W(1 - (2p)^{m+1})(1 - p) + (1 - 2p)[(1 - p^{R+1}) + W2^m p^{m+1}(1 - p^{R-m})]} \tag{6}$$

where R is the Retry Limit, $W = CW_{min} + 1$, m is such that $CW_{max} = 2^m CW_{min}$ and p is the probability that a transmission attempt collides.

The first approximation upon which we base our analysis is the same as [8]: we assume that a station other than the tagged station transmits at each slot time with a constant and independent probability τ. With this assumption, the probability that the tagged station suffers i collisions before transmitting successfully can be computed according to

$$P(i \text{ col}) = P_c^i P_s = \left(1 - (1 - \tau)^{N-1}\right)^i (1 - \tau)^{N-1} \tag{7}$$

where P_s corresponds to the probability that a transmission of the tagged station is successful (i.e. none of the other $N - 1$ stations transmits) and P_c to the probability that it collides (i.e. some other station transmits).

Our second approximation[2] is to assume that the saturation delay given i collisions and j slot times is a gaussian random variable, which we denote by d_{ij}. Note that, assuming independence between different slot times (which is given by the first approximation) and a number of slot times large enough (which is the typical case), the Central Limit Theorem assures that this approximation is accurate.

With the above approximation, it is enough to know the average and the typical deviation of d_{ij} (which we denote by m_{ij} and σ_{ij}, respectively) to compute $P(d < D/i \text{ col}, j \text{ slots})$,

$$P(d < D/i \text{ col}, j \text{ slots}) = \begin{cases} 0.5 + 0.5 \, erf\left(\frac{D-m_{ij}}{\sqrt{2}\sigma_{ij}}\right), & \frac{D-m_{ij}}{\sigma_{ij}} \geq 0 \\ 0.5 \, erfc\left(-\frac{D-m_{ij}}{\sqrt{2}\sigma_{ij}}\right), & \frac{D-m_{ij}}{\sigma_{ij}} < 0 \end{cases} \tag{8}$$

Given the assumption of independence between different slot times, m_{ij} can be computed as the sum of the average duration all slot times in d_{ij},

$$m_{ij} = j \, m_n + i \, T_c + T_s \tag{9}$$

---

[2] This approximation is the key difference between our model and the analyses of [4,5]; while, with our approximation, we only need to compute the average and typical deviation values of d_{ij}, which can be done efficiently, [4,5] compute all the possible values of d_{ij} and their probability, which, as d_{ij} can take a very large number of different values, is very costly computationally.

where m_n is the average duration of a slot time in which the tagged station does not transmit, T_c is the duration of a slot time that contains a collisions and T_s is the duration of a slot time that contains a successful transmission.

The duration of a slot time that contains a successful transmission is equal to [9]

$$T_s = T_{PLCP} + \frac{H+l}{C} + SIFS + \frac{ACK}{C} + DIFS \qquad (10)$$

where T_{PLCP} is the PLCP (Physical Layer Convergence Protocol) preamble and header transmission time, H is the MAC overhead (header and FCS), ACK is the length of an ACK frame and C is the channel bit rate.

Similarly, the duration of a slot time that contains a collision is equal to

$$T_c = T_{PLCP} + \frac{H+l}{C} + EIFS \qquad (11)$$

The average duration of a slot time in which the tagged station does not transmit, m_n, is computed as

$$m_n = P_{s,n} T_s + P_{c,n} T_c + P_{e,n} T_e \qquad (12)$$

where $P_{s,n}$ represents the probability that a slot time in which the tagged station does not transmit contains a successful transmission, $P_{c,n}$ the probability that it contains a collision and $P_{e,n}$ the probability that it is empty.

$P_{s,n}$, $P_{e,n}$ and $P_{c,n}$ can be computed from τ and N as

$$P_{s,n} = (N-1)\tau(1-\tau)^{N-2}, \qquad P_{e,n} = (1-\tau)^{N-1} \qquad (13)$$

and

$$P_{c,n} = 1 - P_{s,n} - P_{e,n} \qquad (14)$$

With the assumption of independence between different slot times, the typical deviation σ_{ij} can be computed from

$$\sigma_{ij}^2 = j \, \sigma_n^2 \qquad (15)$$

with

$$\sigma_n^2 = P_{s,n} T_s^2 + P_{c,n} T_c^2 + P_{e,n} T_e^2 - m_n^2 \qquad (16)$$

which closes the analysis.

3.2 RTS/CTS

In case the RTS/CTS option is used, successful packets are preceded by a RTS/CTS exchange, while collisions occur with RTS frames instead of data packets. Accordingly, the durations of the slot times containing a successful transmission and a collision have to be computed as in [9] for the RTS/CTS case. With this only modification, the analysis of the previous clause can be used to compute the saturation delay distribution for the RTS/CTS case.

3.3 Non Fixed Packet Lengths

Next, we extend our basic model to the case when packet lengths are not fixed but follow a certain distribution. Specifically, we consider that a packet length takes a value l of the set L with probability P_l, being L the set of all possible packet lengths. For simplicity, we assume that all stations transmit the same packet length distribution; however, the analysis would be very similar in the case when this condition does not hold.

In order to account for non fixed packet lengths, we have to modify the expressions to obtain the m_{ij} and σ_{ij} values. m_{ij} is computed as

$$m_{ij} = j\,m_n + i\,m_c + m_s \tag{17}$$

where m_n is the average duration of a slot time in which the tagged station does not transmit, m_c is the average duration of a slot time in which the tagged station collides and m_s is the average duration of a slot time in which the tagged station transmits a packet successfully.

The average duration of a slot time in which the tagged station does not transmit, m_n, is computed as

$$m_n = \sum_{l \in L} P_{s,l,n}\,T_{s,l} + \sum_{l \in L} P_{c,l,n}\,T_{c,l} + P_{e,n}\,T_e \tag{18}$$

where $P_{s,l,n}$ represents the probability that a slot time in which the tagged station does not transmit contains a successful transmission of a packet of length l, $P_{c,l,n}$ the probability that it contains a collision with the longest packet involved of length l and $T_{s,l}$ and $T_{c,l}$ are the slot time durations in each case.

$P_{s,l,n}$ and $P_{c,l,n}$ are computed as

$$P_{s,l,n} = (N-1)\tau(1-\tau)^{N-2}P_l \quad \text{and} \quad P_{c,l,n} = (1 - P_{s,l,n} - P_{e,n})\,P_{c,l} \tag{19}$$

where $P_{c,l}$ is the probability that the longest packet involved in a collision is of length l. Neglecting the collisions of more than two stations,

$$P_{c,l} = 2\,P_l \sum_{k \in L_l} P_k - P_l^2 \tag{20}$$

where L_l is the set of all the packet lengths smaller than or equal to l.

The duration of a slot time that contains a successful transmission of a packet of length l, $T_{s,l}$, and the duration of a slot time that contains a collision of two packets, the longest of length l, $T_{c,l}$, can be computed following Eqs. (10) and (11).

Finally, the typical deviation σ_{ij} for the non fixed packet length case can be computed from

$$\sigma_{ij}^2 = j\,\sigma_n^2 + i\,\sigma_c^2 + \sigma_s^2 \tag{21}$$

with

$$\sigma_n^2 = \sum_{l \in L} P_{s,l,n}\,T_{s,l}^2 + \sum_{l \in L} P_{c,l,n}\,T_{c,l}^2 + P_{e,n}\,T_e^2 - m_n^2, \tag{22}$$

$$\sigma_c^2 = \sum_{l \in L} P_{c,l}\,T_{c,l}^{*2} - m_c^2 \quad \text{and} \quad \sigma_s^2 = \sum_{l \in L} P_l\,T_{s,l}^2 - m_s^2 \tag{23}$$

4   Performance Evaluation

Next, we evaluate the accuracy and computational efficiency of the model proposed. The values of the system parameters used to obtain the results, both for the analytical model and the simulation runs, have been taken from the 802.11b physical layer. The packet length has been taken equal to 1000 bytes for the fixed packet length case, and derived from the measurements of Internet traffic presented in [10] for the non-fixed packet length case. Simulations are performed with an event-driven simulator developed by us, that closely follows the 802.11 DCF protocol details for each independently transmitting station.

Figs. 2, 3 and 4 illustrate the cumulative distribution function (cdf) of the saturation delay –i.e. $P(d < D)$ as a function of D– for our basic model, RTS/CTS extension and non-fixed packet lengths extension, respectively. Analytical results are represented with lines and simulations with points. Simulation results are given with a 95% confidence interval below 0.1%. Results show that our analysis is very accurate; in all cases, and for all values of D and N, simulations coincide almost exactly with analytical results. In addition, results corroborate the intuition that delays are smaller for the RTS/CTS and non-fixed packet lengths cases (the latter due to smaller packets being transmitted).

In order to evaluate the computational efficiency of our method, we measured the times required to compute the cdf values given in Figs. 2, 3 and 4. Measurements have been taken in a Pentium 4 PC with 2.66 GHz of CPU speed and 192 MB of RAM, running under the Linux operating system. We obtained that, for all models (basic, RTS/CTS and non fixed packet lengths) and different values of N (2, 10, 30 and 100), the time required to compute the 20 cdf values given in each of the graphs, ranged from 0.37 to 0.45 seconds. These results show that, with the model proposed, the times required to compute the $P(d < D)$ values keep very low (in all cases below 0.5 seconds for 20 points) and, moreover, are practically constant (almost independent of the model and N). We believe that these results, even though taken in a single platform and running not necessarily optimized code, do proof the low computational cost of our algorithm. Note that the times measured (in the order of 0.5 seconds) are fully acceptable to take an admission control decision; moreover, as (following the discussion of the next section)

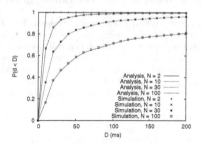

Fig. 2. Saturation delay cdf: Basic Model.

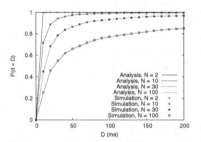

Fig. 3. Saturation delay cdf: RTS/CTS.

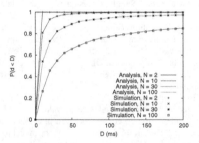

Fig. 4. Saturation delay cdf: Non fixed packet lengths.

in some situations one $P(d < D)$ value may be enough for admission control, the time involved in taking an admission control decision may even be much smaller.

5   Discussion on End-to-End Delay Guarantees in WLANs

The method we have proposed in this paper allows computing the distribution of the backoff delay under saturation conditions. The backoff delay is one of the main components of the end-to-end delay, but not necessarily the only one. Real-time applications require end-to-end delay (i.e. the sum of all the delay components) to be below a certain threshold (at least for most of the packets), or otherwise their performance is unsatisfactory. In this section we discuss how our method can be used to derive the worst-case distribution of the end-to-end delay, and thus allow providing end-to-end delay guarantees by means of admission control.

The fact that our model assumes saturation conditions represents the worst possible case for a tagged station, as this station will experience the largest delays when all the other stations have always packets to transmit. Therefore, this is the case that should be considered if our goal is to provide end-to-end delay guarantees by limiting the number of stations in the WLAN by performing admission control. Many of the previous delay analyses of DCF (namely, [1,2,3]) also assume saturation conditions.

If we consider an end-to-end communication between two WLAN stations, or a WLAN station and the Access Point, then the end-to-end delay consists of two main components: the backoff and the queuing delays. The first is the time elapsed since a packet starts its backoff process until it is successfully transmitted, while the second is the time elapsed since the generation of a packet until it reaches the first position of the transmission buffer. The backoff component of the delay is accurately characterized in the present paper. An open issue is the computation the queuing delay.

The problem of computing the queuing delay in the above case can be seen as analyzing a classical G/G/1 queue, in which the arrivals follow the process given by the packet arrivals at the station, and the queue service time follows the distribution of the backoff delay (which has been characterized in this paper). This problem can be dealt with classical queuing theory [11] – this is the approach taken by [4,5].

The 802.11 standard allows that a station, once it gets access to the channel, sends not only one but multiple packets separated by SIFS times. This option is appropriate e.g. for voice sources, because of the stringent delay requirements of their packets, and also because the short length of voice packets would make the protocol overhead very high otherwise. For a tagged station using this option, and sending all the packets waiting for transmission in its buffer every time it gets access to the channel, the end-to-end delay consists of the backoff delay only, and therefore the model presented in this paper can be used to characterize the end-to-end delay.

6   Summary and Final Remarks

As the capacity of WLANs and their use by real-time applications increases, there arises the need for better understanding and predicting the delay behavior in this type of networks. In this paper we have proposed a method to compute accurately and efficiently the distribution of the backoff delay in 802.11 DCF under saturation conditions. The method proposed is a first step towards an admission control algorithm that, by limiting the number of stations in the WLAN, ensures end-to-end delays low enough for real-time applications.

The backoff delay experienced by a station can be interpreted as the service time seen by its internal queue. Then, classical queuing theory can be used to derive the queuing delay, given the characterization of the backoff delay obtained in this paper. If a station sends all its waiting packets when it accesses the channel, the backoff delay derived here is the only component of the end-to-end delay.

Our model to analyze the backoff delay of a tagged station assumes that all other stations always have packets to transmit. As this corresponds to the worst case for the delay of the tagged station, the results obtained represent an upper bound and are therefore appropriate for providing the tagged station with delay guarantees. However, our analysis could also be reused for non-saturation conditions, if the τ probabilities under non-saturation conditions were given (a rough approximation to compute them is proposed in [4]).

In the literature, there have been many protocol proposals for WLAN that, unlike DCF, have been designed specifically to satisfy the delay requirements of real-time applications (see e.g. [12,13,14]). The PCF scheme of 802.11 [6] was also designed with a similar intention. However, none of these (including PCF) is widely deployed today, which leaves DCF as the only option to provide real-time traffic communication in today's WLANs.

The IEEE 802.11 WG is currently undergoing a standardization activity to extend the 802.11 protocol with QoS support, leading to the upcoming 802.11e standard. The EDCA access mechanism of 802.11e is an extension of the DCF protocol. We believe that our analysis here provides a basis that can be extended to analyze the delay of 802.11e EDCA.

References

1. E. Ziouva and T. Antonkopoulos, "CSMA/CA Performance under high traffic conditions: throughput and delay analysis," *Computer Communications*, vol. 25, no. 1, pp. 313–321, January 2002.
2. P. Chatzimisios, A.C. Boucouvalas, and V. Vitsas, "Packet delay analysis of IEEE 802.11 MAC protocol," *IEE Electronics Letters*, vol. 39, no. 18, pp. 1358–1359, September 2003.
3. B. Li and R. Battiti, "Performance Analysis of An Enhanced IEEE 802.11 Distributed Coordination Function Supporting Service Differentiation," in *Proceedings of QofIS'03*, Stockholm, Sweden, October 2003.
4. O. Tickoo and B. Sikdar, "Queueing Analysis and Delay Mitigation in IEEE 802.11 Random Access MAC based Wireless Networks," in *Proceedings of IEEE INFOCOM'04*, Hong Kong, China, March 2004.
5. H. Zhai and Y. Fang, "Performance of Wireless LANs Based on IEEE 802.11 MAC Protocols," in *Proceedings of IEEE PIMRC'03*, 2003.
6. IEEE 802.11, *Wireless LAN Medium Access Control (MAC) and Physical Layer (PHY) specifications*, Standard, IEEE, August 1999.
7. H. Wu, Y. Peng, K. Long, S. Cheng, and J. Ma, "Performance of Reliable Transport Protocol over IEEE 802.11 Wireless LAN: Analysis and Enhancement," in *Proceedings of IEEE INFOCOM'02*, New York City, New York, June 2002.
8. F. Cali, M. Conti, and E. Gregori, "Dynamic Tuning of the IEEE 802.11 Protocol to Achieve a Theoretical Throughput Limit," *IEEE/ACM Transactions on Networking*, vol. 8, no. 6, pp. 785–799, December 2000.
9. G. Bianchi, "Performance Analysis of the IEEE 802.11 Distributed Coordination Function," *IEEE Journal on Selected Areas in Communications*, vol. 18, no. 3, pp. 535–547, March 2000.
10. K. Claffy, G. Miller, and K. Thompson, "The nature of the beast: Recent traffic measurements from an internet backbone," in *Proceedings of INET'98*, Geneve, Switzerland, July 1998.
11. H. Bruneel and B. Kim, *Discrete-Time Models for Communication Systems Including ATM*, Kluwer Academic Publishers, 1993.
12. V. Kanodia, C. Li, B. Sadeghi, A. Sabharwal, and E. Knightly, "Distributed Multi-Hop with Delay and Throughput Constraints," in *Proceedings of MOBICOM'01*, Rome, Italy, July 2001.
13. J. L. Sobrinho and A.S. Krishnakumar, "Real-Time Traffic over the IEEE 802.11 Medium Access Control Layer," *Bell Labs Technical Journal*, 1996.
14. S. Chevrel et al., "Analysis and optimisation of the HIPERLAN Channel Access Contention Scheme," *Wireless Personal Communications*, vol. 4, pp. 27–39, 1997.

Analysis of the IEEE 802.11 DCF with Service Differentiation Support in Non-saturation Conditions[*]

Bo Li and Roberto Battiti

Department of Computer Science and Telecommunications, University of Trento,
38050 POVO Trento, Italy
{li,battiti}@dit.unitn.it

Abstract. Although the performance analysis of the IEEE 802.11 Distributed Coordination Function (DCF) in saturation state has been extensively studied in the literature, little work is present on performance analysis in non-saturation state. In this paper, a simple model is proposed to analyze the performance of IEEE 802.11 DCF with service differentiation support in non-saturation states, which helps to obtain a deeper insight into the IEEE 802.11 DCF. Based on the proposed model, we can approximately evaluate the most important system performance measures, such as packet delays, which provide one with an important tool to predict and optimize the system performance. Moreover, a practical method to meet packet delay requirements is presented based on our theoretical results. Comparisons with simulations show that this method achieves the specified packet delay requirements with good accuracy.

Keywords: Wireless LAN, IEEE 802.11, Quality of Service Guarantee, Service Differentiation

1 Introduction

In recent years, IEEE 802.11 has become one of the most important international standards for Wireless Local Area Networks (WLAN's) [1]. In the IEEE 802.11 protocol, the fundamental mechanism to access the medium is the Distributed Coordination Function (DCF), which is a random access scheme based on the carrier sense multiple access with collision avoidance (CSMA/CA) protocol. Many performance analyses of 802.11 have been proposed, such as those in [2]-[5]. However, the previous papers consider the assumption of saturation state. That is, it is assumed that the transmission queue for each station is always nonempty, which is not realistic in real-world systems. In [6] and [7], more practical queuing models for IEEE 802.11 DCF are proposed which incorporate practical packet arrival processes. However, the service rate for each node is still based on the results obtained in [4], where saturation state is assumed. The limitation is overcome in [8], where performance analysis in non-saturation state is considered by introducing probability generating functions, which allow the computation of the probability distribution

---

[*] This work is supported by the project of WILMA funded by Provincia Autonoma di Trento (www.wilmaproject.org)

J. Solé-Pareta et al. (Eds.): QofIS 2004, LNCS 3266, pp. 64–73, 2004.
© Springer-Verlag Berlin Heidelberg 2004

function (pdf) of the delay. However, computing pdf values with the proposed method has a high computational cost and therefore the approach is of limited practical use. The other drawback is that the complex analysis method in [8] is of little help to obtain deeper insight into relationships among different system parameters. Moreover, service differentiation support is not considered. In this paper, based on our former work in [9]-[10], a simple analysis model is proposed to analyze the performance of an enhanced 802.11 DCF with service differentiation support in non-saturation state. We considered the following objectives when defining the model.

1. The analysis model should be simple enough to obtain a clear insight into relationships among the most important system parameters.
2. The analysis model should be as practical as possible, so that it can be implemented in real-world systems.
3. Service differentiation must be considered.

2   Performance Analysis

We consider a single-hop wireless LAN, where stations can "hear" each other well. It is assumed that the channel conditions are ideal (i.e., no hidden terminals and capture). M types of traffic are considered with n_i type i ($i = 1,...,M$) stations, and, for simplicity, each station bears only one traffic flow. If the station is busy on the arrival of a packet, the packet must wait in the corresponding transmission queue. The buffer size is assumed to be infinite. It is assumed that the packet arrival processes for type i traffic flows follow independent and identical distributions (i.i.d.), with mean packet inter-arrival duration $T_{p,i}$. The model can consider different arrival processes. Moreover, it is assumed that all packets have the same payload length, which is transmitted in the duration of P_L. It is also assumed that a backoff process starts immediately when the current packet arrives at the head of the queue.

In the following, a type i traffic flow is considered. Let $b_i(t)$ be the stochastic process representing its backoff time counter. Moreover, let us define $W_i = CW_{min,i}$ as its minimum contention window. Denote m_i, "maximum backoff stage" as the value such that $CW_{max,i} = 2^{m_i} \cdot W_i$. $s_i(t)$ is the stochastic process representing its backoff stage $(0,1,...,m_i)$. A two-dimensional discrete-time Markov chain (shown in Fig. 1) is used to model the behavior of the traffic flow. The states are defined as combinations of two integers $\{s_i(t), b_i(t)\}$. It should be noted that apart from using states $\{s_i(t), b_i(t)\}$, a state VTSS (Virtual Time Slot State) is used to model the case that a traffic flow has finished sending a packet and is waiting for the next one. In order to make the system tractable by using a discrete-event Markov chain, the VTSS is sub-divided into different VTS (Virtual Time Slots), whose duration is the same as the time slot in the backoff process. We assume that the station checks if there is a packet available for transmission only at the end of a VTS. In this way, the behavior of the traffic flow in VTSS can be modeled in the same way as the actual backoff processes. For clarity, the above approximated version of DCF is called ADCF. This

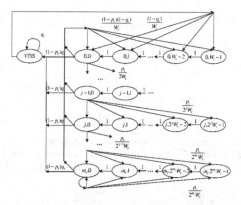

Fig.1. Markov model for a type i traffic flow in ADCF

approximation has very little influence to the final system performance, as verified by extensive simulations. If it is found that the packet transmission queue is not empty after sending the current packet, the state of the traffic flow transits from VTSS to some backoff state. Otherwise, the traffic flow still needs to wait for the arrival of the next packet in VTSS. From Fig. 1, it can be seen that after a packet has been successfully sent or the current VTS has finished, the traffic flow steps into another VTSS with probability q_i. Moreover, parameter p_i is referred to as conditional collision probability, the probability of a collision seen by a packet belonging to a type i traffic flow at the time of its being transmitted on the channel. For simplicity, both q_i and p_i are regarded as constant, which is validated through extensive simulations.

In steady state, $d_{j,k}(i) \equiv \lim_{t \to \infty} P\{s_i(t) = j, b_i(t) = k\}$ ($i = 1,...,M$, $j \in [0,m_i]$, $k \in [0, 2^j W_i - 1]$) is the stationary distribution of backoff states of a type i traffic flow. $P_{VTSS,i}$ is defined as the probability for the traffic flow being at VTSS. Therefore, based on the Markov chain, we have

$$P_{VTSS,i} = d_{0,0} \cdot q_i / (1 - q_i) \qquad (1.1)$$

$$\begin{cases} d_{j,0}(i) = p_i^j \cdot d_{0,0}(i) & (0 < j < m_i) \\ d_{m_i,0}(i) = p_i^{m_i} \cdot d_{0,0}(i)/(1 - p_i) \end{cases} \qquad (1.2)$$

$$d_{j,k}(i) = (2^j W_i - k) \cdot d_{j,0}(i) / 2^j W_i \qquad (1.3)$$

τ_i is the probability that a type i traffic flow transmits in a randomly chosen time slot. It can be given as

$$\tau_i = \sum_{j=0}^{m_i} d_{j,0}(i) = d_{0,0}(i)/(1 - p_i) \qquad (2)$$

Since $P_{VTSS,i} + \sum_{j=0}^{m_i} \sum_{k=0}^{2^j W_i - 1} d_{j,k}(i) = 1$, combing equations 1 and 2, we have

$$\left\{ (1-2p_i)(W_i+1) + p_i W_i [1-(2p_i)^{m_i}] \right\} \cdot \tau_i \Big/ [2(1-2p_i)] + P_{VTSS,i} = 1 \qquad (3)$$

Extensive simulations show that, even if the packet arrival of each traffic flow are assumed to be independent, in some cases there are obvious correlations between behaviors of different traffic flows. Therefore, by introducing compensation factors $\alpha_i > 0$ $(i=1,...,M)$, packet collision rates can be expressed as

$$p_i = \alpha_i \cdot [1-(1-\tau_i)^{n_i-1} \prod_{j=1, j \neq i}^{M} (1-\tau_j)^{n_j}] \qquad (4)$$

In non-saturation state, the system total throughput S and throughputs S_i $(i=1,...,M)$ contributed by type i traffic flows can be expressed as follows with the assumption that all the arrived packets are finally transmitted successfully

$$S = \sum_{i=1}^{M} S_i = \sum_{i=1}^{M} \frac{n_i P_L}{T_{p,i}} = \cfrac{P_L \cdot \sum_{i=1}^{M} n_i \cdot \tau_i \cdot (1-p_i)}{\left(\begin{array}{c} \beta \cdot \sigma \cdot \prod_{i=1}^{M} (1-\tau_i)^{n_i} + P_s \cdot \sum_{i=1}^{M} n_i \cdot \tau_i \cdot (1-p_i) + \\[2mm] [1-\beta \cdot \prod_{i=1}^{M} (1-\tau_i)^{n_i} - \sum_{i=1}^{M} n_i \cdot \tau_i \cdot (1-p_i)] \cdot P_c \end{array} \right)} \qquad (5)$$

where $\beta > 0$ is another compensation factor. It should be noted that the purpose for the introduction of α_i and β is to make our mathematical expressions more rigorous. Extensive experiments show that α_i and β can be approximated as one under the case that the system operates in stable states. Moreover, in equation 5, σ is the duration of an empty time slot (it is also the duration of an empty VTS). P_s is the average time of a slot because of a successful transmission of a packet. And P_c is the average time the channel is sensed busy by each station during a packet collision. We have:

$$P_s = PHY_{header} + MAC_{header} + P_L + SIFS + \delta + ACK + DIFS + \delta \qquad (6)$$

$$P_c = PHY_{header} + MAC_{header} + P_L + DIFS + \delta \qquad (7)$$

where δ is the propagation delay.

We assume that behaviors of all the traffic flows are independent (simulations show that this assumption approximately holds in the case that minimum contention window sizes W_i s are not very small). In this case, the above introduced α_i and β can be approximated as 1. Therefore, given the corresponding offered traffic load (hence, the system throughput S is also given), based on equations 4 and 5, packet sending rates τ_i s and the corresponding packet collision rates p_i s can be determined. Although two sets of solutions can be obtained, only one is preferred, which corresponds to smaller packet collision rates. We denote the preferred solution as $\Phi(\tau_{1...M}^*, p_{1...M}^*)$. Considering the stability, the system should operate close to this solution. It can be seen that $\Phi(\tau_{1...M}^*, p_{1...M}^*)$ can be completely determined without relying on the measurements of other system parameters, such as, packet collision rate p_i s.

Next, we make an analysis on packet delay $T_{d,i}$ $(i = 1,...,M)$, which is defined as the average duration between the beginning of a backoff procedure and the instant that the corresponding packet has been successfully sent. Let us consider a type i traffic flow. $\bar{n}_{VTS,i}$ is the average number of successive VTS following the successful sending of a packet. It can be given as

$$\bar{n}_{VTS,i} = \sum_{j=1}^{\infty} j \cdot q_i^{\ j} \cdot (1 - q_i) = \frac{q_i}{1 - q_i} = \frac{P_{VTSS,i}}{d_{0,0}(i)} = \frac{P_{VTSS,i}}{\tau_i (1 - p_i)} \tag{8}$$

Considering the case that $P_s \approx P_c$ and $\tau_i \ll 1$, the average duration $T_{VTS,i}$ of a VTS can be approximated as

$$T_{VTS,i} \approx \beta \cdot \sigma \cdot \prod_{j=1}^{M} (1 - \tau_j)^{n_j} + [1 - \beta \cdot \prod_{j=1}^{M} (1 - \tau_j)^{n_j}] \cdot P_s - \tau_i P_s \tag{9}$$

According to equation 5, we have

$$T_{VTS,i} \approx n_i \tau_i (1 - p_i) P_L / S_i - \tau_i P_s \tag{10}$$

Therefore, $T_{d,i}$ can be approximated as

$$T_{d,i} = T_{p,i} - \bar{n}_{VTS,i} \cdot T_{VTS,i} \approx \frac{n_i P_L}{S_i} (1 - P_{VTSS,i}) + \frac{P_s}{1 - p_i} \cdot P_{VTSS,i} \tag{11}$$

It should be noted that the above estimated $T_{d,i}$ can be approximated as the average service time for a packet in its transmission queue. Therefore, it can be directly applied to evaluate the average packet queuing delay (waiting time in transmission queues) by using G/G/1 queuing model [11], which is omitted here because of space limitation.

3 Approximation Analysis

Assume that the system operation point can be approximated as $\Phi(\tau_{1...,M}^*, p_{1,...,M}^*)$. Theoretical analysis shows that if the number of traffic flows n_i $(i = 1,...,M)$ and the packet payload P_L is not too small, it is reasonable to assume that $\tau_i^* \ll 1$ [10]. From equation 5, it can be obtained that

$$\tau_i^* \cdot (1 - p_i^*) \cdot T_{p,i} = \tau_j^* \cdot (1 - p_j^*) \cdot T_{p,j} \tag{12}$$

Under the assumption that $\tau_i^* \ll 1$, from equation 4, we have

$$p_i^* \approx p_j^* \tag{13}$$

Therefore, we can make the following approximation,

$$\tau_i^* \cdot T_{p,i} \approx \tau_j^* \cdot T_{p,j} \tag{14}$$

After substituting $\Phi(\tau_{1...,M}^*, p_{1,...,M}^*)$ into equation 3, we have

$$\frac{1-P_{VTSS,i}}{1-P_{VTSS,j}} \approx \frac{\tau_i^* \cdot W_i}{\tau_j^* \cdot W_j} \tag{15}$$

Obviously, when the minimum contention window sizes W_i s are large, both $P_{VTSS,i}$ and p_i are small. Therefore, according to equation 11, the packet delay can be approximated as

$$T_{d,i} \approx \frac{n_i P_L}{S_i} \cdot (1 - P_{VTSS,i}) \tag{16}$$

Based on equations 14, 15 and 16, it can be obtained that

$$\frac{T_{d,i}}{T_{d,j}} \approx \frac{W_i}{W_j} \tag{17}$$

Note that the above approximation only holds in the case that $P_{VTSS,i}$ s are very small. Equation 17 are exactly the same as the approximated results given in [9], which shows that saturation states can be regarded as an extreme case for non-saturation state with $P_{VTSS,i}=0$.

In the following part of this section, we try to find out how to properly set the minimum contention window sizes W_i s so as to achieve the target packet delay requirements, that is, $T_{d,i} < \hat{T}_{d,i}$. Combing equations 3 and 11, we have

$$T_{d,i} \approx \frac{n_i P_L}{S_i} + \left(\frac{P_s}{1-p_i} - \frac{n_i P_L}{S_i} \right) \left(1 - \frac{(1-2p_i)+p_i[1-(2p_i)^{m_i}]}{2(1-2p_i)} W_i \tau_i \right) \tag{18}$$

As we have already mentioned before, for stability, the system should operate near $\Phi(\tau_{1,...,M}^*, p_{1,...,M}^*)$. Therefore, if it is required that $T_{d,i} < \hat{T}_{d,i}$, based on above equation

$$W_i < \left(\hat{T}_{d,i} - \frac{P_s}{1-p_i^*} \right) \Big/ \gamma^* \tag{19}$$

where $\gamma^* = \left(\frac{n_i P_L}{S_i} - \frac{P_s}{1-p_i^*} \right) \cdot \frac{(1-2p_i^*)+p_i^* \cdot [1-(2p_i^*)^{m_i}]}{2(1-2p_i^*)} \tau_i^*$. Equation 19 tells one the upper bounds for W_i s so as to meet the packet delay requirements $\hat{T}_{d,i}$ s.

4 Results and Discussions

In this section, both numerical and simulation results are shown to validate our proposed analysis model. In our experiments, the parameters for the system, which are based on IEEE 802.11b, are summarized as follows: MAC Header = 272 bits; PHY Header = 192 μs; ACK = 112bits + PHY Header; Channel Bit Rate = 11Mbps; Propagation Delay = 1 μs; Slot Time = 20 μs; SIFS = 10 μs; and DIFS = 50 μs. In our discrete-event simulation, a single-hop wireless LAN is considered. In the system, there are n_1 and n_2 type-1 and type-2 sending stations, respectively. Each of them carries only one traffic flow. It is assumed that the channel conditions are ideal (i.e., no hidden terminals and capture).

In the first experiments, two types of traffic flows are considered. Type-1 traffic has priority over type-2 traffic. Therefore, a smaller minimum contention window size W_1 is assigned to type-1 traffic, and a larger minimum contention window size $W_2 = 5W_1$ is allocated to type-2 traffic. Equations 3 and 5 are fundamental in this paper, they are validated in Fig. 2 and Fig. 3, respectively. In Fig. 2, the virtual time slot rates $P_{VTSS,i}$ s versus W_1 are shown. $P_{VTSS,i}$ s are obtained in two ways: one is by simulations. In the second way, packet collision rates p_i s and packet sending rates τ_i s, which are obtained by using simulations, are substituted into equation (3) to calculate the corresponding $P_{VTSS,i}$ s. System parameters are shown in the figure. It can be seen that the $P_{VTSS,i}$ s obtained by using equation 3 are very close to the simulated values, which validates the Markov model shown in Fig. 1. It can also be seen that when the minimum contention window sizes W_i s are very small, the differences between the simulated values and the estimated ones are larger. This is because in this case the packet collision rates increase dramatically and the behavior of the system is unstable.

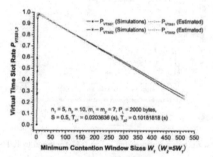

Fig. 2. Virtual time slot rates $P_{VTSS,i}$ versus W_1

In Fig. 3, the throughput S_i s are obtained by using two ways: one is by simulation. In the second way the throughputs are obtained by substituting p_i s and τ_i s, which are obtained from simulations, into equation 5. Again, it can be seen that when W_i s are very small, equation 5 can not describe the behavior of the system well, which is caused by the instability of the system behavior in this case.

In Fig. 4, packet delays $T_{d,i}$ s versus W_i s are shown. Two ways are used to obtain $T_{d,i}$ s. One is through simulations. The other way is that we estimate $T_{d,i}$ s based on equation 11. In order to use equation 11, the solution $\Phi(\tau_1^*, \tau_2^*, p_1^*, p_2^*)$ is calculated by

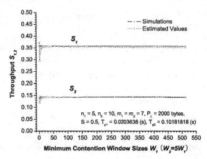

Fig. 3. Throughput S_i versus W_1

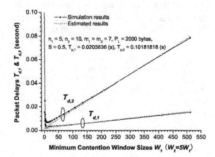

Fig. 4. Packet delays $T_{d,i}$ versus W_1

using equation 5 with the assumption that a_i and β are equal to 1. Then the obtained solution is substituted into equation 11. From the figure, it can be seen that when W_i s are not very small, packet delays can be successfully estimated. In this case, packet delays decrease linearly with the decrease of W_i s (we say that the system operates in "Stable State"). In this case, with the decrease of W_i s, time wasted in backoff processes can be directly converted into VTS without causing significant increase in packet collision rates and packet sending rates. When W_i s are very small, the behavior of the system is unstable (packet collision rates and packet sending rates increase drastically with the slight decrease of W_i s). Therefore, packet delays tend to increase drastically. In this case, the estimations of packet delays are not accurate. However, equation 11 is useful, because one does not want the system to operate far from the "Stable State". However, an interesting future research topic is to guarantee that the system operates under the "Stable State".

Table 1. Guarantee Packet Delay Requirements

$\hat{T}_{d,1}$	$T_{d,2}$	W_1	W_2	$T_{d,1}$	$T_{d,2}$
0.005	0.030	109	924	0.005052	0.030827
0.005	0.025	109	759	0.005085	0.025859
0.005	0.020	109	594	0.005126	0.020966
0.005	0.015	109	429	0.005204	0.016293
0.005	0.010	109	264	0.005313	0.011022
0.005	0.005	109	99	0.005504	0.005506

System parameters: $P_{Len,1} = P_{Len,2} = 2000$ bytes, $n_1 = 5$, $n_2 = 10$, $m_1=m_2=7$, $T_{P,1} = 0.020363636$ s, $T_{P,2} = 0.10181818$ s

In Table 1, we demonstrate a possible application for our analysis model. In equation 19, we propose a way to estimate the upper bounds for the minimum contention window sizes W_i s to meet the required packet delays $T_{d,i} < \hat{T}_{d,i}$. In this example, we first estimate the upper bounds for W_i s based on equation 13. Then, to better understand the performance of the estimated upper bounds, actual packet delays $T_{d,i}$ s are obtained from simulations with the corresponding W_i s being set to be equal to the corresponding estimated upper bounds. Finally, comparisons can be easily made by comparing the obtained packet delays $T_{d,i}$ s and the required packet delays $\hat{T}_{d,i}$ s. In Table 1, the first two columns are the packet delay requirements. The third and fourth columns are estimated minimum contention window sizes by using equation 19. The last two columns are the achieved packet delays obtained from simulations. It can be seen that the packet delay requirements can be approximately met, which suggests a promising application for our proposed model.

5 Conclusions

In this paper, a simple model has been proposed to analyze the performance of IEEE 802.11 DCF with service differentiation support in non-saturation states, which helps one to obtain deeper insight into the IEEE 802.11 DCF. Under the case that the system operates in stable states, we can approximately evaluate the most important system performance measures, such as packet delays, which provide one with an important tool to predict the system performance. Moreover, in order to meet certain packet delay requirements, a practical method has been given based on our theoretical results. Comparisons with simulation results show that this method does achieve the specified packet delay requirements with good accuracy. Possible extensions of this work to consider practical schemes capable to rapidly adapt to changing traffic loads are now being considered.

References

1. Wireless LAN Medium Access Control (MAC) and Physical Layer (PHY) specifications, IEEE Standard 802.11, Aug. 1999.
2. T. S. Ho and K. C. Chen, "Performance evaluation and enhancement of the CSMA/CA MAC protocol for 802.11 wireless LAN's," Proceedings of IEEE PIMRC, Taipei, Taiwan, Oct. 1996, pp.392-396.
3. F. Cali, M. Conti, and E. Gregori, "IEEE 802.11 wireless LAN: Capacity analysis and protocol enhancement," Proceedings of INFOCOM'98, San Francisco, CA, March 1998, vol. 1, pp. 142 -149.
4. G. Bianchi, "Performance analysis of the IEEE 802.11 distributed coordination function," IEEE Journal on Selected Areas In Communications, vol. 18, no. 3, March 2000.
5. E. Ziouva and T. Antonakopoulos, "CSMA/CA performance under high traffic conditions: Throughput and delay analysis," Computer Communications, vol. 25, pp. 313–321, 2002.
6. Foh, C.H., Zukerman, M., "Performance Analysis of the IEEE 802.11 MAC Protocol," Proc. European Wireless, 2002.
7. E.M.M. Winands, T.J.J. Denteneer, J.A.C. Resing and R. Rietman, "A finite-source feedback queueing network as a model for the IEEE 802.11 Distributed Coordination Function," Proc. European Wireless, 2004.
8. O. Tickoo and B. Sikdar, "Queueing Analysis and Delay Mitigation in IEEE 802.11 Random Access MAC based Wireless Networks," in Proceedings of IEEE INFOCOM'04, Hong Kong, China, March 2004.
9. LI Bo, Roberto Battiti, "Performance Analysis of An Enhanced IEEE 802.11 Distributed Coordination Function Supporting Service Differentiation," QoFIS 2003, Sweden, Springer LNCS volume 2811, pp. 152-161.
10. LI Bo and Roberto Battiti, "Achieving Maximum Throughput and Service Differentiation by Enhancing the IEEE 802.11 MAC Protocol," WONS 2004 (Wireless On-demand Network Systems, Trento, Italy), Springer Lecture Notes on Computer Science LNCS volume 2928, pp. 285-301.
11. W. G. Marchal, "An approximate formula for waiting time in single server queues," AIIE Transactions, Dec. 1976, 473-474.

An Interaction Model and Routing Scheme for QoS Support in Ad Hoc Networks Connected to Fixed Networks

Mari Carmen Domingo and David Remondo

Telematics Eng. Dep., Catalonia Univ. of Technology (UPC)
Av del Canal Olímpic s/n. 08860 Castelldefels (Barcelona), SPAIN
+34 93 413 70 51
{cdomingo, remondo}@mat.upc.es

Abstract. We propose a new protocol, named DS-SWAN (Differentiated Services-Stateless Wireless Ad Hoc Networks), to support end-to-end QoS in ad hoc networks connected to one fixed DiffServ domain. DS-SWAN warns nodes in the ad hoc network when congestion is excessive for the correct functioning of real-time applications. These nodes react by slowing down best-effort traffic. Furthermore, we present a routing protocol for the ad hoc network, named SD-AODV (Service Differentiation- Ad Hoc On-demand Distance Vector), where new route requests are suppressed to maintain the desired QoS requirements for real-time flows. Simulation results indicate that DS-SWAN and SD-AODV significantly improve end-to-end delays for real-time flows without starvation of background traffic.

1 Introduction

There has been little research on the support of QoS when a wireless ad hoc network is attached to a fixed IP network. In this context, co-operation between the ad hoc network and the fixed network can facilitate the end-to-end QoS support [1]. In the present work, we propose a new protocol, named DS-SWAN, that is based on the co-operation between a QoS model named SWAN within the ad hoc network and a Differentiated Services (DiffServ) [2] domain in the fixed network.

The authors in [3] study the behavior of voice traffic in an isolated ad hoc network that uses SWAN. However, there has been no prior work on analyzing the transmission of real-time traffic that shares resources with background traffic between a mobile ad hoc network and a fixed IP network.

The routing protocols play an important role in support of delivering QoS because the network performance relies on the speed at which routing protocols can recompute new routes between source-destination pairs after topology changes. Therefore in the present work we present a new routing protocol for the ad hoc network, named SD-AODV, that interoperates with the QoS scheme (DS-SWAN) to maintain the desired QoS between the wired and the wireless network.

J. Solé-Pareta et al. (Eds.): QofIS 2004, LNCS 3266, pp. 74–83, 2004.
© Springer-Verlag Berlin Heidelberg 2004

The paper is structured as follows: Section 2 describes the SWAN model. Section 3 presents the new protocol, named DS-SWAN (DiffServ-SWAN). Section 4 introduces the new routing protocol SD-AODV. Section 5 explores the dynamics of the system. Finally, Section 6 concludes the paper.

2 SWAN

SWAN is a stateless scheme designed to provide end-to-end service differentiation in ad hoc networks employing a best-effort distributed wireless MAC [3]. It distinguishes between two traffic classes: real time and best effort.

When best-effort packets arrive at a node, they enter a leaky-bucket traffic shaper that has a previously calculated rate, derived from an AIMD (Additive Increase Multiplicative Decrease) rate control algorithm. Every node measures the MAC delays continuously and this information is used as feedback to the rate controller. Every T seconds, each device increases its transmission rate gradually (additive increase with increment rate of c bit/s) until the packet delays at the MAC layer become excessive. As soon as the rate controller detects excessive delays, it reduces the rate of the shaper with a decrement rate (multiplicative decrease of r %).

Rate control restricts the bandwidth of best-effort traffic so that real-time applications can use the required bandwidth. On the other hand, the bandwidth not used by real-time applications can be efficiently used by best-effort traffic.

For the real-time traffic, SWAN uses sender-based admission control. This mechanism works by sending an end-to-end request/response probe along the existing route to estimate the bandwidth availability at each node and then determines whether a new real-time session should be admitted or not.

3 DS-SWAN

We consider a scenario where background traffic and real-time VBR (Variable Bit Rate) traffic are transmitted as the mobile nodes in the ad hoc network communicate with one of the fixed hosts located in the fixed network (see Fig. 2). For the sake of simplicity, we include a single DiffServ domain and we assume that the traffic goes from the ad hoc network to corresponding nodes in the fixed network.

For the real-time traffic, the DiffServ service class is the EF (Expedited Forwarding) PHB (Per-Hop Behavior), which provides low loss, low latency, low jitter and end-to-end assured bandwidth service. The EF aggregates are policed with a token bucket at the ingress edge router. The traffic that exceeds the profile is dropped.

The number of dropped packets at the ingress edge router and the end-to-end delay of the real-time connections are associated with the QoS parameters of the SWAN model in the ad hoc network. We observe that if the rate of the best-effort leaky bucket traffic shaper is lower then best-effort traffic is more efficiently rate controlled and real-time traffic is not so much influenced by best-effort traffic and it is able to maintain the required QoS parameters. We propose a new protocol that enables the co-operation between the DiffServ architecture at the fixed network and the explained SWAN scheme in the ad hoc network to improve end-to-end QoS support. In the proposed protocol, DS-SWAN, the ingress edge router periodically monitors the number of EF packets that are dropped by its token bucket meter. On the other hand, the corresponding nodes in the fixed IP network periodically monitor the average end-to-end delays of the real-time flows.

We have selected a specific type of real-time application that implies burstiness and that contains end-to-end delay information: VBR Voice-over-IP (VoIP) [4]. The ITU-T recommends in its standard G.114 that the end-to-end delay should be kept below 150 ms to maintain an acceptable conversation quality [5]. Also, for Pulse Code Modulation encoding with the G. 711 codec, the packet loss rate should never be larger than 5% [6].

In DS-SWAN, a destination node sends a QoS_LOST warning message to the ingress edge router when the end-to-end delay of one VoIP flow becomes greater than 140 ms.

We have observed from initial simulation runs that the number of dropped VoIP packets in the ad hoc network is usually well below 1% when SWAN is used. Therefore, we establish that if the number of dropped VoIP packets at the ingress edge router is less than 4% and this router has received a QoS_LOST message, then it sends the QoS_LOST message to the ad-hoc network to inform that the system is too congested to maintain the desired QoS (due to excessive delays at the ad hoc network). When the number of lost packets at the edge router is higher than 4% the packet loss rate will be larger than 5% so that the VoIP quality will be degraded and it does not have any sense to send QoS_LOST messages to act over the end-to-end delays because the packet loss rate will not be diminished.

When a node in the ad hoc network receives the QoS_LOST message, it will react by modifying the parameter values in the AIMD rate control algorithm mentioned above. In DS-SWAN, every time that a QoS_LOST message is received, the node decreases the value of c by Δc- bit/s with a certain minimum value. When no QoS_LOST message is received during T seconds, the node increases the value of c by Δc+ bits/s unless the initial value has been reached.

When a wireless node receives a QoS_LOST message, it also increases the value of r by Δr+ up to a maximum value. When no QoS _LOST message has been received in the period T, the value of r is decreased by Δr- until the initial value is reached.

SWAN has a minimum rate m for the best-effort leaky bucket traffic shaper. In DS-SWAN nodes are also allowed to reduce m. When a node receives a QoS_LOST message, it reduces the m by Δm- bit/s. However, this parameter value is kept above a minimum value of m_0 bit/s and is increased Δm+ bits/s every T seconds up to the initial value when the mobile nodes do not receive a warning message in T seconds. Thus, we change the parameter values of SWAN dynamically according to the traffic conditions not only in the ad hoc network but also in the fixed network.

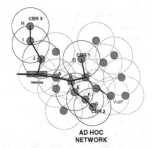

Fig. 1. Example network

We have designed two different versions of DS-SWAN:

- When all the CBR sources and the intermediate nodes along the routes are advised ("DS-SWAN – CBR sources") so that they throttle their best-effort traffic.

- When the edge router sends a QoS_LOST message only to the VoIP sources generating flows that have problems to keep their end-to-end delays under 150 ms and to the intermediate nodes along the routes. Then these nodes forward the QoS_LOST message as a broadcast packet to all their neighbours because they may be contending with them for medium access. Only when a node receives a QoS_LOST message as a broadcast packet, it throttles its best-effort traffic. ("DS-SWAN – VoIP sources + neighbours").

The different functioning of both versions of DS-SWAN is illustrated in Fig. 1. It shows an example of an ad hoc network where one VoIP real-time flow and three CBR best-effort flows have been established so that packets are sent toward Internet through the gateway. First we apply the version ("DS-SWAN – CBR sources"). If we consider that the VoIP flow has problems to keep its end-to-end delay under 150 ms QoS_LOST messages will be sent to the CBR sources and the intermediate nodes along the routes in the ad hoc network so that nodes A, B, C, D, E, F, G, H, I, J and the gateway will throttle their best-effort traffic.

Although nodes H and I do not compete for medium access with the nodes along the route towards the VoIP problematic flow they will slow down their CBR traffic either because these nodes have been warned in this DS-SWAN version. On the other hand, if the second version ("DS-SWAN – VoIP sources and neighbours") is applied, then the CBR sources and intermediate nodes that are not neighbours of the VoIP problematic source and its intermediate nodes along the route are not warned so that in the example nodes H and I will not throttle their best-effort traffic. It is important to notice, however, that with this DS-SWAN version some nodes like node E in the example will receive the QoS_LOST broadcast packet more than once because they are neighbours from many nodes (node E has nodes B and C as neighbours) so that they will act over the Leaky Bucket parameters to rate control best-effort traffic several times.

4   SD-AODV (Service Differentiation-AODV)

We have used the Internet draft "Global Connectivity for IPv6 Mobile Ad-Hoc Networks" [7] to provide Internet access to mobile ad-hoc networks modifying the Ad-Hoc On-demand Distance Vector (AODV) routing protocol [8]. AODV is a best-effort routing protocol that does not provide QoS to real-time traffic.

AODV uses the shortest number of wireless hops towards a destination as the primary metric for selecting a route. with independence of the traffic congestion. However, real-world experiments have shown that the shortest hop count metric often chooses routes poorly, thus leading to low throughput, unreliable paths. One reason for the shortest hop count metric performing below par is that the metric does not account for traffic load during the route selection process.

It is perfectly possible that AODV selects as shortest route between source and destination a route with congested nodes or with nodes with real-time flows having problems to keep their end-to-end delays. Consequently, there is a clear need that this protocol is modified so that it would be able to interwork with the QoS interaction model to maintain the average end-to-end delays of real-time flows.

We have considered that the QoS interaction model "DS-SWAN – VoIP sources + neighbours" is applied to our system. In this version a node receives a QoS_LOST message because it is either a problematic VoIP source or node along the route towards the source that has problems to keep its end-to-end delay under 150ms or because it is a neighbour from these nodes and it is contending with them for medium access producing congestion. Under such conditions nodes consume more energy and their packet loss is increased together with their end-to-end delays as we have observed. With the aid of DS-SWAN it is possible to mitigate the bad effects of congestion in order to maintain the desired quality for VoIP flows delaying the access of best-effort traffic to the MAC layer and consequently to the medium. Furthermore, new actions should be taken in conjunction with the explained DS-SWAN scheme not only to reduce the existing congestion in some problematic regions but also to avoid that new traffic load can increase it.

We present a simple and scalable routing protocol named SD-AODV that takes advantage of the co-operation between SWAN and DiffServ to avoid that the degree of congestion in the network grows.

The simple goal of SD-AODV is to redirect new "routes" away from nodes that have received a QoS_LOST broadcast message in the case that these nodes have congestion problems (with exception of the gateways). We consider that a node has congestion problems if its average MAC delays during the RTS-CTS-DATA-ACK cycle exceed a predefined value D_{MAX}. The MAC delay [9] can be estimated by the total defer time accumulated plus the time to acknowledge the packet if no collision occurs For example, if RTS/CTS is enabled:

$$d = t_{defer} + t_{RTS} + t_{CTS} + t_{packet} + t_{ACK} + 3t_{SIFS} + 3\tau, \tag{1}$$

where τ estimates the maximum propagation delay. Each node independently monitors and computes its average MAC delays and can declare itself as a congested node if these delays are excessive. SD-AODV acts in a fully distributed manner suppressing new route requests to these congested nodes to ensure that new routed traffic does not increase the congestion in the bottlenecks in order to continue maintaining the desired QoS parameters for real-time traffic. The naïve suppression of route creation may prevent the use of the only possible path between two hosts but we argue that further one can not offer a priori a new VoIP or CBR flow a route towards the destination at the expense of on going real-time flows.

The congested zones are motivated due to the excessive contention of the shared medium and are transient in nature because flows are rerouted due to changes in the network topology and changes in the traffic load. When there is a link failure (for example due to node mobility) and a route towards a destination is broken, the routing protocol has to find a new route towards the destination (or a stored route in a routing table can be used). Nevertheless, if some or all nodes along the old route were marked as congested nodes now these nodes are unmarked because the traffic and topology conditions have changed and it is not possible to know a priori if these nodes will still continue experiencing congestion or not. The Differentiated Services architecture will continue co-operating with the SWAN model so that DS-SWAN will receive updated information dynamically about the QoS parameter values in the ad hoc network and the unmarked nodes could be marked as congested nodes again in the future if it is necessary.

With SD-AODV it is possible to reduce the energy consumption from the congested nodes and thus these nodes will not exhaust their energy resources prematurely and the problem that the network suffers a premature partition can be reduced.

We observe that the node selection of the congested nodes depends on the interaction between the Differentiated Services architecture and the SWAN model and a co-operation between the DS-SWAN model and SD-AODV is absolute necessary so that SD-AODV may be able to avoid that the traffic is concentrated at certain nodes. SD-AODV interworks with the existing QoS model to maintain the desired QoS parameters for real-time flows. SD-AODV is a modification of the AODV routing protocol, but it would have been possible to modify any routing protocol working in the ad hoc network. SD-AODV can not be considered a QoS routing protocol because it does not try to provide QoS support or QoS routes; however it contributes with the aid of DS-SWAN to maintain the QoS in the ad hoc network

5   Simulations

We have run simulations with the NS-2 tool [10] to investigate the performance of DS-SWAN with a relatively realistic physical layer model.

The system framework is shown in Fig. 2. We consider a single DiffServ domain (DS-domain) covering the whole network between the corresponding hosts and the two wireless gateways. The chosen scenario consists of 20 mobile nodes, 2 gateways, 3 fixed routers and 3 corresponding hosts. In this work we consider a hybrid gateway discovery method [11] to find a gateway. The mobile nodes are uniformly distributed in a rectangular region of 700 m by 500 m. Each mobile node selects a random destination within the area and moves toward it at a velocity uniformly distributed between 0 and

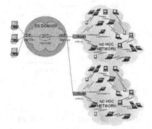

Fig. 2. Simulation framework.

3 m/s. Upon reaching the destination, the node pauses for 20 s, selects another destination and repeats the process. The dynamic routing algorithm is AODV [8] and the wireless links are IEEE 802.11b.

Background traffic is generated by 13 of the mobile hosts, while VBR VoIP traffic is generated by 15 of the mobile hosts. The destinations of each of the background and VoIP flows are chosen randomly among the three hosts in the wired network.

For the voice calls, we use the ITU G711 a-Law codec [4]. The VoIP traffic is modelled as a source with exponentially distributed on and off periods with 1.004 s and 1.587 s average each. Packets have a constant and are generated at a constant inter-arrival time during the on period. The VoIP connections are activated at a starting time chosen from a uniform distribution in [10 s, 15 s]. Background traffic is Constant Bit Rate (CBR) with a rate of 48 Kbit/s and a packet size of 120 bytes. To avoid synchronization, the CBR flows have starting times chosen randomly from the interval [15 s, 20 s] for the first source, [20 s, 25 s] for the second source and so on.

Shaping of EF and BE traffic at the edge router is done in two different drop tail queues. The EF and BE aggregates are policed with a token bucket meter with CBS = 1000 bytes and CIR = 200 Kbit/s.

We have run 40 simulations to assess the end-to-end delay and packet loss of VoIP traffic and the throughput of CBR traffic. We have evaluated and compared the performance of SWAN (Case 1) with the two different implementations of DS-SWAN discussed in Section 3: "DS-SWAN – CBR sources" (Case 2) and "DS SWAN – VoIP sources + neighbours" (Case 3).

Fig. 3 shows the average end-to-end delay for VoIP traffic in the three cases. Using SWAN the end-to-end delays increase progressively because the system is congested with VoIP flows and background traffic. From the second 115 until the end of the simulation the end-to-end delays are too high for an acceptable conversation quality [5]. In DS-SWAN the end-to-end delays of the VoIP flows are reduced because some nodes in the ad-hoc network are advised and react throttling their best-effort traffic. For this reason it is possible that the VoIP flows are able to maintain their QoS parameters achieving an acceptable conversation quality. In Case 3 the average end-to-end delays are lower than in Case 2 because some nodes receive more than once the

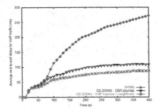

Fig. 3. Average end-to-end delay for VoIP traffic: SWAN (Case 1) vs. DS-SWAN (Cases 2 and 3).

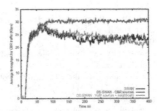

Fig. 4. Average throughput for background traffic: SWAN (Case 1) vs. DS-SWAN (Cases 2 and 3).

same QoS_LOST broadcast message. On the other hand, in Case 2 some nodes with CBR packets slow down this kind of traffic unnecessarily because they may be not contending with VoIP flows having problems to keep their delays although they receive the QoS_LOST message. Case 3 shows the best results because this DS-SWAN version acts only when and where it is needed.

Fig. 4 shows the average throughput for background traffic. In DS-SWAN, the average throughput for this kind of traffic is lower than in SWAN because some nodes in the ad hoc network react by decreasing the rate of the best-effort traffic shaper when they receive a warning.

In Case 2, the average throughput is sometimes smaller than in Case 3 because all nodes with best-effort traffic rate control their flows. However, in other time intervals the average throughput is higher in Case 2 in comparison with Case 3 because some nodes receive the same QoS_LOST message as broadcast packet more than once. Therefore the reduction of the average throughput for CBR traffic of the DS-SWAN model in comparison with the SWAN model will depend on the number of CBR sources in Case 2 and on the number of nodes that receive a QoS_LOST broadcast message more than once and on how many times they receive it. In any case, there is not starvation of background traffic.

The packet loss rate for VoIP was well below the required 5% in all simulations.

Now we have run 40 simulations to assess the end-to-end delay of VoIP traffic and the throughput of CBR traffic. We have evaluated and compared the performance of a

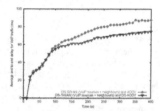

Fig. 5. Average end-to-end delay for VoIP traffic: DS-SWAN (VoIP sources + neighbours) and AODV (Case 1) vs. DS-SWAN (VoIP sources + neighbours) and SD-AODV (Case 2).

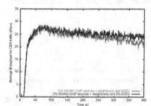

Fig. 6. Average throughput for background traffic: DS-SWAN (VoIP sources + neighbours) and AODV (Case 1) vs. DS-SWAN (VoIP sources + neighbours) and SD-AODV (Case 2).

system using ("DS-SWAN – VoIP sources + neighbours") as QoS scheme and AODV as routing protocol (Case1) with a system using the same QoS scheme but SD-AODV as routing protocol (Case 2). The parameter value D_{MAX} has a great impact on the number of congested nodes that is detected and in our simulations it is set to 20 ms. However, each node calculates the average MAC delays during the simulation.

Fig. 5 represents the average end-to-end delays for VoIP. We can appreciate that there is a significant improvement in the average end-to-end delays for VoIP traffic when SD-AODV is used as routing protocol (Case 2) in comparison with Case 1. The reason for this behaviour is that the VoIP traffic sources having problems to keep their end-to-end delays and nodes along the routes as well as their neighbours that compete for medium access are not overloaded with more CBR or VoIP traffic flows once it is checked that these nodes are congested. and the new route requests towards these nodes are suppressed. Besides, the probability that new VoIP flows experience more congestion conditions is reduced because the new routes for these flows avoid selecting previously declared congested nodes.

Fig. 6 represents the average throughput for CBR best-effort traffic. We can appreciate that in both cases there is not starvation of best-effort traffic. However, using SD-AODV as routing protocol the new CBR traffic sources select routes avoiding congested nodes and as a result the average throughput is increased with respect to Case 1 because it is less probable that these sources would have to throttle their flows.

6 Conclusions

The combination of DS-SWAN and SD-AODV reduces the average end-to-end delays of VoIP flows and improves the average throughput for best-effort traffic. For this reason, it is recommended to combine a QoS interaction model just as DS-SWAN with such a routing scheme in order to improve the network performance. As future work we want to show how the DS-SWAN QoS model and the SD-AODV routing protocol perform in a range of configurations under a variety of representative traffic loads and in the case of a change of traffic mix.

Acknowledgements. This work was partially supported by the "Ministerio de Ciencia y Tecnología" of Spain under the project TIC2003-08129-C02 and under Ramón y Cajal programme.

References

[1] Y.L. Morgan and T. Kunz, "PYLON: An Architectural Framework for Ad-hoc QoS Interconnectivity with Access Domains," in *Proc. 36th Annual Hawaii Int. Conf. on System Sciences*, Jan. 2003, pp. 309–318.

[2] S. Blake, D. Black, M. Carlson, E. Davies, Z. Wang and W. Weiss, "An architecture for differentiated service," *Request for Comments (Informational) 2475*, Internet Engineering Task Force, Dec. 1998.

[3] G.-S. Ahn, A.T. Campbell, A. Veres and L.-H. Sun, "SWAN," draft-ahn-swan-manet-00.txt, *Work in Progress*, Feb. 2003.

[4] D. Chen, S. Garg, M. Kappes and K.S. Trivedi, "Supporting VBR Traffic in IEEE 802.11 WLAN in PCF Mode," in *Proc. OPNETWORK'02*, Washington D.C., Aug. 2002.

[5] ITU-T Recommendation G.114, "One way transmission time," May 2000.

[6] P.B. Velloso, M.G. Rubinstein and M.B. Duarte, "Analyzing Voice Transmission Capacity on Ad Hoc Networks," in *Proc. Int. Conf. on Communications Technology - ICCT*, Beijing, China, Apr. 2003.

[7] R. Wakikawa, J. T. Malinen, C. E. Perkins, A. Nilsson, and A. J. Tuominen, "Global connectivity for IPv6 mobile ad-hoc networks", Internet Engineering Task Force, Internet Draft (Work in Progress), July 2002.

[8] C.E. Perkins, E.M. Royer, "Ad-hoc On-demand Distance Vector routing," in *Proc. of the 2nd IEEE Workshop on Mobile Computing Systems and Applications*, New Orleans, U.S.A., Feb. 1999.

[9] A. Veres, A.T. Campbell, M. Barry, and L.-H. Sun, "Supporting Service Differentiation in Wireless Packet Networks Using Distributed Control," *IEEE J. Selected Areas in Comm., special issue on mobility and resource management in next-generation wireless systems*, vol. 19, no. 10, pp. 2094-2104, Oct. 2001.

[10] NS-2: Network Simulator, *http://www.isi.edu/nsnam/ns*.

[11] J. Xi and C. Bettstetter, "Wireless Multi-Hop Internet Access: Gateway Discovery, Routing, and Addressing", In Proc. International Conference on Third Generation Wireless and Beyond (3Gwireless'02), San Francisco, CA, USA, May 28-21, 2002.

Load Analysis of Topology-Unaware TDMA MAC Policies for Ad Hoc Networks

Konstantinos Oikonomou[1] and Ioannis Stavrakakis[2]

[1] INTRACOM S.A., Emerging Technologies & Markets Department,
19.5 Km Markopoulou Avenue, Paiania 190 02, Athens, Greece
Tel: +30 210 6677023, Fax: +30 210 6671312
okon@intracom.gr
[2] University of Athens, Department of Informatics & Telecommunications
Panepistimiopolis, Ilissia 15 784, Athens, Greece
Tel: +30 210 7275343, Fax: +30 210 7275333
ioannis@di.uoa.gr

Abstract. Medium Access Control (MAC) policies in which the
scheduling time slots are allocated irrespectively of the underline topology are suitable for ad-hoc networks, where nodes can enter, leave or
move inside the network at any time. *Topology-unaware* MAC policies,
that allocate slots deterministically or probabilistically have been proposed in the past and evaluated under heavy traffic assumptions. In this
paper, the heavy traffic assumption is relaxed and the *system throughput*
achieved by these policies is derived as a function of the traffic load. The
presented analysis establishes the conditions and determines the values of
the access probability for which the system throughput under the probabilistic policy is not only higher than that under the deterministic policy
but it is also close to the *maximum* achievable, provided that the *traffic load* and the *topology density* of the network are known. sub-optimal
solutions are also provided. Simulation results for a variety of topologies
with different characteristics support the claims and the expectations of
the analysis and show the comparative advantage of the Probabilistic
Policy over the Deterministic Policy.

1 Introduction

The design of an efficient Medium Access Control (MAC) protocol in ad-hoc
networks is challenging. The idiosyncrasies of these networks, where no infrastructure is present and nodes are free to enter, leave or move inside the network
without prior configuration, allow for certain choices on the MAC design depending on the particular environment and therefore, it is not surprising that
several MAC protocols have been proposed so far. Various MAC protocols are
widely employed in ad-hoc networks, such as [1], [2], [3], [4], [5], [7]. In general,
optimal solutions to the problem of time slot assignment often result in NP-hard
problems, [6], which are similar to the n-coloring problem in graph theory.

Topology-unaware: TDMA scheduling schemes determine the scheduling time
slots irrespectively of the underlying topology and in particular, irrespectively

J. Solé-Pareta et al. (Eds.): QofIS 2004, LNCS 3266, pp. 84–93, 2004.
© Springer-Verlag Berlin Heidelberg 2004

of the scheduling time slots assigned to neighbor nodes. The topology-unaware scheme presented in [8], exploits the mathematical properties of polynomials with coefficients from finite Galois fields to randomly assign scheduling time slot sets to each node of the network. For each node it is guaranteed that at least one time slot in a frame would be collision-free, [8]. Another scheme, proposed in [9], maximizes the minimum guaranteed throughput. However, both schemes employ a deterministic policy (to be referred to as the *Deterministic Policy*) for the utilization of the assigned time slots that fails to utilize non-assigned time slots that could result in successful transmissions, [10]. Therefore, a new policy was proposed that probabilistically utilizes the non-assigned time slots according to a common *access probability p* (to be referred to as the *Probabilistic Policy*).

All the aforementioned works ([8], [9], [10]) have focused on heavy traffic conditions and no work has been conducted for the non-heavy traffic case. In this paper, the probability λ, $0 \leq \lambda \leq 1$, that a node has data for transmission during one time slot (probability λ is also referred to as the *traffic load*) is considered.

In Section 2 both transmission policies are described. In Section 3, analytical expressions for the system throughput as a function of the traffic load are derived for both policies. In Section 4, the conditions for the existence of an *efficient range* of values for the access probability p (values of the access probability p under which the Probabilistic Policy outperforms the Deterministic Policy) are established through an approximate analysis. Furthermore, this analysis determines the value of the access probability that *maximizes* the system throughput provided that the traffic load and the topology density are known. In Section 5 the case when the traffic load and/or the topology density are not known is considered and expressions are derived such that the system throughput under the Probabilistic Policy, even though not maximized, is higher than that under the Deterministic Policy. Simulation results, presented in Section 6 for a variety of topologies with different characteristics, support the claims and expectations addressed by the results of the analysis in the previous sections. Section 7 presents the conclusions.

2   Transmission Policies

An ad-hoc network may be viewed as a time varying multihop network and may be described in terms of a graph $G(V, E)$, where V denotes the set of nodes and E the set of links between the nodes at a given time instance. Let $|X|$ denote the number of elements in set X and let $N = |V|$ denote the number of nodes in the network. Let S_u denote the set of neighbors of node u, $u \in V$. These are the nodes v to which a direct transmission from node u (*transmission* $u \to v$) is possible. Let D denote the maximum number of neighbors for a node; clearly $|S_u| \leq D$, $\forall u \in V$. Time is divided into time slots with fixed duration and collisions with other transmissions is considered to be the only reason for a transmission not to be successful (*corrupted*). It has been shown, [10], that transmission $u \to v$ is corrupted in time slot i if at least one transmission $\chi \to \psi$, $\chi \in S_v \cup \{v\} - \{u\}$ and $\psi \in S_\chi$, takes place in time slot i.

According to the transmission policy proposed in [8] and [9], each node $u \in V$ is randomly assigned a unique polynomial f_u of degree k with coefficients from a finite Galois field of order q $(GF(q))$. Polynomial f_u is represented as

$$f_u(x) = \sum_{i=0}^{k} a_i x^i (\text{mod } q), [9], \text{ where } a_i \in \{0, 1, 2, ..., q-1\}; \text{ parameters } q \text{ and } k$$

are calculated based on N and D, according to the algorithm presented either in [8] or [9].

The access scheme considered is a TDMA scheme with a frame consisted of q^2 time slots. If the frame is divided into q subframes s of size q, then the time slot assigned to node u in subframe s, $(s = 0, 1, ..., q-1)$ is given by $f_u(s) \text{mod } q$, [9]. Consequently, one time slot is assigned for each node in each subframe. Let Ω_u be the the set of time slots assigned to node u. Given that the number of subframes is q, $|\Omega_u| = q$.

The deterministic transmission policy, proposed in [8] and [9], is the following.

The Deterministic Policy: Each node u transmits in a slot i only if $i \in \Omega_u$, provided that it has data to transmit.

Let $O_{i,u \to v}$ be that set of nodes χ whose transmissions corrupt a particular transmission $u \to v$ (i.e. $\chi \in S_v \cup \{v\} - \{u\}$) and which are also allowed to transmit in time slots i (i.e., they are assigned slot i or $i \in \Omega_\chi$). Let $O_{i,u \to v}{}^c$ be the complementary set of nodes in $S_v \cup \{v\} - \{u\}$ for which $i \notin \Omega_\chi$.

$$O_{i,u \to v} = \left\{ \chi : \chi \in S_v \cup \{v\} - \{u\}, i \in \Omega_\chi \right\}, \tag{1}$$

$$O_{i,u \to v}{}^c = \left\{ \chi : \chi \in S_v \cup \{v\} - \{u\}, i \notin \Omega_\chi \right\}. \tag{2}$$

Obviously, $|O_{i,u \to v}| + |O_{i,u \to v}{}^c| = |S_v|$.

Depending on the particular random assignment of the polynomials, it is possible that two nodes be assigned overlapping time slots (i.e., $\Omega_u \cap \Omega_v \neq \emptyset$). Let $C_{u \to v}$ be the set of overlapping time slots between those assigned to node u and those assigned to any node $\chi \in S_v \cup \{v\} - \{u\}$.

$$C_{u \to v} = \Omega_u \cap \left(\bigcup_{\chi \in S_v \cup \{v\} - \{u\}} \Omega_\chi \right). \tag{3}$$

Obviously, if $i \in C_{u \to v}$ ($|O_{i,u \to v}| > 0$ when $i \in C_{u \to v}$), it is possible that another transmission will corrupt transmission $u \to v$ in time slot i, provided that there are data for transmission. If $i \in \Omega_u - C_{u \to v}$ ($|O_{i,u \to v}| = 0$), no transmission corrupts transmission $u \to v$ in time slot i.

Let $R_{u \to v}$ denote the set of time slots i, $i \notin \Omega_u$, over which transmission $u \to v$ would be successful even for heavy traffic conditions ($\lambda = 1$). Equivalently,

$R_{u \to v}$ contains those slots not included in set $\bigcup_{\chi \in S_v \cup \{v\}} \Omega_\chi$. Consequently,

$$|R_{u \to v}| = q^2 - \left| \bigcup_{\chi \in S_v \cup \{v\}} \Omega_\chi \right|. \tag{4}$$

In order to use all non-assigned time slots, $R_{u \to v}$, without the need for further coordination among the nodes, the following probabilistic transmission policy was introduced in [10].

The Probabilistic Policy: Each node u always transmits in slot i if $i \in \Omega_u$ and transmits with probability p in slot i if $i \notin \Omega_u$, provided it has data to transmit.

3   System Throughput

In order to derive the expressions for the system throughput under both the Deterministic Policy and the Probabilistic Policy, it is necessary to derive the expressions for the *probability of success of a specific transmission* (throughput) under both policies. Let $P_{i,D,u \to v}$ $(P_{i,P,u \to v})$ denote the probability that transmission $u \to v$ in slot i is successful, under the Deterministic (Probabilistic) Policy. Let $P_{D,u \to v}$ $(P_{P,u \to v})$ be the average probability over a frame for transmission $u \to v$ to be successful during a time slot, under the Deterministic (Probabilistic) Policy. That is, $P_{D,u \to v} = \frac{1}{q^2} \sum_{i=1}^{q^2} P_{i,D,u \to v}$ $(P_{P,u \to v} = \frac{1}{q^2} \sum_{i=1}^{q^2} P_{i,P,u \to v})$, where q^2 is the frame size, in time slots.

Under the Deterministic Policy each node u transmits over a time slot i with probability λ, if $i \in \Omega_u$. Any other node χ, for which $i \in \Omega_\chi$, also transmits with probability λ. Consequently, $P_{i,D,u \to v} = 0$, for $i \notin \Omega_u$, and $P_{i,D,u \to v} = \lambda(1-\lambda)^{|O_{i,u \to v}|}$, for $i \in \Omega_u$. Given that $|O_{i,u \to v}| = 0$ for those slots $i \in \Omega_u - C_{u \to v}$ $(|O_{i,u \to v}| > 0$ when $i \in C_{u \to v})$, it is concluded that $P_{i,D,u \to v} = \lambda$, for $i \in \Omega_u - C_{u \to v}$. As a result,

$$P_{D,u \to v} = \frac{q - |C_{u \to v}| + \sum_{i \in C_{u \to v}} (1 - \lambda)^{|O_{i,u \to v}|}}{q^2} \lambda, \tag{5}$$

where $|\Omega_u| = q$.

Under the Probabilistic Policy, if $i \in \Omega_u$, each node u transmits over a time slot i with probability λ, while if $i \notin \Omega_u$ each node u transmits with probability $p\lambda$. Any other node χ, for which $i \in \Omega_\chi$, transmits in the same time slot i with probability λ, whereas if $i \notin \Omega_\chi$ it transmits with probability $p\lambda$. Consequently, for $i \in \Omega_u$, $P_{i,P,u \to v} = \lambda(1 - \lambda)^{|O_{i,u \to v}|}(1 - p\lambda)^{|O_{i,u \to v}^c|}$, while for $i \notin \Omega_u$, $P_{i,P,u \to v} = p\lambda(1 - \lambda)^{|O_{i,u \to v}|}(1 - p\lambda)^{|O_{i,u \to v}^c|}$. Given that $|O_{i,u \to v}| + |O_{i,u \to v}^c| = |S_v|$, for $i \in \Omega_u$, $P_{i,P,u \to v} = \lambda \left(\frac{1-\lambda}{1-p\lambda} \right)^{|O_{i,u \to v}|} (1 - p\lambda)^{|S_v|}$, while for $i \notin \Omega_u$, $P_{i,P,u \to v} = p\lambda \left(\frac{1-\lambda}{1-p\lambda} \right)^{|O_{i,u \to v}|} (1 - p\lambda)^{|S_v|}$. By definition, $|O_{i,u \to v}| = 0$, if $i \in$

$(\Omega_u - C_{u \to v}) \cup R_{u \to v}$, and $|O_{i,u \to v}| > 0$ if $i \in C_{u \to v}$ or $i \notin \Omega_u \cup R_{u \to v}$. As a result,

$$
\begin{aligned}
P_{P,u \to v} = & \frac{\sum_{i \in C_{u \to v}} \left(\frac{1-\lambda}{1-p\lambda} \right)^{|O_{i,u \to v}|}}{q^2} \lambda (1 - p\lambda)^{|S_v|} \\
& + \frac{q - |C_{u \to v}| + p|R_{u \to v}|}{q^2} \lambda (1 - p\lambda)^{|S_v|} \\
& + \frac{p \sum_{i \notin R_{u \to v} \cup \Omega_u} \left(\frac{1-\lambda}{1-p\lambda} \right)^{|O_{i,u \to v}|}}{q^2} \lambda (1 - p\lambda)^{|S_v|}.
\end{aligned}
\tag{6}
$$

The destination node v of a particular transmission $u \to v$, depends on the destination of the data and consequently, on the *network application* as well as on the *routing protocol*. For the rest of this work it will be assumed that a node u transmits to only one v, $v \in S_u$. For any of the numerical or simulation results presented in the sequel, node v will be a node randomly selected out of S_u.

The probability of success of a transmission (averaged over all nodes in the network) under each corresponding policy, is referred to as the *system throughput* and is denoted by P_D (P_P) under the Deterministic (Probabilistic) Policy. According to equations (5) and (6), the system throughput P_D and P_P is derived by $P_D = \frac{1}{N} \sum_{\forall u \in V} P_{D,u \to v}$ and $P_P = \frac{1}{N} \sum_{\forall u \in V} P_{P,u \to v}$, respectively.

4   System Throughput Maximization

As it is difficult to derive analytical expressions and establish the conditions for the system throughput maximization, a more tractable form of P_P is considered.

Let $\hat{P}_P$ denote that value of P_P when $\frac{1-\lambda}{1-p\lambda}$ is replaced by 1. Then,

$$
\hat{P}_P = \frac{1}{N} \sum_{\forall u \in V} \frac{1 + p(q-1)}{q} \lambda (1 - p\lambda)^{|S_v|}.
\tag{7}
$$

Since $\frac{1-\lambda}{1-p\lambda} \leq 1$, it is clear that $\hat{P}_P \geq P_P$.

Even though $\hat{P}_P$ has a more tractable form than that of P_P, it still cannot be easily analyzed further. It can be seen from Equation (7) that $\hat{P}_P$ corresponds to a polynomial of $D + 1$ degree with respect to p, which is difficult or impossible to be solved in the general case ($D > 1$). A more tractable form of $\hat{P}_P$, $\tilde{P}_P$, is analyzed instead. $\tilde{P}_P$ is equal to $\frac{1}{N} \sum_{\forall u \in V} \frac{1+p(q-1)}{q} \lambda (1 - p\lambda)^{\overline{|S|}}$, where $\overline{|S|} = \frac{1}{N} \sum_{\forall v \in V} |S_v|$. Finally, $\tilde{P}_P$ is given by the following equation.

$$
\tilde{P}_P = \frac{1 + p(q-1)}{q} \lambda (1 - p\lambda)^{\overline{|S|}}.
\tag{8}
$$

$\overline{|S|}$ is the average number of neighbor nodes of each node in the network. Let $\overline{|S|}/D$ be referred to as the *topology density* for a particular network. Given that D is known (and constant), knowledge of $\overline{|S|}$ is enough to determine the topology density $\overline{|S|}/D$ and vice versa. It is clear that $\overline{|S|}$ (or the topology density $\overline{|S|}/D$) influences exponentially the system throughput as it can be concluded from Equation (8).

The following two theorems show that there exists a range of values of p of the form $[0, p_{max}]$ (where p_{max}, $0 < p_{max} \le 1$, corresponds to that value of p for which $\tilde{P}_P = P_D$) such that $\tilde{P}_P \ge P_D$, irrespectively of the value of λ.

Theorem 1. *There exists an efficient range of values for p such that $\tilde{P}_P \ge P_D$, irrespectively of the value of λ.*

Proof. Under the Deterministic Policy, q is the maximum number of time slots during which a transmission is successful and therefore, $q\lambda$ is the maximum (average) number of successful transmissions in one frame. For any transmission $u \to v$, according to Equation (5), $q - |C_{u \to v}| + \sum_{i \in C_{u \to v}} (1 - \lambda)^{|O_{i,u \to v}|} \lambda \le q\lambda$. Consequently, $P_D \le \frac{\lambda}{q}$. It thus suffices to identify the range of values of p for which $\tilde{P}_P \ge \frac{\lambda}{q}$.

It can be shown that $\frac{d\tilde{P}_P}{dp} = \frac{p\lambda(\overline{|S|}+1)(1-q)+q-1-\lambda\overline{|S|}}{q} \lambda(1-p\lambda)^{\overline{|S|}-1}$ and therefore, $\frac{d\tilde{P}_P}{dp} = 0$ for $p = \frac{q-1-\lambda\overline{|S|}}{\lambda(\overline{|S|}+1)(q-1)}$ ($\equiv p_0$). Given that $\tilde{P}_P = P_D$ for $p = 0$, the required range of values for p does not exist if $\frac{d\tilde{P}_P}{dp} < 0$ for all $p \in (0, 1]$. $\frac{d\tilde{P}_P}{dp} < 0$ is satisfied if $p_0 < 0$ and as it may also be shown, $p_0 < 0$, if $\lambda > \frac{q-1}{\overline{|S|}}$. Given that $\overline{|S|} \le D$ (equality holds if all nodes have D neighbor nodes) and $q \ge kD + 1 \ge D + 1$ (equality holds for $k = 1$), it is concluded that $\overline{|S|} \le q - 1$ or $\frac{q-1}{\overline{|S|}} \ge 1$. Given that $\lambda \le 1$, $\lambda > \frac{q-1}{\overline{|S|}}$ is not satisfied and consequently, it is not possible that $\frac{d\tilde{P}_P}{dp} < 0$ for all $p \in (0, 1]$. Consequently, there exists a range of values for p such that $\tilde{P}_P \ge P_D$, irrespectively of the value of λ. $\square$

Theorem 2. *The efficient range of values for p of Theorem 1 is of the from $[0, p_{max}]$, for some $0 < p_{max} \le 1$.*

Proof. If $\lambda \ge \frac{q-1}{q\overline{|S|}+q-1}$ is satisfied then, it may be shown that $p_0 \le 1$ is satisfied, where $p_0 = \frac{q-1-\lambda\overline{|S|}}{\lambda(\overline{|S|}+1)(q-1)}$. It can be shown that for p close to 0, $\frac{d\tilde{P}_P}{dp} > 0$. Given that $\frac{d\tilde{P}_P}{dp} = 0$ for $p = p_0$, $\frac{d\tilde{P}_P}{dp} > 0$ for $p < p_0$, and $\frac{d\tilde{P}_P}{dp} < 0$ for $1 \ge p > p_0$ ($\frac{d\tilde{P}_P}{dp} = 0$ for only one value of $p \in (0, 1)$. As a result, when $\lambda \ge \frac{q-1}{q\overline{|S|}+q-1}$, $\tilde{P}_P = \frac{\lambda}{q}$ for $p = 0$, and as p increases, $\tilde{P}_P$ increases, until $p = p_0$. As p increases and $p > p_0$, $\tilde{P}_P$ decreases until $p = 1$, where $\tilde{P}_P = \lambda(1-\lambda)^{\overline{|S|}}$. If $\lambda(1-\lambda)^{\overline{|S|}} > \frac{\lambda}{q}$, then $p_{max} = 1$. If $\lambda(1-\lambda)^{\overline{|S|}} \le \frac{\lambda}{q}$, there exists a value of $p = p_{max} \le 1$, such that $\tilde{P}_P = \frac{\lambda}{q}$.

If $\lambda < \frac{q-1}{q\overline{|S|}+q-1}$, then $\frac{d\tilde{P}_P}{dp} > 0$, for any value of p. Consequently, $\tilde{P}_P$ constantly increases and therefore, $p_{max} = 1$ and the maximum is assumed for $p = 1$. $\square$

Theorem 3. *If* $\lambda < \frac{q-1}{q|S|+q-1}$, $\tilde{P}_P$ *is maximized for* $p = 1$. *If* $\lambda \geq \frac{q-1}{q|S|+q-1}$, $\tilde{P}_P$ *is maximized for* $p = \frac{q-1-\lambda|S|}{\lambda(|S|+1)(q-1)}$.

Proof. For $\lambda < \frac{q-1}{q|S|+q-1}$, as it was shown in the proof of Theorem 2, $\frac{d\tilde{P}_P}{dp} > 0$. Consequently, the maximum value for $\tilde{P}_P$ is assumed for $p = 1$.

For $\lambda \geq \frac{q-1}{q|S|+q-1}$, as it was shown in the proof of Theorem 2, $\frac{d\tilde{P}_P}{dp} \geq 0$ for $p \in [0, p_0]$ and $\frac{d\tilde{P}_P}{dp} \leq 0$ for $p \in [p_0, 1]$. Consequently, the maximum value for $\tilde{P}_P$ is assumed for $p = p_0 = \frac{q-1-\lambda|S|}{\lambda(|S|+1)(q-1)}$. □

Based on the results of Theorem 3, the value of the access probability p that maximizes $\tilde{P}_P$, denoted by $\tilde{p}_{\lambda,\overline{|S|}}$, is given by the following equation.

$$\tilde{p}_{\lambda,\overline{|S|}} = \begin{cases} 1, \text{ if } \lambda < \frac{q-1}{q|S|+q-1}; \\ \frac{q-1-\lambda\overline{|S|}}{\lambda(\overline{|S|}+1)(q-1)}, \text{ if } \lambda \geq \frac{q-1}{q|S|+q-1}. \end{cases} \quad (9)$$

For $p = \tilde{p}_{\lambda,\overline{|S|}}$, $\tilde{P}_P$ is maximized but it is also expected that P_P will have a better performance than that achieved by a fixed value of p when λ is not constant.

5   Unawareness of λ and $\overline{|S|}$

Knowledge of both λ and $\overline{|S|}$ is required in order to utilize the results of the analysis presented in Section 4 and in particular, by Equation (9). If the topology density $\overline{|S|}/D$ is known, $\overline{|S|}$ can be calculated (D is known). In the general case, it is possible that λ and/or $\overline{|S|}$ are not known and therefore, it would be useful to derive analytical expressions for those values of p that the system throughput under the Probabilistic Policy even though it is not maximized, at least is higher than that under the Deterministic Policy. For the case for which λ is known but $\overline{|S|}$ is not, the maximum topology density value (corresponding to $\overline{|S|} = D$) is considered and the corresponding value for p, denoted by $\tilde{p}_\lambda$, is given by Equation (10). For the case for which $\overline{|S|}$ is known but λ is not, a heavy traffic scenario (corresponding to $\lambda = 1$) is considered and the corresponding value for p, denoted by $\tilde{p}_{\overline{|S|}}$, is given by Equation (11). Finally, for the case for which both λ and $\overline{|S|}$ are not known, the corresponding value for p ($\lambda = 1$ and $\overline{|S|} = D$), denoted by $\tilde{p}$, is given by Equation (12).

$$\tilde{p}_\lambda = \begin{cases} 1, \text{ if } \lambda < \frac{q-1}{qD+q-1}; \\ \frac{q-1-\lambda D}{\lambda(D+1)(q-1)}, \text{ if } \lambda \geq \frac{q-1}{qD+q-1}. \end{cases} \quad (10)$$

$$\tilde{p}_{\overline{|S|}} = \frac{q-1-\overline{|S|}}{(\overline{|S|}+1)(q-1)}. \quad (11)$$

$$\tilde{p} = \frac{q-1-D}{(D+1)(q-1)}. \quad (12)$$

6   Simulation Results

In this section, networks of 100 nodes are considered for various values of D and the topology density $\overline{|S|}/D$. The aim is to demonstrate the applicability of the analytical results for a variety of topologies with different characteristics. In particular, four different topology categories are considered. The number of nodes in each topology category is set to $N = 100$, while D is set to 5, 10, 15 and 20. These four topology categories are denoted as D5N100, D10N100, D15N100 and D20N100, respectively, and correspond to a certain value of topology density $\overline{|S|}/D$. 1 depicts simulation results for the system throughput P_P, for different topology density values close to 0.6. For all cases the number of neighbor nodes for each node is not the same; this leads to nonzero values for the topology density variation $Var\{|S|\}$. The algorithm presented in [9] is used to derive the sets of scheduling slots and the system throughput is calculated averaging the simulation results over 100 frames. Time slot sets Ω_χ are assigned randomly to each node χ, for each particular topology. The particular assignment is kept the same for each topology category throughout the simulations.

Figure 1 depicts simulation results for the system throughput P_P as a function of λ, for different values of p ($p = 0$, $p = \tilde{p}_{\lambda,\overline{|S|}}$, $p = \tilde{p}_\lambda$, $p = \tilde{p}_{\overline{|S|}}$, $p = \tilde{p}$ and $p = 1.0$). For $p = 0$ the system throughput under the Probabilistic Policy is identical to that under the Deterministic Policy ($P_P = P_D$). This case is depicted throughout the simulation results for comparison reasons. For $p = 1.0$, it can be seen that as λ increases, P_P increases rather fast until a certain maximum value and then decreases until $\lambda = 1.0$, where $P_P = 0$. For $p = \tilde{p}_{\lambda,\overline{|S|}}$ it is evident that the system throughput under the Probabilistic Policy is not only higher than that under the Deterministic Policy but it is also close to the maximum for all different topologies.

For $p = \tilde{p}_\lambda$ and for small values of λ, as it can be seen in Figure 1, the system throughput is identical to that obtained for $p = \tilde{p}_{\lambda,\overline{|S|}}$ and $p = 1.0$. This can be concluded from Equation (9) and Equation (10) where for $\lambda < \frac{q-1}{qD+q-1}$, $\tilde{p}_{\lambda,\overline{|S|}} = \tilde{p}_\lambda = 1.0$. For $\lambda > \frac{q-1}{qD+q-1}$ the system throughput, even though not close to that obtained for $p = \tilde{p}_{\lambda,\overline{|S|}}$, is higher than that obtained under the Probabilistic Policy. For $p = \tilde{p}_{\overline{|S|}}$ and for small values of λ, the system throughput is not close to that obtained for $p = \tilde{p}_{\lambda,\overline{|S|}}$ but is higher than that obtained under the Deterministic Policy, irrespectively of the topology density value. As λ increases the system throughput increases and for large values of λ it is close to the maximum. For $p = \tilde{p}$, the system throughput curve is similar to that for $p = \tilde{p}_{\overline{|S|}}$, except that the obtained system throughput is smaller. As the topology density increases, the system throughput for both cases ($p = \tilde{p}$ and $p = \tilde{p}_{\overline{|S|}}$) converges. Again, the obtained system throughout is higher than that obtained under the Deterministic Policy. This is an important observation since $\tilde{p}$ is calculated without knowledge of either λ or $\overline{|S|}$.

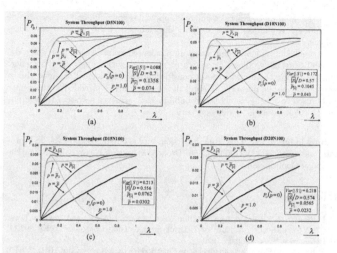

Fig. 1. System throughput simulation results as a function of λ for different values of p ($p = 0$, $p = \tilde{p}_{\lambda,\overline{|S|}}$, $p = \tilde{p}_\lambda$, $p = \tilde{p}_{\overline{|S|}}$, $p = \tilde{p}$ and $p = 1.0$) and medium topology density values $\overline{|S|}/D$.

7 Conclusions

The system throughput under both the Deterministic Policy and the Probabilistic Policy has been investigated previously for heavy traffic conditions, [8], [9], [10]. In this work, expressions have been derived and analyzed for the general non-heavy traffic case. The results of this analysis established conditions and provided expressions in order for the system throughput under the Probabilistic Policy to be not only higher than that under the Deterministic Policy, but also be maximized, for various traffic loads.

Expressions for the system throughput were derived for both policies as a function of the traffic load. These expressions determine the values of the access probability p for which the system throughput under the Probabilistic Policy is higher than that under the Deterministic Policy. For the particular case for which both the traffic load (λ) and the topology density ($\overline{|S|}/D$) are known, the system throughput is maximized for the derived value of access probability, $p = \tilde{p}_{\lambda,\overline{|S|}}$. For the cases for which λ and/or $\overline{|S|}/D$ are not known, analytical expressions for the appropriate values of p were also derived leading to a system throughput which, even though not maximized, is higher than that achieved under the Deterministic Policy ($p = \tilde{p}_\lambda$, $p = \tilde{p}_{\overline{|S|}}$ and $p = \tilde{p}$). Simulations have been conducted for a variety of different topologies with different characteristics and different traffic loads that support the claims and the expectations of the analysis.

In conclusion, a simple and easily implemented transmission policy like the Probabilistic Policy, allows for nodes to transmit without any need for coordination among them and without considering the topological changes; this is essential for ad-hoc networks. The results of the analysis provided in this paper can be used to specify an access probability that achieves a system throughput close to the maximum for different traffic loads.

Acknowledgement. This work has been supported in part by the IST program under contract IST-2001-32686 (BroadWay) which is partly funded by the European Commission.

References

1. IEEE 802.11, "Wireless LAN Medium Access Control (MAC) and Physical Layer (PHY) specifications," Nov. 1997. Draft Supplement to Standard IEEE 802.11, IEEE, New York, January 1999.
2. P. Karn, "MACA- A new channel access method for packet radio," in ARRL/CRRL Amateur Radio 9th Computer Networking Conference, pp. 134-140, 1990.
3. V. Bharghavan, A. Demers, S. Shenker, and L. Zhang, "MACAW: A Media Access Protocol for Wireless LAN's," Proceedings of ACM SIGCOMM'94, pp. 212-225, 1994.
4. C.L. Fullmer, J.J. Garcia-Luna-Aceves, "Floor Acquisition Multiple Access (FAMA) for Packet-Radio Networks," Proceedings of ACM SIGCOMM'95, pp. 262-273, 1995.
5. J. Deng and Z. J. Haas, "Busy Tone Multiple Access (DBTMA): A New Medium Access Control for Packet Radio Networks," in IEEE ICUPC'98, Florence, Italy, October 5-9, 1998.
6. A. Ephremides and T. V. Truong, "Scheduling Broadcasts in Multihop Radio Networks," IEEE Transactions on Communications, 38(4):456-60, April 1990.
7. L. Bao and J. J. Garcia-Luna-Aceves, "A new approach to channel access scheduling for ad hoc networks," ACM Mobicom 2001, July 2001.
8. I. Chlamtac and A. Farago, "Making Transmission Schedules Immune to Topology Changes in Multi-Hop Packet Radio Networks," IEEE/ACM Trans. on Networking, 2:23-29, 1994.
9. J.-H. Ju and V. O. K. Li, "An Optimal Topology-Transparent Scheduling Method in Multihop Packet Radio Networks," IEEE/ACM Trans. on Networking, 6:298-306, 1998.
10. K. Oikonomou and I. Stavrakakis, "Analysis of a Probabilistic Topology-Unaware TDMA MAC Policy for Ad-Hoc Networks," IEEE Journal on Selected Areas in Communications (JSAC), Special Issue on Quality-of-Service Delivery in Variable Topology Networks. Accepted for publication. To appear 3rd-4th Quarter 2004.

Performance of TCP/IP with MEDF Scheduling

Ruediger Martin, Michael Menth, and Phan-Gia Vu

Department of Distributed Systems, Institute of Computer Science
University of Würzburg, Am Hubland, D-97074 Würzburg, Germany
{martin|menth|phan}@informatik.uni-wuerzburg.de

Abstract. To achieve Quality of Service (QoS) in Next Generation Networks
(NGNs), the Differentiated Services architecture implements appropriate Per Hop
Behavior (PHB) for service differentiation. Common recommendations to enforce
appropriate PHB include Weighted Round Robin (WRR), Deficit Round-Robin
(DRR) and similar algorithms. They assign a fixed bandwidth share to Transport
Service Classes (TSCs) of different priority. This is a viable approach if the ratio
of high priority traffic TSC_{high} over low priority traffic TSC_{low} is known in
advance. If TSC_{high} holds more and TSC_{low} less users than expected, the QoS
for TSC_{high} can be worse than for TSC_{low}. As shown in preceding work, the
Modified Earliest Deadline First (MEDF) algorithm heals this problem on the
packet level. Therefore, we investigate its impact in congested TCP/IP networks
by simulations and show its attractiveness as a powerful service differentiation
mechanism.

1 Introduction

Current research for multi-service Next Generation Networks (NGNs) focuses amongst
others on the provision of Quality of Service (QoS) for different service classes. The
Differentiated Services architecture [1], [2] achieves QoS by implementing appropriate Per Hop Behavior (PHB) for different Transport Service Classes (TSCs). Flows
of different TSCs compete for the resources buffer space and forwarding speed in the
routers. Mechanisms that assign those resources divide buffer space among different
TSCs (buffer management) and control the order in which packets are dequeued and
forwarded (scheduling). Therefore, those mechanisms can be characterized along two
dimensions: space and time.

Common examples and recommendations [3] [4] to enforce appropriate PHB are
algorithms like Weighted Round Robin (WRR), Class Based Queueing (CBQ) [5], and
Deficit Round-Robin (DRR) [6]. The common goal is to assign a fair share of network
resources to different TSCs. The share is set in advance and fixed independently of the
actual traffic mix. This behavior is desirable in many situations. In a network where a
ratio q of high priority TSC (TSC_{high}) traffic over low priority TSC (TSC_{low}) traffic is
expected, e.g. due to network admission control, the algorithms can be used to assign this
share. The low priority TSC (TSC_{low}) uses the remaining bandwidth where a fraction of
$1-q$ is guaranteed. However, if resources are scarce and buffers always contain packets
of both classes, these algorithms enforce the share q regardless of the current traffic mix.
Particularly, if the TSC_{high} traffic exceeds the limit set by the control parameters, it
suffers from QoS degradation.

J. Solé-Pareta et al. (Eds.): QofIS 2004, LNCS 3266, pp. 94–103, 2004.
© Springer-Verlag Berlin Heidelberg 2004

The authors of [7] introduced the priority algorithm Modified Earliest Deadline First (MEDF). They showed that MEDF prefers TSC_{high} over TSC_{low} on the packet level equally regardless of the traffic mix ratios. This is a clear advantage of MEDF compared to the previously mentioned algorithms that assign a fixed share for the whole TSC_{high} aggregate.

In this paper we focus on the impact of MEDF in TCP/IP networks. For saturated TCP sources conventional algorithms are problematic because of the fixed share of bandwidth assigned to each traffic class regardless of the current number of flows. We bring the MEDF algorithm into play to achieve a relative traffic-mix-independent per-flow-prioritization among TSCs. But still, this behavior is easily configurable by per-class relative delay factors.

The algorithms that work on the IP packet level impact the performance of adaptive TCP flows by packet loss and delay (round trip time). Packet loss is influenced by space priority mechanisms, delay by time priority mechanisms. In this work we combine MEDF with space priority mechanisms like Full Buffer Sharing (FBS) and Buffer Sharing with Space Priority (BSSP) and contrast it to time priority mechanisms like First In First Out (FIFO) and Static Priority (SP).

This work is structured as follows. In Section 2 we present the algorithms under study in detail. Section 3 discusses the simulation environment, the respective parameters used for our performance evaluation study, and presents the results obtained from our simulations. Sections 4 and 5 finally conclude this work with a short summary and outlook on future research.

2   Space and Time Priority Mechanisms

Network congestion arises where different flows compete for resources at routers in the network. To avoid this problem at least for a certain subset of high priority flows, flows of higher priority should receive preferential service as opposed to low priority flows. Basically, if packet arrivals exceed the router forwarding speed temporarily or permanently, congestion arises and buffers fill up. This leads to longer network delays and high packet loss rates, to degraded Quality of Service. Buffer sizes and forwarding speed are fixed parameters for given networks. To assign these scarce resources, we can limit the space available to the respective flows (buffer management) or we can dequeue the packets depending on their priority (scheduling). Thus, mechanisms to achieve service differentiation can be divided along two dimensions: space and time. Combinations of both are also possible.

2.1   Space Priority Mechanisms

We use two kinds of space priority mechanisms for our performance evaluation: Full Buffer Sharing and Shared Buffers with Space Priority. In [10] we compare a third space priority mechanism Random Early Detection gateways [11]. RED was originally designed to detect incipient congestion by measuring the average queue length. Several improvements have been suggested for instance in [12] and [13] to achieve fairness in

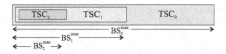

Fig. 1. Buffer Sharing with Space Priority for $i = 3$ TSCs

the presence of non-adaptive connections and to introduce TSC priorities. We omit this section for lack of space here and refer to our technical report [10].

In the following sections, we denote the router buffer by B and packets by P. The function $S(B)$ refers to the maximum buffer size and $F(B)$ to the current fill level of the buffer. The function $enqueueTail(P, B)$ enqueues the packet P into the buffer B. The function $drop(P)$ drops the packet P if the algorithms cannot accept the packet.

Full Buffer Sharing (FBS). The FBS strategy allows all flows to share the same buffer irrespective of their priority. If not mentioned differently, we use this mechanism as default in our simulations.

Buffer Sharing with Space Priority (BSSP). The BSSP queueing strategy (cf. Alg. 1) is threshold based and allows packets to occupy buffer space available for their TSC and for all TSCs of lower priority. Let TSC_i, $i \in \{0, \dots, n-1\}$ be TSCs of different priority, 0 being the highest priority. TSC_i can at most demand space BS_i^{max} in the buffer, where $BS_i^{max} \geq BS_{i+1}^{max}$ and BS_0^{max} is set to the actual buffer size. The concept is illustrated in Fig. 1 for three TSCs and fully described in Alg. 1 with $F(B, TSC_i)$ denoting the space in the buffer B that is currently filled by TSC_i. There is a guaranteed amount of buffer space for the highest priority class only, lower priority classes possibly find their share taken by classes of higher priority. This concept resembles the Russian dolls bandwidth constraints model (RDM) suggested by the IETF traffic engineering working group (TEWG) in [14].

Require: Packet P, Buffer B, max TSC Buffer Size BS_i^{max} for $i = 0 \dots n-1$
  { max Buffer Size $S(B) = BS_0^{max}$ }
  i = TSC(P)
  if $\sum_{j=i}^{j=(n-1)} F(P, TSC_j) \leq BS_i^{max}$ **then**
    $enqueTail(P, B)$
  else {space limit exceeded for TSC i}
    $drop(P)$
  end if

Algorithm 1: Buffer Sharing with Space Priority ENQUEUE

2.2   Time Priority Mechanisms

Once packets arrive at the queue and the space priority mechanism assigns available buffer space, i.e., it decides whether the packet is accepted or dropped, the time priority mechanism decides which packet to dequeue next. This decision on the packet level influences the delay and therefore the TCP sending rate via RTT. We contrast two time priority mechanisms to Modified Earliest Deadline First (MEDF).

First in First Out (FIFO). FIFO leaves the prioritization to the enqueueing option and is used as the performance baseline to compare with. Packets proceed in the order they arrive and are accepted by the space priority mechanism.

Static Priority (SP). The Static Priority concept chooses TSC_{high} packets in FIFO order as long as packets of that class are in the buffer. TSC_{low} packets wait in the router queue until low priority packets only are available. Then they are also dequeued in a FIFO manner until new TSC_{high} packets arrive.

Modified Earliest Deadline First (MEDF). In the context of the UMTS Terrestrial Radio Access Network, the authors of [7] introduced a modified version of the Earliest Deadline First (EDF) algorithm called Modified Earliest Deadline First (MEDF). It supports n only different TSCs, but in contrast to EDF it is easier to implement. Packets are stored in n TSC specific queues in FIFO manner. They are stamped with a modified deadline that is their arrival time plus an offset $M_i, 0 \le i < n$, which is characteristic for each TSC. The MEDF scheduler selects the packet for transmission that has the earliest due date among the packets in the front positions of all queues. For only two TSCs, this is the choice between two packets and sorting according to ascending deadlines is not required. The difference $|M_i - M_j|$ between two TSCs i and j is a relative delay advantage that influences the behavior of the scheduler. We are interested in the performance of this scheduling algorithm in the presence of adaptive traffic, here TCP.

For our simulations we use two TSCs whose queues are implemented as shared buffers such that the space priority mechanisms are applicable. With two TSCs we set the MEDF parameters to $M_{high} = 0$ and $M_{low} = x, x \in \{0s, 0.1s, 0.5s, 1.0s, 1.5s\}$. Thus, TSC_{high} obtains no additional delay. The deadline for TSC_{low} packets is increased by the M_{low} parameter.

3   MEDF Performance Evaluation

In this section we describe the general goals and approach of our performance evaluation study and present the results. We used the network simulator (NS) version 2 [15] to run the experiments deploying the RENO TCP implementation [16]. Standard simulation methods as replicate-delete were applied to obtain statistically reliable results of the non-ergodic random processes. In the following sections we only give average values as the simulated time was chosen to yield very narrow confidence intervals. Our goal is the measurement of the prioritization of TSC_{high} traffic. For that purpose, we define the normalized bandwidth ratio. Let n_{high} be the number of TSC_{high} flows, n_{low} the number of TSC_{low} flows. The functions $B(TSC_{high})$ and $B(TSC_{low})$ denote the bandwidth used by all TSC_{high} and TSC_{low} flows, respectively. The normalized bandwidth ratio $\overline{B}_{ratio}(TSC_{high}, TSC_{low})$ is the amount of bandwidth used by TSC_{high} per flow divided by the amount of bandwidth used by TSC_{low} per flow:

$$\overline{B}_{ratio}(TSC_{high}, TSC_{low}) = \frac{\frac{B(TSC_{high})}{n_{high}}}{\frac{B(TSC_{low})}{n_{low}}}$$

A mechanism with traffic-mix-independent per-flow-prioritization among TSCs exhibits the same normalized bandwidth ratio regardless of the traffic mix. The number of saturated TCP sources is the same for both TSCs in the following if not mentioned otherwise.

3.1 MEDF Characteristics

To isolate the general behavior more easily and to eliminate unpredictable side effects, we start with single link simulations and extend it to multiple links.

MEDF Single Link Scenario

Simulation environment. We use the classical dumbell topology for our single link simulation environment. A number of TSC_{high} TCP traffic sources and a number of TSC_{low} TCP traffic sources connect to Router A. Router A uses a space and a time priority mechanism described above and sends the packets over a single link to router B. Router B has sufficient capacity to serve the link and its single task is to distribute the arriving packets to the corresponding destinations.

We choose the number of simultaneous active TCP connections n as $n_{min} \cdot 2^i, i \in \{0, \dots, 8\}$, n_{min} being the minimum number of TCP connections to get a theoretical load of 100% on the link. Otherwise there is no overload, space and time priorities do not have effect, and the flow control is not active. Here $n_{min} = 2$. The packet size $S(P)$ is a common standard value of 500 Bytes including headers. Regarding the link parameters, with the link bandwidth being $C_l = 1.28Mbit/s$, we set the link propagation delay D_{prop} to 46.875 ms so that the theoretical round trip time RTT sums up to $RTT = 2 \cdot (n_{links} \cdot D_{prop} + (n_{links} + 1) \cdot D_{TX}) = 2 \cdot (1 \cdot 46.875ms + 2 \cdot 3.125ms) = 100ms$, where $D_{TX} = \frac{S(P)}{C_l}$ is the transmission delay to send a packet and n_{links} the number of links between routers A and B.

The default value for the buffer size S_{Buffer} is 160 packets so that a router is able to store packets for 0.5 seconds transmission. We use the parameters mentioned here as default parameters and write down the respective values in the following text only if they are set differently. Other parameters like algorithm specific settings are subject to the analysis and we indicate their values appropriately.

Simulation. Figure 2 shows the normalized bandwidth ratio $\overline{B}_{ratio}(TSC_{high}, TSC_{low})$ for traffic mixes $n_{high} : n_{low}$ of $1:3, 1:1$, and $3:1$ with MEDF parameter $M_{low} = 0.5s$, i.e., one buffer size. The link bandwidth is the x-axis parameter. The value $n_{min} = 2$ is omitted here and in the following figures as there is virtually no priority for the minimum number of users. The link capacity is fully shared between the single user of each class, thus, they both reach the maximum rate. This behavior – as expected – is sound for lack of competition on the link.

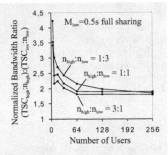

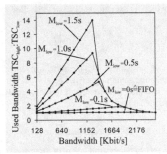

Fig. 2. MEDF prioritization independent of the traffic mix

Fig. 3. MEDF prioritization for two TSCs

The figure shows the traffic-mix-independent per-flow-prioritization property of MEDF. The small differences for low congestion are due to the slight influence of the buffer space. Few high priority flows can occupy relatively more buffer space per flow in contrast to many high priority flows under low congestion. However, the difference is negligibly small and the normalized bandwidth ratio converges very quickly. Opposed to that, conventional algorithms like WRR are insensitive to the traffic mix and therefore the normalized bandwidth ratio would decrease severely with the ratio of TSC_{high} traffic over TSC_{low}. We further emphasize that this property is achieved by a single parameter per class and originates from the relative delay advantage controlled by MEDF.

The traffic-mix-independent per-flow-prioritization property of MEDF was already shown in [7] on the packet level. Due to this property on the TCP flow level as well, we use the same number of saturated TCP sources for both TCSs in the following.

For the minimum number of users $n_{min} = 2$, there is virtually no prioritization for lack of competition. Prioritization of TSC_{high} traffic reaches its maximum at $n = 4$ users (2 users per TSC) and degrades with a rising number of users. As we cannot simulate any value between two and four users – one and two users per TSC – we vary the bandwidth while keeping the number of users fixed at a value of 4 to derive the basal characteristics of the algorithm by having a more continuous range in Fig. 3. This demonstrates the behavior at various levels of slight congestion in real networks.

At a bandwidth of 1.280 Mbit/s this experiment corresponds to a simulation with default values and 4 users, at a bandwidth of 2.560 Mbit/s it is equivalent to 2 users. Higher offset values M_{low} lead directly to a higher prioritization of TSC_{high} packets. The throughput ratio rises with the bandwidth which is inversely proportional to the number of users. Low bandwidth (same holds for many users) limits the rate that connections for TSC_{high} can achieve dramatically. Besides, the actually measured round trip time increases and shortens the maximum obtainable rate. Thus, TSC_{low} connections are able to grasp a higher relative share of the bandwidth. The bandwidth ratio rises until it reaches a maximum. Here, slowly sufficient capacity becomes available for both TSCs and low priority packets can use more of the additional bandwidth. At 2.560 Mbit/s there is virtually no competition for bandwidth anymore.

Another important aspect that can be seen here to understand MEDF is the interaction of M_{low} and the round trip time. The round trip time rises both with increasing traffic on the link and with decreasing bandwidth. Low bandwidth results in a longer transmission delay. The delay advantage becomes smaller relative to the round trip time and the prioritization decreases.

The MEDF parameter M_{low} can be used to adjust the priority ratio for the anticipated level of competition for network resources. If sufficient resources are available, the MEDF algorithm does not influence normal network operation. For very scarce resources – here large numbers of users and low bandwidth, respectively – the network is under heavy overload and anticipatory action like admission control to block some of the connections must be taken to prevent such situations. Otherwise, only a very small portion of the overall bandwidth remains for each TSC_{high} flow anyway — no matter whether they receive preferential service or not. For low and medium overload, MEDF shows a very clear and easy adjustable behavior.

MEDF Multi Link. We now extend our single link experiment to multiple links to assess the influence of MEDF on TSC priority if applied multiple times.

Simulation environment. Figure 4 shows the simulation topology for the multi link experiment in the case of two links. If we simply add additional links and routers, the first router receives the packets from the TCP sources in an unordered way and applies the priority algorithm. Thus, the packets arrive at the router serving the next link one by one and the priority algorithm has no additional effect. To overcome this problem, we introduce cross traffic. Additional TCP sources connect to the interior routers and generate traffic that crosses the way of the measured traffic.

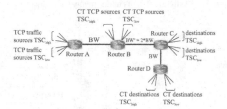

Fig. 4. Multi link simulation topology

It is important to send the cross traffic over the same number of links to account for comparable round trip times for the measured traffic and the cross traffic. Furthermore, the round trip time for both the single link and the multi link experiment should be the same. Otherwise, significant parameters that depend on the round trip time such as the maximum rate that can be achieved by a TCP connection are different and the

experiments are not comparable. Therefore, we calculate the new link propagation delay
$D_{prop} = \frac{46.875-(n_{link}-1) \cdot D_{TX}}{n_{link}} ms$.

The TCP connections need the same bandwidth per flow on all links. If the bandwidth differs from link to link, the link with the lowest capacity becomes the bottleneck and dominates the observable effects. However, doubling the bandwidth of the links with cross traffic solves this problem.

Simulation. Figure 5 shows the effect of MEDF over multiple links. We used the standard parameters with $M_{low} = 1s$, i.e., twice the buffer size, and the default Full Buffer Sharing mechanism as buffer management.

In general, the degree of prioritization of TSC_{high} increases with the number of links on the path, hence, with the number of applications of MEDF scheduling instances. However, when the competition for network resources is low, the increase in priority is much more obvious. The reason behind this is similar to the situation for the single link experiment. The bandwidth theoretically available to a single connection is higher, hence, the actually measured round trip time is lower. Therefore, few TSC_{high} connections achieve higher rates in contrast to the situation when the network is highly overloaded. Rising competition for network resources makes the conditions for TSC_{high} more disadvantageous. TSC_{low} now obtains a larger share of the bandwidth. The priority does not increase linearly if additional links are added. The overall bandwidth ratio can be controlled by setting the MEDF parameter appropriately.

MEDF and Space Priority. We now consider the MEDF characteristics with the usage of space priority mechanisms. Figure 6 shows the influence of the buffer sharing option. FIFO with FBS leads to an even division of available bandwidth between both TSCs as no packet preferences exist. FIFO with BSSP spreads the bandwidth equally as long as there is enough buffer space available ($n \leq 2$). Then it reaches its maximum when router buffers fill completely and slightly flattens under heavy traffic load.

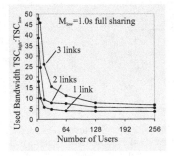

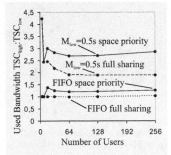

Fig. 5. MEDF prioritization in a multi link topology

Fig. 6. MEDF and the impact of space priority

MEDF with parameter $M_{low} = 0.5s$ and FBS clearly outperforms both FIFO experiments and exhibits the behavior characterized in the preceding sections. If we add BSSP, we observe a superposition of the MEDF curve and the curve for FIFO with BSSP. For few users we clearly identify the typical MEDF characteristics, for more users the router buffers fill completely and the space priority comes into play. Thus, space priority prohibits the typical decrease of the bandwidth ratio.

3.2 MEDF in Comparison to Other Priority Mechanisms

We used FIFO as the comparison baseline in the previous experiments. FIFO does not prioritize the traffic in time and therefore is one extreme of the spectrum of time priority mechanisms. Another extreme is Static Priority (SP).

Static Priority (SP). Under network congestion, the time priority mechanism Static Priority leads to starvation of TSC_{low} regardless of the buffer management in use. There are always TSC_{high} packets waiting in the router queues. SP dequeues those packets and even though the TSC_{low} packets occupy most of the buffer space, their chance to leave the buffer is very low and, thus, the TCP timers for those connections expire. Accordingly, the TCP source tries to re-establish the connection but will suffer from starvation again. As a consequence, SP is completely inadequate for severely congested networks. In contrast to MEDF it does not consider a maximum delay for low priority traffic to prevent this effect.

For a comparison to RED, a pure space priority algorithm, we refer to our technical report [10].

4   Conclusion

In this work we examined the impact of the pure time priority (packet scheduling) mechanism Modified Earliest Deadline First (MEDF) in congested TCP/IP networks. Conventional algorithms like Weighted Round Robin (WRR), Deficit Round Robin (DRR) or Class Based Queuing (CBQ) assign fixed bandwidth shares among Transport Service Classes (TSCs) of different priorities. This is problematic with a varying number of users per TSC and saturated TCP sources. If the TSC of high priority (TSC_{high}) holds more users than expected and the TSC of low priority (TSC_{low}) holds fewer users, then the Quality of Service (QoS) for TSC_{high} can be worse then for TSC_{low}. MEDF, however, achieves a relative traffic-mix-independent per-flow-prioritization among all TSCs.

In contrast to MEDF, Static Priority (SP) leads to starvation of low priority traffic while First In First Out (FIFO) effects no prioritization at all. MEDF achieves the desired priority ratio of the high priority TSC over the low priority TSC in realistic overload situations by its adjustable parameter M_{low} which reflects a relative delay advantage.

Full Buffer Sharing (FBS) was the default buffer management scheme in our experiments. To estimate the influence of the buffer management algorithms, we combined MEDF with Buffer Sharing with Space Priority (BSSP). The results showed an increased prioritization of TSC_{high}.

In conclusion, MEDF has powerful service differentiation capabilities and our performance study revealed that it is an attractive mechanism to achieve service differentiation for TCP flows in congested networks. MEDF is especially interesting since it does not require per-class bandwidth configuration which might be problematic in the presence of unknown traffic mix.

5 Outlook

The practical adaptation of the relative delay parameter M_{low}, especially its dependence on the propagation delay, is an interesting field of further research.

Acknowledgment. The authors would like to thank Prof. Tran-Gia for the stimulating environment which was a prerequisite for that work.

References

1. Blake, S., Black, D.L., Carlson, M.A., Davies, E., Wang, Z., Weiss, W.: RFC2475: An Architecture for Differentiated Services. ftp://ftp.rfc-editor.org/in-notes/rfc2475.txt (1998)
2. Grossman, D.: RFC3260: New Terminology and Clarifications for Diffser. ftp://ftp.rfc-editor.org/in-notes/rfc3260.txt (2002)
3. Davie, B., et al.: An Expedited Forwarding PHB (Per-Hop Behavior). ftp://ftp.rfc-editor.org/in-notes/rfc3246.txt (2002)
4. Charny, A., et al.: Supplemental Information for the New Definition of the EF PHB (Expedited Forwarding Per-Hop Behavior). ftp://ftp.rfc-editor.org/in-notes/rfc3247.txt (2002)
5. Floyd, S., Jacobson, V.: Link-sharing and Resource Management Models for Packet Networks. IEEE/ACM Transactions on Networking **3** (1995)
6. Shreedhar, M., Varghese, G.: Efficient Fair Queueing Using Deficit Round-Robin. IEEE/ACM Transactions on Networking **4** (1996)
7. Menth, M., Schmid, M., Heiß, H., Reim, T.: MEDF - A Simple Scheduling Algorithm for Two Real-Time Transport Service Classes with Application in the UTRAN. IEEE INFOCOM 2003 (2003)
8. Ramabhadran, S., Pasquale, J.: Stratified round Robin: a low complexity packet scheduler with bandwidth fairness and bounded delay. Proceedings of the 2003 conference on Applications, technologies, architectures, and protocols for computer communications Karlsruhe, Germany (2003) 239–250
9. Stiliadis, D., Varma, A.: Efficient fair queueing algorithms for packet-switched networks. IEEE/ACM Transactions on Networking (TON) **6** (1998) 175–185
10. Martin, R., Menth, M.: Performance of TCP/IP with MEDF Scheduling. Technical Report 323, University of Würzburg (2004)
11. Floyd, S., Jacobson, V.: Random Early Detection Gateways for Congestion Avoidance. IEEE/ACM Transactions on Networking **1** (1993) 397–413
12. Anjum, F., Tassiulas, L.: Balanced-RED: An Algorithm to Achieve Fairness in the Internet. IEEE INFOCOM 1999 (1999)
13. Bodin, U., Schelén, O., Pink, S.: Load-tolerant Differentiation with Active Queue Management. SIGCOMM Computer Communication Review (2000)
14. Le Faucheur, F.: Russian Dolls Bandwidth Constraints Model for Diff-Serv-aware MPLS Traffic Engineering. Internet Draft TEWG (2003)
15. Fall, K., Varadhan, K.: The ns Manual. http://www.isi.edu/nsnam/ns/doc/ns_doc.pdf (2003)
16. Stevens, W.R.: TCP/IP Illustrated, Volume 1. Addison-Wesley Longman (1994)

Ping Trunking: A Vegas-Like Congestion Control Mechanism for Aggregated Traffic[*]

Sergio Herrería-Alonso, Manuel Fernández-Veiga, Miguel Rodríguez-Pérez,
Andrés Suárez-González, and Cándido López-García

Departamento de Enxeñería Telemática, Universidade de Vigo
ETSE Telecomunicación, Campus universitario s/n, 36310 Vigo, Spain
sha@det.uvigo.es

Abstract. We present a new edge-to-edge management technique, called *Ping Trunking*, that can provide soft service guarantees to aggregated traffic. A Ping trunk is an aggregate traffic stream that flows between two nodes in a network at a rate dynamically determined by a Vegas-like congestion control mechanism. A management connection associated to each trunk is in charge of regulating the user data transmission rate. Due to this managing, trunks are able to probe for available bandwidth adapting themselves to changing network conditions in accordance with their subscribed target rates. We demonstrate analytically and through simulation experiments the effectiveness of our proposal.

Keywords: Traffic management, aggregated traffic, TCP Vegas.

1 Introduction

Only two simple key principles, namely the best effort paradigm for data transport and the end-to-end philosophy for traffic control and management, have prevented the current Internet to collapse while it underwent an exponential growth. However, the best effort model with no service guarantee is no longer acceptable in view of the proliferation of interactive applications such as Internet telephony or video conferencing. Along the past years, the IETF has standardized several frameworks to meet the demand for Quality of Service (QoS) support as, for instance, RSVP [1] in the control plane or Differentiated Services [2] in the architecture area.

For scalability reasons, it is impractical to enforce performance guarantees at a fine-grained level (e.g. per flow) and so the QoS requirements will most likely be applied to aggregate traffic streams rather than to individual flows. An aggregate traffic stream bundles a number of flows for common treatment between two nodes in a network. Such form of aggregation clearly simplifies the allocation of network resources and promotes the deployment of QoS frameworks notably.

---

[*] This work was supported by the project TIC2000-1126 of the "Plan Nacional de Investigación Científica, Desarrollo e Innovación Tecnológica" and by the "Secretaría Xeral de I+D da Xunta de Galicia" through the grant PGIDT01PX132202PN.

J. Solé-Pareta et al. (Eds.): QofIS 2004, LNCS 3266, pp. 104–113, 2004.
© Springer-Verlag Berlin Heidelberg 2004

For similar reasons, service providers are likely to offer performance commitments at the aggregate level, either to end users or to other peer providers. Within this context, and irrespective of the end-to-end transport-level protocols used by individual flows, it is essential that aggregated traffic is responsive to network congestion at an aggregate level. Several approaches proposed to apply congestion control to aggregates such as *Congestion Manager* [3] and *Coordination Protocol* [4] architectures require the modification of user applications at the endpoints and, therefore, they are not suitable to be used in the Internet backbone. In contrast, *TCP Trunking* [5,6] is an interesting method for providing the management of aggregate traffic streams without changing neither user protocols, nor applications. This technique employs some control TCP connections to regulate the flow of each aggregate traffic stream into the network.

In this paper we introduce *Ping Trunking*, an enhancement of the *TCP Trunking* technique that improves the original one by changing the control overlay. In particular, instead of using several TCP connections to manage each aggregate, a new preventive Vegas-like connection is used. *Ping Trunking* has several advantages over *TCP Trunking*. Foremost, it does not cause sharp variations in the transmission rate of the trunks and reduces the size of the queues at the core of the network, due to the dynamics of its Vegas-like congestion response. Additionally, it introduces far less overhead and makes trunks operation simpler.

The remainder of this paper is organized as follows. In Sect. 2, we give a brief overview of *TCP Trunking*, the traffic management technique which *Ping Trunking* is based on. Section 3 describes the *Ping Trunking* mechanism. In Sect. 4, we present a simple analysis of its performance. Section 5 contains some ns-2 simulation experiments to validate the theoretical analysis. *TCP* and *Ping Trunking* techniques are compared in Sect. 6. We end the paper with some concluding remarks and future lines in Sect. 7.

2   TCP Trunking

A TCP trunk [5,6] is an aggregate traffic stream where data packets are transmitted at a rate dynamically determined by the TCP congestion control algorithm [7]. Each trunk carries a varying number of user flows between two nodes of the network. The flow of the aggregated traffic is regulated by a control TCP connection established between the two edges of the trunk. This control connection injects control TCP packets into the network to probe its congestion level (Fig. 1). *TCP Trunking* fully decouples control from data transmission: the introduction of control packets is not conditioned by the user data protocols, but it is based on control packet drops as in usual TCP connections. In addition, trunks will not retransmit user packets if they are lost. If it is required, retransmissions should be handled by the user applications on the end hosts.

TCP Trunking is implemented in the following manner. User packets arriving at the sender edge are temporarily queued into the buffer of the trunk. After a control packet is transmitted, user packets can be forwarded, totalling at most *vmss* (virtual maximum segment size) bytes. When *vmss* user bytes

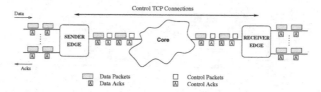

Fig. 1. TCP Trunking basics

have been transmitted, the control connection will generate and send a control packet if its control TCP congestion window allows it. This way, control packets regulate the transmission of user packets and then, the loss of control packets in the network not only slows down the transmission rate of control packets but also reduces the transmission rate of user data.

Multiple control TCP connections are employed with each trunk. If each trunk were regulated by a single control connection, a control packet loss would cause the entire trunk to halve its sending rate, as dictated by the TCP dynamics. When dealing with traffic aggregates, this abrupt reduction in transmission rate on packet losses is undesirable. In [6], the authors argue that four control connections per trunk are enough to produce smooth bandwidth transitions.

To conclude with this overview, it is important to point out that both user and control packets must follow the same path between the edges of the trunk to ensure that control connections are probing the proper available bandwidth. This assumption can be absolutely guaranteed if trunks are run on top of ATM virtual circuits or MPLS label-switched paths [8].

3   Ping Trunking

Our proposal, named *Ping Trunking*, borrows from *TCP Trunking* the concept of decoupling control from data in IP networks, but the original technique has been improved by changing the control overlay. In *Ping Trunking*, only a single control connection is established between the two network edges of each trunk. This connection controls the flow of user packets into the core of the network using a Vegas-like congestion control mechanism that adapts its congestion window (*cwnd*) based on the observed changes in the round-trip time (RTT), and not only on the loss of control packets. Figure 2 provides greater detail on the operation of this mechanism. Incoming user packets at the sender edge are classified as belonging to a particular trunk and queued in the corresponding buffer.[1] User packets can only be forwarded when credit for their trunk is available. The credit value represents the amount of user data allowed to be forwarded. When

---

[1] The identification method for classifying packets from different aggregates could be based on the value of several header fields, such as source or destination addresses, source or destination port numbers, ATM virtual circuit identifier or MPLS label.

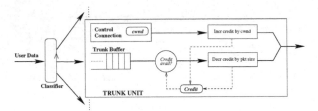

Fig. 2. Ping Trunking diagram

an user packet is sent, the credit is decremented by the size of the packet. When a control packet is sent, the credit is incremented by *cwnd* bytes. Therefore, the transmission of user data is regulated by both the forwarding of control packets and the *cwnd* value.

3.1   Control Connection

Control TCP connections have been substituted by a new control connection whose main task consists of measuring the RTT between the edges of the trunk accurately. The designed Vegas-like congestion control mechanism will employ this RTT estimate when adapting the transmission rates of the trunks. Let us start by giving a brief description of the operations accomplished by the control connection. When the first user packet arrives to the trunk, a control packet is generated and sent. For each control packet that reaches the receiver edge of the trunk, its corresponding acknowledgment (ack) is generated. The arrival of an ack back at the sender edge triggers the transmission of a new control packet. Therefore, the control connection only sends control packets on reception of acks. To avoid the starvation of control connections, a waiting time-out timer is needed. This timer is started every time a control packet is sent. If the ack of the last control packet sent does not arrive to the sender edge before the timer expires, the control connection will consider that the packet has been lost and it will send a new control packet.[2]

Control connections use a method similar to the one used in the TCP estimation of RTT. To carry out this task, they timestamp with its local time every control packet and this timestamp is echoed in the acks. The value of the last RTT sample observed is computed as the difference between the current time and the timestamp field in the ack packet. The RTT is eventually estimated using an exponential moving average taken over RTT samples.

---

[2] The control connection proposed is analogous to the *ping* command used to send ICMP ECHO_REQUEST/REPLY packets to network hosts. This explains why we refer to our mechanism as *Ping Trunking*.

3.2   Vegas-Like Congestion Control Mechanism

The transmission rate of each trunk should be able to update dynamically according to the current network conditions. We propose the use of a Vegas-like congestion control mechanism to discover the available bandwidth that each trunk should obtain. TCP Vegas [9] is an implementation of TCP that employs proactive techniques to increase throughput and decrease packet losses. The congestion control mechanism introduced by Vegas gives TCP the ability to detect incipient congestion before losses are likely to occur. We devise a similar mechanism adapted to trunks.

Upon receiving each ack, control connections calculate the expected throughput and the actual throughput as in TCP Vegas. If it is assumed that active trunks are not overflowing the path, the expected throughput can be calculated as $cwnd/d$, where d is the round-trip propagation delay and can be estimated as the minimum of all measured RTTs. On the other hand, the actual throughput is given by $cwnd/D$, where D is the RTT estimation. These throughputs are compared and then, control connections adjust their congestion windows accordingly. Let $Diff$ be the difference between the expected and the actual throughput:

$$Diff = Expected - Actual = \left(\frac{cwnd}{d} - \frac{cwnd}{D} \right) d . \tag{1}$$

The $Diff$ value has been scaled with the minimum RTT so that $Diff$ can be seen as the amount of user data in transit.

There are two thresholds defined: α, β, with $\alpha \leq \beta$. When $Diff < \alpha$, a trunk is allowed to increment its amount of user data in transit, and therefore, the assigned control connection increases its congestion window linearly. If $Diff > \beta$, the control connection is forced to decrease its congestion window linearly. In any other case, the congestion window remains unchanged. This mechanism stabilizes the value of the congestion window and reduces packet drops. If a control packet loss is detected, the congestion window will be halved, but this should happen sporadically.

3.3   Avoiding Burstiness

Our mechanism as described so far may introduce long bursts of data packets into the network of length $cwnd$ bytes. Nevertheless, we can easily avoid this undesired effect fixing a maximum burst length (max_burst_length) after which trunks have to wait $burst\_delay$ seconds before sending the next packet. This value is computed in the following manner:

$$burst\_delay = D \cdot \frac{max\_burst\_length}{cwnd} . \tag{2}$$

4   Performance Analysis

Consider a network shared by a set T of trunks. Let C denote the network capacity. Each trunk $i \in T$ has associated a subscribed target rate r_i. The overall aggregated demand R is the sum of the subscribed target rates for all active trunks. If $R < C$, the excess unsubscribed bandwidth should be distributed among trunks in proportion to the contracted target rates. Therefore, each trunk should receive ideally the following share of bandwidth:

$$r_i + (C - R)\frac{r_i}{R} = \frac{r_i C}{R} \quad \text{where} \quad R = \sum_i r_i . \tag{3}$$

According to one interpretation of Vegas [10], congestion windows of control connections must satisfy the following equation in the equilibrium (we assume for simplicity that $\alpha_i = \beta_i$):

$$\left(\frac{cwnd_i}{d_i} - \frac{cwnd_i}{D_i} \right) d_i = \alpha_i . \tag{4}$$

On the other hand, the transmission rate x of each trunk is determined by the $cwnd$ value of its corresponding control connection ($cwnd = xD$). Substituting this in (4), we have

$$\left(\frac{x_i D_i}{d_i} - \frac{x_i D_i}{D_i} \right) d_i = \alpha_i , \tag{5}$$

and, from (5), it follows that

$$x_i = \frac{\alpha_i}{D_i - d_i} . \tag{6}$$

The RTT can be calculated as the sum of two delays: the round-trip propagation delay (d) and the queueing delay (B/C), where B denotes the total backlog buffered in the network. Then, from (6), and using $D_i = d_i + B/C$, the transmission rate of each trunk can be expressed as

$$x_i = \frac{\alpha_i C}{B} . \tag{7}$$

Finally, equating (3) and (7) yields the suitable value of α threshold that permits to allocate to each trunk its desired share of bandwidth:

$$\alpha_i = \frac{r_i B}{R} . \tag{8}$$

Therefore, to compute the α threshold, each trunk must know both B and R parameters. The B parameter should have a fixed low value set by the network manager to encounter small queues at the core. However, the necessity of determining the overall aggregated demand in all edge nodes may complicate our proposal substantially.

Fortunately, we can demonstrate that it is not required to know the value of the overall demand very accurately. Assume $\hat{R} \neq R$ was used as the aggregated demand. The total backlog $\hat{B}$ actually buffered in the network is obtained as the sum of α thresholds of all competing trunks. Then,

$$\hat{B} = \sum_i \alpha_i = \sum_i \frac{r_i B}{\hat{R}} = \frac{B}{\hat{R}} \sum_i r_i = \frac{BR}{\hat{R}} . \tag{9}$$

From (7), and using $\hat{B}\hat{R} = BR$ derived from (9), we can conclude that the fairness condition is still satisfied although the $\hat{R}$ value employed is false:

$$x_i = \frac{\alpha_i C}{\hat{B}} = \frac{r_i BC}{\hat{R}\hat{B}} = \frac{r_i C}{R} \ . \tag{10}$$

5   Performance Evaluation

We have implemented *Ping Trunking* in the ns-2 simulator [11]. The following experiments have been conducted to validate the analysis performed in the previous section. Figure 3 shows the network topology employed. It consists of three edge nodes and one core node belonging to a particular domain. Each edge node is connected to a client or traffic source. We consider two competing aggregates: aggregate 1 running from client 1 to client 3 and aggregate 2 running between clients 2 and 3. Therefore, both aggregates pass through a single bottleneck (link C-E3). Both aggregates comprise 20 TCP flows. All TCP connections established are modeled as eager FTP flows that always have data to send and last for the entire simulation time. We use the TCP New Reno implementation and the size of user packets is set to 1 000 bytes.

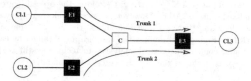

Fig. 3. Network topology. Every link has a 10 Mbps capacity and a propagation delay of 10 ms. All queues are FIFO buffers and are limited to 50 packets

A Ping trunk is used to manage each aggregate traffic stream: trunk 1 manages aggregate 1 and trunk 2 manages aggregate 2. The sender and receiver of trunk 1 are E1 and E3, respectively, while E2 and E3 are the sender and receiver of trunk 2, respectively. Control connections send 44-byte packets.[3] Each trunk buffer is a simple FIFO queue with capacity for 50 packets. A total backlog of 10 000 bytes (10 packets) is chosen to be buffered in the network. The maximum burst length is set to 5 000 bytes (5 packets).

We run the simulations for 50 seconds and measure the average throughput achieved by each aggregate over the whole simulation period. Each simulation experiment is repeated 10 times changing slightly the initial transmission time of each TCP flow and then, an average is taken over all runs.

---

[3] Control connections do not actually transmit any real data, so control packets only consist of the TCP/IP header with no payload (40 bytes plus 4 additional bytes required to estimate RTT).

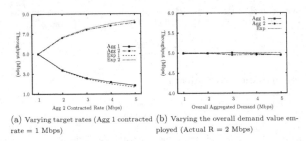

(a) Varying target rates (Agg 1 contracted rate = 1 Mbps)

(b) Varying the overall demand value employed (Actual R = 2 Mbps)

Fig. 4. Performance evaluation results

In the first experiment, aggregate 1 contracts a fixed throughput of 1 Mbps while the subscribed target rate of aggregate 2 varies from 1 to 5 Mbps. Figure 4(a) shows both the obtained and the expected results. As required, the network bandwidth is fairly distributed between the two competing aggregates according to their contracted rates. In the second experiment, both aggregates contract a throughput of 1 Mbps setting, therefore, the overall aggregated demand R to a fixed value of 2 Mbps. However, we force trunks to employ a different value when calculating their α thresholds. Thus, the $\hat{R}$ value employed varies from 1 to 5 Mbps. Figure 4(b) depicts the throughput obtained by each trunk. Simulation results confirm theoretical analysis: our technique still works despite of using an incorrect value of the overall aggregated demand.

6   Comparison with TCP Trunking

In this section, we compare *TCP* and *Ping Trunking* techniques. Foremost, *Ping Trunking* facilitates the operation of trunks: while *TCP Trunking* assigns four control TCP connections to each aggregate to produce a smoother behavior, each Ping trunk is regulated by a unique control connection. Therefore, our technique is clearly simpler, easier to understand and it operates more efficiently.

In addition, despite of managing each aggregate with a single control connection, trunks do not suffer from sharp variations on their sending rate. The applied Vegas-like congestion control mechanism reduces control packet drops and stabilizes the transmission rate while maintaining the ability to adapt to changing conditions. To confirm this interesting feature, we have conducted an extra simulation experiment using the same network topology with the same parameters as in the previous section. In this experiment, we compare the variations in the transmission rate of TCP and Ping trunks. Firstly, we use TCP trunks to regulate the two competing aggregates. In this case, each aggregate will be managed by four control TCP connections. Then, we use Ping trunks to manage them. The

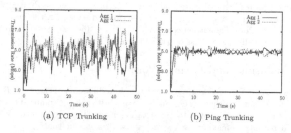

(a) TCP Trunking   (b) Ping Trunking

Fig. 5. Transmission rate comparison

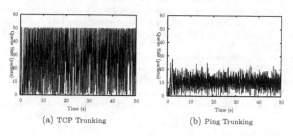

(a) TCP Trunking   (b) Ping Trunking

Fig. 6. Core queue size comparison

obtained results are shown in Fig. 5.[4] It can be seen how *Ping Trunking* in effect causes smoother variations in the sending rate of the trunks than *TCP Trunking*. We have also compared the core queue size for the two techniques. As shown in Fig. 6, the core queue size is greatly decreased with *Ping Trunking*. Moreover, with our proposal, no packet loss takes place at the core node. This result can be explained by the different congestion control mechanisms used in each scheme. While the Reno algorithm induces losses to learn the available network capacity, Vegas sources adjust their rates so as to maintain α packets buffered in their paths. Additionally, a smaller queue size helps to reduce both the latency and the jitter suffered by the packets.

The last issue to take into account is that both techniques add some overhead due to the introduction of control packets. *TCP Trunking* technique injects a control packet into the network each *vmss* data bytes. In the experiments conducted to study *TCP Trunking* performance [6], the *vmss* value has been set to 1.5 or 3.0 kBytes to obtain good results. In *Ping Trunking*, a control packet is injected into the network each *cwnd* data bytes. This value is not fixed and

---

[4] We performed exhaustive simulations to verify that the examples shown are representative.

depends on both the trunk rate and the RTT, but, as can be derived from the previous experiments, the *cwnd* is considerably greater than the *vmss* used in TCP trunks for common networks and contracts (5-45 kBytes). Therefore, Ping trunks usually add far less overhead than TCP trunks.

7   Conclusions and Future Work

We have presented *Ping Trunking*, an enhanced modification of the *TCP Trunking* technique, able to share the available bandwidth among competing aggregates in proportion to their subscribed target rates. The new control overlay proposed permits to manage each aggregate with a unique control connection that does not introduce large variations in the transmission rate of the handled aggregates. In addition, the size of the shared queues at the core of the network is reduced significantly helping to support lower latency and jitter times. Lastly, the overhead added due to the introduction of control packets is also greatly decreased. Thus, the bandwidth used by control packets can be considered practically negligible.

There are some issues requiring further investigation. Firstly, we should consider the stability problems shown in Vegas as delay increases [12]. This could be specially serious when we use *Ping Trunking* in the core of networks with large delays. Also, our current technique is applied to one-to-one network topologies. In future work, we plan to extend this technique to one-to-many topologies.

References

1. Braden, B., Zhang, L., Berson, S., Herzog, S., Jamin, S.: Resource reSerVation Protocol (RSVP) – Version 1 Functional Specification. RFC 2205 (1997)
2. Blake, S., Black, D., Carlson, M., Davis, E., Wang, Z., Weiss, W.: An architecture for differentiated services. RFC 2475 (1998)
3. Balakrishnan, H., Rahul, H.S., Seshan, S.: An integrated congestion manager architecture for internet hosts. In: Proc. of ACM SIGCOMM. (1999)
4. Ott, D.E., Sparks, T., Mayer-Patel, K.: Aggregate congestion control for distributed multimedia applications. In: Proc. of IEEE INFOCOM. (2004)
5. Chapman, A., Kung, H.T.: Traffic management for aggregate IP streams. In: Proc. of 3rd Canadian Conference on Broadband Research. (1999)
6. Kung, H.T., Wang, S.Y.: TCP Trunking: Design, implementation and performance. In: Proc. of 7th Int. Conference on Network Protocols. (1999)
7. Jacobson, V.: Congestion avoidance and control. In: Proc. of ACM SIGCOMM. (1988)
8. Rosen, E., Viswanathan, A., Callon, R.: Multiprotocol label switching architecture. RFC 3031 (2001)
9. Brakmo, L., O'Malley, S., Peterson, L.: TCP Vegas: New techniques for congestion detection and avoidance. In: Proc. of ACM SIGCOMM. (1994)
10. Low, S., Peterson, L., Wang, L.: Understanding Vegas: A duality model. In: Proc. of ACM SIGMETRICS. (2001)
11. NS: Network Simulator, ns. http://www.isi.edu/nsnam/ns/ (2003)
12. Choe, H., Low, S.: Stabilized Vegas. In: Proc. of IEEE INFOCOM. (2003)

An Aggregate Flow Marker for Improving TCP Fairness in Multiple Domain DiffServ Networks

Ji-Hoon Park[1], Kyeong Hur[1], Sung-Jun Lee[1], Choon-Geun Cho[1], NhoKyung Park[2], and Doo-Seop Eom[1]

[1] Dept. of Electronics and Computer Engineering, Korea University, Seoul, Korea
{jhpak, sjlee, ckcho}@final.korea.ac.kr, {hkyeong, eomds}@korea.ac.kr
[2] Dept. of Information and Telecommunication Eng., Hoseo University,
Asan-Si Chungnam, Korea
nkpark@office.hoseo.ac.kr

Abstract. The differentiate services (DiffServ) is proposed to provide packet level service differentiations in a scalable manner. To provide end-to-end service differentiation to users having a connection over multiple domains, an intermediate marker is necessary at the edge routers. The intermediate marker has a fairness problem among the TCP flows due to the TCP's congestion control. In this paper, we propose an aggregate fairness marker (AFM) as an intermediate marker which works with a user flow three color marker (uf-TCM) operating as a flow marker for a user flow. Through the simulations, we show that the proposed AFM can improve fairness to the TCP flows with different RTTs.

1   Introduction

To support the quality of service (QoS) [1] on IP based networks, the Internet Engineering Task Force (IETF) has proposed the Differentiated Services (Diff-Serv) [2]. DiffServ provides simple and predefined per-hop behavior (PHB) level service differentiation. The IETF has defined one class for Expedited Forwarding (EF) PHB and four classes for Assured Forwarding (AF) PHB [3][4]. AF PHB allows an Internet service provider (ISP) to provide different levels of forwarding assurances according to the user profile.

Recent works on DiffServ deal with an aggregate flow at edge routers [5]. The random early demotion promotion (REDP) was proposed as an aggregate flow management [6]. We think that the main contribution of REDP is the introduction of the packet demotion and promotion concept in DiffServ networks. REDP achieves good UDP fairness in demotion and promotion. However, it fails to give good fairness to the TCP flows with different RTTs. In the case of the TCP flows, their transmission rates highly depend on the round trip time (RTT) due to the TCP's congestion control algorithm, which brings about the unfairness of throughput among the TCP flows with different RTTs. Note that it is very difficult to resolve this problem at the aggregate flow level without the individual state information of each TCP flow such as RTT and the sending rate.

J. Solé-Pareta et al. (Eds.): QofIS 2004, LNCS 3266, pp. 114–123, 2004.
© Springer-Verlag Berlin Heidelberg 2004

In this paper, to resolve the TCP fairness problem of an intermediate marker such as REDP, we propose an aggregate fairness marker (AFM) as an intermediate marker. Also, it works with a user flow three color marker (uf-TCM), which is proposed as a flow marker for a user flow. The fundamental assumption of the proposed marker is that the relative transmission rates among TCP flows do not remain constant so that at certain times the TCP flow with a longer RTT sends relatively more packets than the TCP flow with a shorter RTT. In that case, if we decrease the demotion probability and increase the promotion probability at those times, the TCP flow with a longer RTT gets more advantages from demotion and promotion than the TCP flow with a shorter RTT. The problem, then, is how to determine the demotion and promotion probabilities for improving TCP fairness using only the aggregate flow state information. We measure the green, yellow, red and total rates at the edge router, and infer the individual TCP flow states from the measurement results. Using the inferred individual TCP flow state information, the demotion and promotion probabilities are determined. Through the simulations, we show that the AFM can improve TCP fairness as well as link utilization without per-flow management in multiple domain environments.

2   A Proposed Scheme

Our proposed scheme is composed of a uf-TCM and an AFM operating independently as shown in Fig. 1. A flow marker is necessary for monitoring the user profile and it initially marks the packets from a user flow according to its profile. The proposed uf-TCM is a flow marker that monitors each user profile and marks the packets from the flow as green, yellow, or red. The proposed AFM is an intermediate marker that monitors the aggregate traffic and according to incoming traffic situation, it fairly performs packet demotion and promotion.

2.1   User Flow Three Color Marker(uf-TCM)

For assured services, in general, the packets of each user flow are initially marked as either green or red according to their profile obedience. A simple method for monitoring the traffic profile obedience of the flow is to use a token bucket

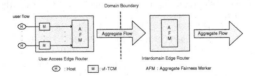

Fig. 1. The structure of the proposed scheme at Edge Router

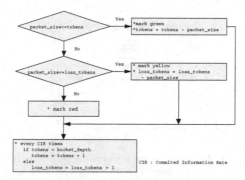

Fig. 2. The proposed uf-TCM algorithm

marker. TCP flow sends packets in a burst within an RTT interval; thus, there are some time intervals during which no traffic is generated from a user source. During these time intervals, token loss is likely to occur from the token bucket; thus, it is hard to guarantee minimum throughput to a user [7]. In order to improve this situation, we use a three color marking process instead of the normal two color marking process at each user flow marker. For this purpose, we propose a user flow three color marker (uf-TCM) based on a simple token bucket algorithm. The operation of the proposed uf-TCM is shown in Fig. 2. A packet is marked as green if there are enough tokens in the token bucket. It is marked as yellow if there are not enough tokens in the token bucket but there are enough loss tokens. Otherwise it is marked as red. Note that if a packet is marked as yellow in this way, the throughput of green packets of a flow cannot grow beyond the contracted rate even when yellow packets are promoted to green packets in the following domains. This is because yellow packets can consume only the loss tokens. That is, the uf-TCM ensures that the packets from a flow never disobey its traffic profile which is agreed upon between a user and an ISP. However, the throughput of a TCP flow can be improved because yellow packets have chance to be promoted to green in the following domains. In our scheme, we assume that yellow packets have the same drop precedence as red packets at the core routers. Thus, we can use the RIO buffer management scheme [8].

2.2   Aggregate Fairness Marker(AFM)

AFM measures the green, yellow, red and total rates using the rate estimator. The marking decider infers the individual TCP flow states from the measurement results and determines the demotion and promotion probabilities using the inferred individual TCP flow state information and the token level of the token bucket (i.e., the number of the remaining tokens in the token bucket). Then, the

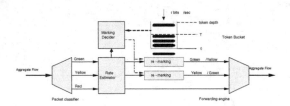

Fig. 3. The functional structure of AFM

marking decider determines whether green and yellow packets are remarked to yellow and green packets, respectively. Figure 3 shows the functional structure of AFM.

AFM has two operation modes. If the rate of green packets is beyond the aggregate contract rate or negotiated rate between two domains, it enters the demotion mode in which the token bucket is divided into the demotion region and the balance region, as shown in Fig. 4. And, whenever the token level of bucket is within the demotion region, AFM fairly demotes green packets to yellow to prevent the phase effect. On the other hand, if the green rate is under the contract rate, it enters the promotion mode in which the token bucket is divided into the balance region and the promotion region. And, whenever the token level is within the promotion region, it fairly promotes yellow packets to green to increase link utilization. The balanced region is needed to prevent unnecessary packet demotion and promotion [6].

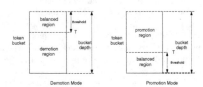

Fig. 4. The operation modes of AFM mode

As previously stated, due to the TCP's congestion control algorithm, the TCP flow with a shorter RTT generates more green and yellow packets than the TCP flow with a longer RTT, which brings about the unfairness of throughput among the TCP flows with different RTTs as well as lowering link utilization. To resolve this problem, AFM measures the green, yellow, red and total rates at an edge router, infers the individual TCP flow states from the measurement results and determines the demotion and promotion probabilities using the inferred individual TCP flow state information. This is done to improve TCP fairness using only the aggregate flow state information. In the following, we explain the reason why this works well.

We first consider the over-provisioning case, i.e., the sum of the user contract rates is lower than the aggregate contract rate. In this case, a promotion situation is dominant. Figure 5(a) shows the ratios of the incoming red traffic rate to the total incoming traffic rate and the incoming yellow traffic rate to the total incoming traffic rate under an over-provisioning situation at an edge router. It is well known that the throughput of TCP oscillates due to its congestion control algorithm. Thus, the ratio of the incoming red traffic rate to the total incoming traffic rate at an edge router also oscillates. If the ratio is close to the maximum value 'A' as shown in Fig. 5(a), we can infer that the sending rates of the TCP flows with a shorter RTT are also close to the maximum value 'A', as shown in Fig. 5(b). That is, the TCP flows with a shorter RTT determine the dynamics of the aggregate flow. Therefore, if we increase the promotion probability in this case, it may bring about unfairness because of the increased probability that the TCP flows with a shorter RTT will receive more promoted packets than the TCP flows with a longer RTT. Therefore, it is desirable to lower the promotion probability in this case.

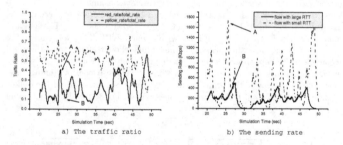

a) The traffic ratio   b) The sending rate

Fig. 5. The traffic ratio in over-provisioning case and TCP behavior

On the other hand, if the ratio is close to the minimum value 'B' as shown in Fig. 5(a), we can infer that the sending rates of the TCP flows with a shorter RTT are also close to the minimum value 'B' as shown in Fig. 5(b). This is because when the sending rates of the TCP flows with a shorter RTT reach maximum values, the packet dropping probability of the TCP flows with a shorter RTT becomes much higher than that of the TCP flows with a longer RTT, and thus the sending rates of the TCP flows with a shorter RTT become lower. But, by this reduction of the sending rates, the available bandwidth for the TCP flows with a longer RTT increases, so that their sending rates become higher until the time corresponding to the minimum value 'B'. After that, the ratio increases again since the TCP flows with a shorter RTT begin recovering their previous transmission window. From the above observation, if we increase the promotion probability when TCP flows with a longer RTT send relatively more packets than TCP flows with a shorter RTT, a TCP flow with a longer RTT gets more advantages from promotion than a TCP flow with a shorter RTT, which means that TCP fairness is improved. Similarly, we decrease the promotion probability when the TCP flow with a shorter RTT sends relatively more packets than the TCP flow with a longer RTT. Therefore, we can improve TCP fairness by using the ratio of the incoming red traffic rate to the total incoming traffic rate without per-flow management. For the over-provisioning case, the packet promotion probability P_{promo} is determined as follows:

$$P_{promo} = 1 - \frac{red\_rate}{total\_rate} \cdot \left(1 - \frac{number\_of\_tokens - threshold}{bucket\_depth - threshold}\right) \qquad (1)$$

We next consider the under-provisioning case, i.e., the sum of the user contract rates is higher than the aggregate contract rate. Even in the under-provisioning case, due to the TCP congestion control algorithm, the incoming green traffic rate is hardly beyond the aggregate contract rate at an edge router. Therefore, a promotion situation is dominant. If the network is under-provisioned, the packets from a TCP flow cannot consume all of the loss tokens in most cases and thus TCP flows can hardly generate any red traffic, as shown in Fig. 6(a). However, the TCP flows with a shorter RTT generate more yellow traffic than the TCP flows with a longer RTT. Therefore, unlike the over-provisioning case, the ratio of the incoming yellow traffic rate to the total incoming traffic rate should be used for determining the promotion probability when the network is under-provisioned. Similar to the over-provisioning case, it is desirable to lower the promotion probability when the ratio of the yellow traffic rate to the total traffic rate becomes higher. For the under-provisioning

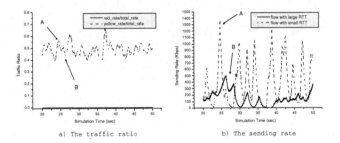

a) The traffic ratio   b) The sending rate

Fig. 6. The traffic ration in under-provisioning case and TCP behavior

case, the packet promotion probability P_{promo} is determined as follows:

$$P_{promo} = 1 - \frac{yellow\_rate}{total\_rate} \cdot \left(1 - \frac{number\_of\_tokens - threshold}{bucket\_depth - threshold}\right) \quad (2)$$

Finally, we consider the case that the incoming green traffic rate is beyond the negotiated contract rate at interdomain edge routers. In this case, green packets should be demoted to yellow. In the demotion situation, we don't need to separately consider the over-provisioning case and the under-provisioning case because the ratio of the incoming green traffic rate to the aggregate contract rate at an interdomain edge router also oscillates, according to input traffic condition. In a demotion mode, the higher the incoming rate of green packet is, the more the demotion probability, P_{demo}, must be increased. For both provisioning cases, the packet demotion probability P_{demo} is determined as follows:

$$P_{demo} = \left(1 - \frac{contract\_rate}{green\_rate}\right) \cdot \left(1 - \frac{number\_of\_tokens - threshold}{bucket\_depth - threshold}\right) \quad (3)$$

Note that P_{promo} and P_{demo} are determined by using not only the traffic ratios but also the token level of the token bucket. This is because the token level reflects the difference between the incoming green traffic rate and the contract rate. For example, if there are enough tokens in the bucket, by making the promotion probability higher, we can increase link utilization in a promotion situation, On the other hand, by making the demotion probability lower, we can prevent lowering of link utilization in a demotion situation.

AFM algorithm

```
if contract_rate => green_rate
    enter Promotion mode
    if threshold => number_of_tokens
        enter Balanced region,
        P_promo = 0
    elseif threshold < number_of_tokens
        enter Promotion region
        if red_rate/total_rate > 0.01
            decision over_provisioning
            R_promo = red_rate/total_rate
        elseif red_rate/total_rate <= 0.01
            R_promo = yellow_rate/total_rate
        P_promo = 1 - R_promo*(1-(number_of_tokens - threshold)
                /(bucket_depth - threshold))
elseif contract_rate < green_rate
    enter Demotion mode
    if threshold => token_depth - number_of_tokens
        enter Balanced region,
        P_demo = 0
    elseif threshold < number_of_tokens
```

```
enter Demotion region
R_demo = 1 - contract_rate/green_rate
P_demo = R_demo*(1-(number_of_tokens)
           /(bucket_depth - threshold))
```

In the proposed AFM algorithm, The operation mode is determined by a comparison between the contract rate and the rate of green packets. When the ratio of the incoming red traffic rate to the total incoming traffic rate at an edge router is under '0.01' in a promotion situation, as shown In the proposed AFM algorithm, we assume that it is an under-provisioning case. This is because there are some possibilities that TCP flows with a shorter RTT may send some red packets in spite of the under-provisioning situation. The AFM uses a time sliding window (TSW) [8] to estimate the incoming traffic rates at edge router.

3   Performance Study

In this section, we analyze the performance of the proposed scheme, in comparison with the REDP combined with a two color token bucket marker acting as a flow marker, and the REDP combined with the proposed uf-TCM. We implemented a simple RIO queue [8] with parameters $(q_{min}/q_{max}/p_{max}) = 45/60/0.02$ for in packets and $20/40/0.2$ for out packets in $ns2$ simulator [9]. We set the bucket depth as one packet for a flow marker and as 60 packets for an intermediate marker. In the case of REDP, T_L is set to 15 packets, T_H is set to 45 packets, MAX_{demo} is set to 0.5, and MAX_{promo} is set to 0.5 [6]. In the case of AFM, we set appropriately the range of the balanced region as 10 packets by the number of simulations

3.1   Multiple Domain Scinario

Figure 7 shows the simulation model for the multiple domain case. There are 10 TCP hosts and each host contracts 400Kbps.

The RTT of each flow is ranged from 30ms to 84ms with 6ms differences and each flow connection spans through domain 1, 2. The contract rate of the aggregate flow is 4Mbps in domain 1. So, there is exact-provisioning in domain 1. In domain 2, the interdomian negotiated contract rate is a Mbps as shown in Fig. 7 and we set a as 6Mbps for the over-provisioning (150% provisioning) case, as 4Mbps for the exact-provisioning (100% provisioning) case, as 3Mbps for the under-provisioning (75% provisioning) case, respectively.

Figures 8 show the average total throughput of each host for the three cases. Our proposed scheme shows the best results for all cases. The REDP combined with the two color token bucket marker corresponds to the original REDP, cannot improve TCP fairness. However, in the case of the REDP combined with the proposed uf-TCM, compared with the above original REDP scheme, as well as link utilization, TCP fairness is improved to some degree because the yellow

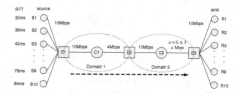

Fig. 7. The simulation model for the multiple domain case

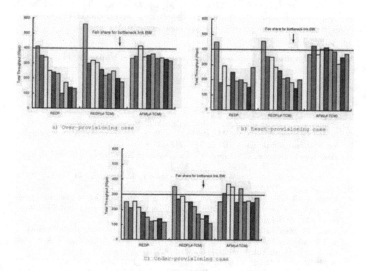

Fig. 8. The average total throughput of each host for the three cases

packet is generated from each flow source by the uf-TCM and has a chance to be promoted to a green packet at the edge router. This means that the throughput of each flow can be improved. However, compared with our proposed scheme, the performance improvement is restricted, due to the simple probability decision of REDP marker that depends only on the token level. For the multiple domain simulation model, our simulation results show that the fairness indexes [10] of REDP/REDP(uf-TCM)/AFM(uf-TCM) are 0.874/ 0.897/ 0.988 for the over-provisioning case, 0.848/ 0.943/ 0.983 for the exact-provisioning case, and 0.928/0.892/0.976 for the under-provisioning case, respectively.

4 Conclusion

To provide end-to-end service differentiation of assured services in DiffServ, as well as a flow marker that initially marks user flow, an intermediate marker is necessary. To the best of our knowledge, there have been no works to resolve the TCP fairness problem of intermediate markers because it is very difficult to resolve this problem without individual TCP flow information. In this paper, to resolve this TCP fairness problem of an intermediate marker, we have proposed an aggregate fairness marker (AFM) as an intermediate marker which works with a user flow three color marker (uf-TCM) operating as a flow marker for a user flow. The proposed AFM improves the fairness among the TCP flows with different RTTs without per-flow management in multiple domains.

Acknowledgements. This research was supported by University IT Research Center Project.

References

1. Ferguson, P., Huston, G.: Quality of Service. Willey, New York (1998)
2. Blake, S., Blake, D., Carlos, M., Davies, E., Wang, Z., Weiss, W.:An architecture for Differentiated Services. IETF, RFC 2475, (1998)
3. Jacobson, V., Nichols, K., Poduri, K.: An Expedited Forwarding PHB. IETF, RFC 2598, (1999)
4. Heinanen, J., Baker, F., Weiss, W., Wroclawski, J.: Assured Forwarding PHB Group. IETF, RFC 2597, (1999)
5. Yeom, I., Reddy, A. L. N.: Impact of marking strategy on aggregate flows in a Differentiated Services network. In Proceedings of IEEE International Workshop on Quality of Service. (1999)
6. Wang, F., Mohapatra, P., Mukherjee, S., Bushmitch, D.: Random Early Demotion and Promotion Marker for Assured Services. IEEE Journal of Selected Areas on Communications. Vol. 18 (2000)
7. Feng. W., et. al.: Understanding and Improving TCP Performance Over Networks with Minimum Rate Guarantee. IEEE/ACM Transactions on Networking. Vol. 7 (1999) 173–187
8. Clark, D., Fang, W.: Explict Allocation of Best Effort Packet Delivery Service. IEEE/ACM Transactions on Networking. Vol. 6 (1998) 362–373
9. UCB/LBNL/VINT.: Network Simulator-ns(version2). http://www-mash.cs. berkeley.edu/ns. (1998)
10. Jain, R.: The art of Computer Systems Performance Analysis. John Wiley and Sons Inc, New York (1991) 36

Combined Use of Prioritized AIMD and Flow-Based Traffic Splitting for Robust TCP Load Balancing*

Onur Alparslan, Nail Akar, and Ezhan Karasan

Electrical and Electronics Engineering Department,
Bilkent University, Ankara 06800, Turkey

Abstract. In this paper, we propose an AIMD-based TCP load balancing architecture in a backbone network where TCP flows are split between two explicitly routed paths, namely the primary and the secondary paths. We propose that primary paths have strict priority over the secondary paths with respect to packet forwarding and both paths are rate-controlled using ECN marking in the core and AIMD rate adjustment at the ingress nodes. We call this technique "prioritized AIMD". The buffers maintained at the ingress nodes for the two alternative paths help us predict the delay difference between the two paths which forms the basis for deciding on which path to forward a new-coming flow. We provide a simulation study for a large mesh network to demonstrate the efficiency of the proposed approach in terms of the average per-flow goodput and byte blocking rates.

Keywords: Traffic engineering; load balancing; multi-path routing; TCP.

1 Introduction

IP Traffic Engineering (TE) controls how traffic flows through an IP network in order to optimize the resource utilisation and network performance [4]. In multi-path routing-based TE, multiple explicitly routed paths with possibly disjoint links and nodes are established between the two end points of a network in order to optimize the resource utilisation by intelligent traffic splitting. These explicitly routed paths are readily implementable using standard-based layer 2 technologies like ATM or MPLS or using source routed IP tunnels. The work in [5] proposes a dynamic multi-path routing algorithm in connection-oriented networks where the shortest path is used under light traffic conditions and as the shortest path becomes congested, multiple paths are used upon their availability in order to balance the load. Recently, there have been a number of multi-path TE proposals specifically for MPLS networks that are amenable to distributed

---

* This work is supported in part by the Scientific and Technical Research Council of Turkey (Tübitak) under project EEEAG-101E048.

J. Solé-Pareta et al. (Eds.): QofIS 2004, LNCS 3266, pp. 124–133, 2004.
© Springer-Verlag Berlin Heidelberg 2004

online implementation. In [7], the ingress node uses a gradient projection algorithm for balancing the load among the Label Switched Paths (LSP) by sending probe packets into the network and collecting congestion status. Additive Increase/Multiplicative Decrease (AIMD) feedback algorithms are used generally for flow and congestion control in computer and communication networks [6]. The multi-path AIMD-based approach of [17] uses binary feedback information for detecting the congestion state of the LSPs and a traffic splitting heuristic using AIMD is proposed in [17] which ensures that source LSRs do not send traffic to secondary paths of longer length before making full use of their primary paths.

Some multi-path routing proposals cause possible de-sequencing (or reordering) of packets of a TCP flow. This is due to sending successive packets of a TCP flow over different paths with different one-way delays. The majority of the traffic in the current Internet is based on TCP and this packet de-sequencing adversely affects the application-layer performance of TCP flows [10]. In order to avoid packet de-sequencing in multi-path routing, a flow-based splitting scheme that operates on a per-flow basis can be used [16]. In [14], flow-based multi-path routing of elastic flows are discussed. Flow-based routing in the QoS routing context in MPLS networks is described in [11], but the flow awareness requirement inside the core network may cause scalability problems with increasing number of instantaneous flows.

Recently, a new scalable flow-based multi-path TE approach for best-effort IP/MPLS networks is first proposed in [2] which employs max-min fair bandwidth sharing using an explicit rate control mechanism. This approach imposes flow awareness only at the edges of an MPLS backbone. This work demonstrates the performance enhancements attained by the flow-based splitting approach using comparisons with packet-based (i.e., non-flow based) multi-path routing and single-path routing when streaming traffic (i.e., UDP) is used. Significant reductions in packet loss rates are obtained relative to single-path routing in all the scenarios tested. This architecture is then studied for load balancing of elastic traffic (i.e., TCP) with AIMD-based rate control (as opposed to explicit rate for the sake of practicality) using a simple three node topology [3]. It is shown in [3] that flow-based multi-path routing method consistently outperforms the case of single-path. In the current paper, we provide an extensive simulation study of the approach proposed in [3] for TCP load balancing in larger and realistically sized mesh networks.

It is well-known that using alternative longer paths by some sources force other sources whose min-hop paths share links with these alternative paths to also use alternative paths [13]. This fact is called the knock-on effect in the literature and is studied in depth for alternately routed circuit switched networks [9]. Precautions should be taken to mitigate the knock-on effect for example the well-known "trunk reservation" concept in circuit switched networks [9]. One of the key ingredients of our proposed architecture is the use of strict priority queuing that favors packets of primary paths (PP) over those of secondary paths (SP) to cope with the knock-on effect. In this paper, we also compare and contrast strict priority queuing with the widely deployed FIFO queuing in their capabilities to deal with the knock-on effect in the TCP load-balancing context.

The remainder of the paper is organized as follows. In Section 2, we present our TE architecture. We provide our simulation results in Section 3. The final section is devoted to conclusions and future work.

2   Architecture

This section is mainly based on [3] but the proposed architecture is outlined here for the sake of completeness. In this study, we envision an IP backbone network which consists of edge and core nodes (i.e., routers) and which has mechanisms for establishing explicitly routed paths. In this network, edge (ingress or egress) nodes are gateways that originate/terminate explicitly routed paths and core nodes carry only transit traffic. Edge nodes are responsible for per-egress and per-class based queuing, flow classification, traffic splitting, and rate control. Core nodes support per-class queuing and Explicit Congestion Notification (ECN) marking. In this architecture, flow awareness requirement is restricted to edge nodes making the overall architecture scale better than some other flow-based architectures.

Our architecture is based on the following building blocks: (i) queuing in network nodes, (ii) path establishment, (iii) feedback mechanism and rate control, and (iv) traffic splitting. As far as queuing is concerned, the core nodes employ per-class queuing with three drop-tail queues, namely the gold, silver, and bronze queues and strict priority queuing with the highest (lowest) priority given to the gold (bronze) queue, The gold queue is used for Resource Management (RM) and TCP ACK . We envision that ACK packets are identified by the ingress node and the encapsulation header for such packets are marked accordingly. Silver and bronze queues are used for TCP data packets according to the selection of paths as explained below. We assume in this study that edge nodes are single-homed, i.e., they have a link to a single core node. We setup one PP and one SP from an ingress node to every other egress node. We impose that the two paths are link-disjoint within the scope of the core network. The PP is first established as the min-hop path. If there are multiple min-hop paths, the one with the minimum propagation delay is chosen as the PP. In order to find the route for the SP, we prune the links used by the PP and compute the min-hop path in the remaining network graph. A tie in this step is broken similarly. If the connectivity is lost after the first step, we do not establish an SP. We prefer to use this simple path selection scheme since we do not assume a-priori knowledge of the traffic demand matrix.

In this paper, we study two queuing models based on the work in [2]. The first one is FIFO (first-in-first-out) queuing in which all the TCP data packets join the silver queue irrespective of the type of path they ride on. However, this queuing policy triggers the knock-on effect due to the lack of preferential treatment to packets using fewer resources (i.e., traversing fewer hops). Using longer secondary paths by some sources may force other sources whose primary

Table 1. The AIMD algorithm

if RM packet marked as CE
$\quad$ ATR := ATR − RDF × ATR
else
$\quad$ ATR := ATR + RIF × PTR
ATR := min(ATR, PTR)
ATR := max(ATR, MTR)

paths share links with these secondary paths to also use secondary paths. In order to mitigate this cascading effect, longer secondary paths should be resorted to only if primary paths can no longer accommodate additional traffic. Based on the work described in [2] and [3], we propose strict priority queuing in which TCP data packets routed over PPs use the silver service and those routed over SPs receive the bronze service.

Another building block of the proposed architecture is the feedback mechanism and rate control. In our proposed architecture, ingress nodes periodically send RM packets to egress nodes, one over the PP (P-RM) and the other over the SP (S-RM). These RM packets are sent in every T_{RM} seconds with the direction bit set to indicate the direction of flow. If strict priority queuing is used and when an P-RM (S-RM) packet arrives at the core node on its forward path, the node compares the percentage queue occupancy of its silver (bronze) queue on the outgoing interface with a predetermined configuration parameter μ and it sets the CE (Congestion Experienced) bit (if not already set) of the P-RM (S-RM) packet accordingly. If FIFO queuing is used then it is the silver queue occupancy that needs to be checked for both P-RM and S-RM packets. When an RM packet arrives at the egress node, it is sent back to the ingress node after resetting the direction bit of the RM packet. RM packets travelling over the reverse path are not marked by the core nodes. When the RM packet arrives back at the ingress node, the CE bit indicates the congestion status of the path it was sent over. According to the information, the ingress node updates the Allowed Transmission Rate (ATR) of the corresponding rate-controlled path by using the AIMD algorithm given in Table 1 [6]. In this algorithm, MTR and PTR denote the Minimum Transmission Rate and Peak Transmission Rate and RDF and RIF denote the Rate Decrease Factor and Rate Increase Factor, respectively. Therefore, an ingress node maintains two per-egress queues, one for the PP and the other for the SP, that are drained using AIMD-based rate control. The proposed architecture is depicted in Fig. 1 for an example 3-node network in which solid lines are for PPs whereas the dotted lines stand for SPs originating at ingress node 0. We also assume that the switching technology in the core network has the necessary fields in the encapsulation header for implementing the above-mentioned mechanisms.

The final ingredient to the proposed approach is the way we split traffic over the PP and the SP. The edge nodes first identify new flows. The delay estimates for the PP and SP queues (denoted by D_{PP} and D_{SP}, respectively) in the edge

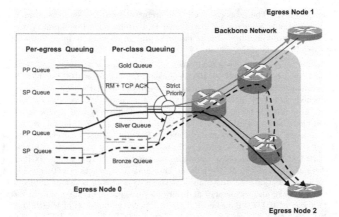

Fig. 1. The proposed architecture for an example 3-node network

nodes are then calculated by dividing the occupancy of the corresponding queue with the current drain rate. Upon the arrival of the first packet of the nth flow (i.e., a TCP SYN segment) a running estimate of the delay difference (denoted by d_n) is calculated as $d_n = \beta(D_{PP} - D_{SP}) + (1-\beta)d_{n-1}$, where β is the smoothing parameter. If $d(n) \leq min_{th}$ $(d(n) \geq max_{th})$ then we forward the flow over the PP (SP). When $min_{th} < d_n < max_{th}$, then the new flow is forwarded over the SP with probability $p_0(d_n - min_{th})/(max_{th} - min_{th})$ where min_{th}, max_{th} and p_0 are the splitting algorithm parameters to be set. In this paper, we use $p_0 = 1$. Once a path decision is made for the first packet of a flow, all the remaining packets of the flow will follow the same path. This traffic splitting mechanism is called Random Early Reroute (RER) which is inspired by the RED (Random Early Detect) algorithm used for active queue management in the Internet [8]; note the similarity in the algorithm parameters. RED is used for controlling the average queue occupancy whereas the average smoothed delay difference of silver and bronze queues is controlled by RER. RER parameters are generally chosen so that the PP is favoured (i.e., $min_{th} \geq 0$) and proportional control (as opposed to on-off control) is used, i.e., $max_{th} > min_{th}$.

3   Simulation Study

In this paper, we present the simulation results of our AIMD-based multi-path TE algorithm for TCP traffic over a mesh network called the hypothetic US topology that has 12 POPs (Point of Presence). This network topology and the traffic demand matrix are given in www.fictitious.org/omp and also described in [2]. The proposed TCP TE architecture is implemented over ns-2 (Network Simulator) version 2.27 [12] and TCP-Reno is used in our simulations. We introduced a number of new modules and modifications in ns-2 that are available in [1].

In our simulations, we scaled down the capacities of all links and the demand matrix by a factor of 45/155 (replace all OC-3 links with DS-3) to reduce the simulation run-times. We assume that each of the POPs has one edge node connected via a very high speed link to one core node. We use a traffic model where flow arrivals occur according to a Poisson process and flow sizes have a bounded Pareto distribution denoted by $BP(k, p, \alpha)$ [15]. The following parameters are used for the bounded Pareto distribution in this study: $k = 4000$ Bytes, $p = 50 \times 10^6$ Bytes, and $\alpha = 1.20$, corresponding to a mean flow size of $m = 20,362$ Bytes. The delay averaging parameter is set to $\beta = 0.3$. TCP data packets are assumed to be 1040 Bytes long and RM packets are 50 Bytes long (after encapsulation). All the buffers at the edge and core nodes including per-egress (primary and secondary) and per-class queues (gold, silver and bronze), have a size of 104,000 Bytes each. The TCP receive buffer is of length 20,000 Bytes. We fix the following parameters for the AIMD algorithm. PTR is chosen as the speed of the slowest link on the corresponding path. We use very small but nonzero MTR in order to eliminate cases causing division by zero in the simulations. If the expected delay of a buffer exceeds 0.36 s, then the packets destined to the corresponding queue are dropped. We use $T_{RM} = 0.02$ s and $\mu = 20\%$. The simulation runtime is selected as 300 s. We report only the statistics related to those flows that have been initiated in the interval [90 s, 250 s].

We compare and contrast three TE policies using simulations. *Shortest path routing* policy uses the minimum-hop path with the AIMD-ECN capability turned on and there is no traffic splitting. The second TE policy is the *Flow-based Multi-path with Shortest Delay (SD) and FIFO queuing*. In this policy, SD refers to the specific RER setting $min_{th} = max_{th} = 0$ and therefore SD forwards each flow to the path with the minimum estimated queuing delay at the ingress edge node and it does not necessarily favour the PP. Moreover, we use SD in conjunction with the FIFO queuing discipline where there is no preferential treatment between the PP and the SP at the core nodes. The third TE policy is the *Flow-based Multi-path with RER and Strict Priority queuing* approach proposed in this paper.

The goodput of the TCP flow i (in bit/s), denoted by G_i, is defined as the service rate received by flow i during its lifetime. Mathematically, $G_i = \Delta_i/T_i$, where Δ_i is the number of bits successfully delivered to the application layer by the TCP receiver for flow i and T_i is the sojourn time of the flow i within the simulation runtime. We note that if flow i terminates before the end of the simulation, then Δ_i will be equal to the flow size S_i. One performance measure we study is the normalized average goodput defined as

$$G = \frac{\sum_i \Delta_i G_i}{\sum_i \Delta_i}.$$

However, we note that some flows are not fully carried due to overloading of certain links in the network. In order to take this effect into account, we introduce a new performance measure, called the net average goodput, denoted by G_{net}

$$G_{net} = \frac{\sum_i \Delta_i G_i}{\sum_i S_i},$$

by means of equating the service rate of un-carried packets to zero. For the same effect, we suggest a new measure, called the Byte Rejection Ratio (BRR), to quantify the portion of data that cannot be delivered within the simulation duration, in percentage. Mathematically,

$$\text{BRR} = \frac{\sum_{s,d} N(s,d) - \sum_{s,d} \Gamma(s,d)}{\sum_{s,d} N(s,d)} * 100,$$

where $N(s,d)$ is the sum of the sizes of flows demanded from node s to node d, and $\Gamma(s,d)$ is the total traffic (in bytes) successfully delivered to the application layer from node s to node d.

We first study the role of AIMD parameterization on the proposed TE in terms of G_{net} and BRR. Figures 2(a) and 2(b) demonstrate the effect of RIF and RDF on G_{net}. Similarly, Figures 2(c) and 2(d) present the effect of these AIMD parameters on BRR. In these simulations, RER parameters are chosen as $min_{th} = 1$ ms and $max_{th} = 15$ ms. We observe that flow-based multi-path with RER and strict priority queuing gives better performance in both measures than shortest-path routing. The choice of RDF= 0.0625 and RIF=0.0625 gives relatively good and robust performance in terms of G_{net} and therefore we use these parameters in the rest of the paper.

The effect of RER parameters on G_{net} and BRR are presented in Figures 3(a) and 3(b), respectively. We observe that the performance of the RER is quite robust except for the choices of RER parameters close to $min_{th} = max_{th} = 0$, i.e., the SD policy. We observe a sharp decline in the performance of the system when we apply the SD policy due to the induced knock-on effect. The simulation results show that G_{net} for the multi-path routing policy with RER and Strict Priority satisfies $G_{net} \geq 5.50$ Mbit/s when the RER parameters are in the range $0 \leq min_{th} \leq 1$ ms and 1 ms $\leq max_{th} \leq 15$ ms. For the same example, G_{net} is given by $G_{net} \approx 5.24$ Mbit/s and $G_{net} \approx 3.90$ Mbit/s for the shortest-path routing policy with and without AIMD, respectively. This shows that for a wide operational range for RER, multi-path routing policy outperforms single-path routing policies and the performance of the RER converges to that of the shortest-path routing policy with AIMD as we increase min_{th} and max_{th}. Based on these observations, we choose the RER parameters as $min_{th} = 1$ ms and $max_{th} = 15$ ms from this wide operational range.

Finally, we scale the incoming traffic by multiplying the flow sizes with a scaling parameter γ where $0.5 \leq \gamma < 1$ while fixing the flow arrival times. We then vary γ to see its impact on network performance. In Fig. 4(a), the multi-path TE with strict-priority and RER is shown to achieve the highest G_{net} for all values of γ. It is also observed from Fig. 4(a) that the proposed TE approach

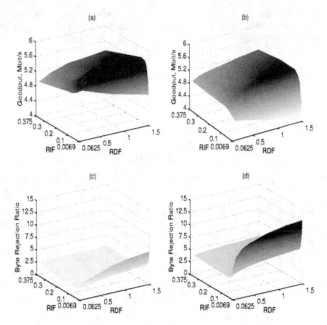

Fig. 2. As a function of RIF and RDF: (a) G_{net} for the multi-path TE with strict-priority and RER, (b) G_{net} for the shortest-path routing, (c) BRR for the multi-path TE with strict-priority and RER (d) BRR for the shortest-path routing

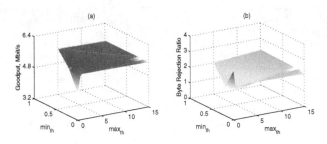

Fig. 3. As a function of min_{th} and max_{th}: (a) G_{net} for the multi-path TE with strict-priority and RER (b) BRR for the multi-path TE with strict-priority and RER

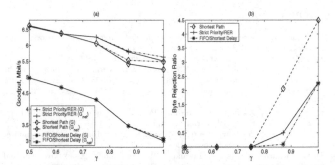

Fig. 4. As a function of traffic scaling parameter γ: (a) G_{net} and G (b) Byte Rejection Ratio

outperforms the other policies in terms of G as well. This shows that the multipath TE with strict-priority and RER not only carries more traffic but also the carried flows are transported faster.

In Fig. 4(b), we observe that the policy of multi-path routing with strict-priority and RER has a BRR which is approximately half of that of the shortest-path routing policy for $\gamma = 1$. As the offered traffic decreases, the gap between the multi-path routing with strict-priority and RER and the shortest-path routing disappears. This is due to the fact that the PP is not congested at light traffic loads and the multi-path routing nearly boils down to shortest-path routing. We also observe that the SD routing with FIFO queuing gives lower BRR than the proposed TE policy for some values of γ. However, the net goodput of the multi-path routing with SD and FIFO queuing is 25-50% lower than the proposed TE approach when γ varies between 0.5 and 1.0, as shown in Fig. 4(a).

4   Conclusions

In this paper, we report our findings on a recently proposed TCP load balancing architecture that uses prioritized AIMD and flow-based multi-path routing with RER. Using a publicly used test network, we show that our proposed architecture consistently outperforms the case of a single path in terms of average normalized goodput and the byte rejection ratio. We show in this paper that the architecture stays robust for relatively large networks, extending our existing results for small topologies. On the other hand, we also show that employing load balancing with conventional FIFO queuing and shortest delay policies does not always produce better results than that of a single path, which can be explained by the knock-on effect. Future work in this area will consist of incorporating a-priori knowledge on the traffic demand matrix into the proposed architecture.

References

1. The multi-path routing project at Bilkent University Information Networks Laboratory (BINLAB). Web page:
   http://www.binlab.bilkent.edu.tr/onur/index.html, July 2004.
2. N. Akar, I. Hokelek, M. Atik, and E. Karasan. A reordering-free multipath traffic engineering architecture for Diffserv/MPLS networks. In *Proceedings of IEEE Workshop on IP Operations and Management*, pp. 107–113, Kansas City, Missouri, USA, 2003.
3. O. Alparslan, N. Akar, and E. Karasan. AIMD-Based Online MPLS Traffic Engineering for TCP Flows via Distributed Multi-Path Routing. www.binlab.bilkent.edu.tr/e/journal/125/annales_final.pdf. Accepted for publication in Annales Des Telecommunications.
4. D. O. Awduche, A. Chiu, A. Elwalid, I. Widjaja, and X. Xiao. Overview and principles of Internet traffic engineering. IETF Informational RFC-3272, May 2002.
5. S. Bahk and M. E. Zarki. Dynamic multi-path routing and how it compares with other dynamic routing algorithms for high speed wide area networks. In *ACM SIGCOMM*, pp. 53–64, 1992.
6. D. M. Chiu and R. Jain. Analysis of the increase/decrease algorithms for congestion avoidance in computer networks. *Computer Networks and ISDN Systems*, 17(1):1–14, June 1989.
7. A. Elwalid, C. Jin, S. Low, and I. Widjaja. MATE: MPLS Adaptive Traffic Engineering. In *Proceedings of INFOCOM*, pp. 1300–1309, 2001.
8. S. Floyd and V. Jacobson. Random early detection gateways for congestion avoidance. *IEEE/ACM Transactions on Networking*, 1(4):397–413, 1993.
9. F. P. Kelly. Routing in circuit switched networks: Optimization, shadow prices and decentralization. *Advances in Applied Probability*, 20:112–144, 1988.
10. M. Laor and L. Gendel. The effect of packet reordering in a backbone link on application throughput. *IEEE Network Magazine*, 16(5):28–36, 2002.
11. Y-D. Lin, N-B. Hsu, and R-H. Hwang. QoS routing granularity in MPLS networks. *IEEE Comm. Mag.*, 46(2):58–65, 2002.
12. S. McCanne and S. Floyd. ns Network Simulator. Web page:
    http://www.isi.edu/nsnam/ns/, July 2002.
13. S. Nelakuditi, Z. L. Zhang, and R. P. Tsang. Adaptive proportional routing: A localized QoS routing approach. In *Proceedings of INFOCOM*, Anchorage, USA, 2000.
14. S. Oueslati-Boulahia and J. W. Roberts. Impact of trunk reservation on elastic flow routing. In *Networking 2000*, March 2000.
15. I. A. Rai, G. Urvoy-Keller, and E. W. Biersack. Analysis of LAS scheduling for job size distributions with high variance. In *Proceedings of ACM Sigmetrics*, pp. 218–228, 2003.
16. A. Shaikh, J. Rexford, and Kang G. Shin. Load-sensitive routing of long-lived IP flows. In *ACM SIGCOMM*, pp. 215–226, 1999.
17. J. Wang, S. Patek, H. Wang, and J. Liebeherr. Traffic engineering with AIMD in MPLS networks. In *7th IFIP/IEEE International Workshop on Protocols for High-Speed Networks*, pp. 192–210, 2002.

Online Traffic Engineering:
A Hybrid IGP+MPLS Routing Approach

Antoine B. Bagula

Department of Computer Science
University of Stellenbosch, 7600 Stellenbosch, South Africa
Phone: +2721 8084070, Fax: +2721 8084416
bagula@cs.sun.ac.za

Abstract. The dualism of IGP and MPLS routing has raised a debate separating
the IP community into divergent groups expressing different opinions concerning
how the future Internet will be engineered. This paper presents an on-line traffic
engineering model which uses a hybrid IGP+MPLS routing approach to achieve
efficient routing of flows in IP networks. The approach referred to as *Hybrid*
assumes a network-design process where an optimal network configuration is built
around optimality, reliability and simplicity. The IGP+MPLS routing approach is
applied to compute paths for the traffic offered to a 50-node network. Simulation
reveals performance improvements compared to both IGP and MPLS routing in
terms of several performance parameters including routing optimality, network
reliability and network simplicity.

1 Introduction

The Internet has developed beyond the best effort service delivered by traditional IP
protocols into a universal communication platform where traffic management methods
such as QoS and Traffic Engineering (TE), once the preserve of telephone networks, will
be re-invented and used to deliver differing resource requirements.

Traffic engineering allows traffic to be efficiently routed through the network by
implementing QoS agreements between the available resources and the current and
expected traffic. The dualism of IGP and MPLS routing has raised a debate separating
the IP community into divergent groups with different views concerning how the future
Internet will be engineered [1]. On one hand, there are proponents of the destination-
based TE model using traditional IGP routing. They point to (1) the ability of the Internet
to support substantial increases of traffic load without the need for TE mechanisms such
as proposed by MPLS and (2) the capability of traditional IGP routing to optimize routing
by using appropriate adjustments to the link weights. On the other hand, the advocates
of the newly proposed MPLS standard argue for a flow-based TE model using source-
or connection-oriented routing. In MPLS packet forwarding and routing are decoupled
to support efficient traffic engineering and VPN services. MPLS networks route traffic
flows over bandwidth-guaranteed tunnels referred to as Label Switched Paths (LSPs)
which are set up and torn down through signaling.

Hybrid routing approaches which either combine the best of destination-based
and flow-based TE or allow a smooth migration from traditional IGP routing to the

J. Solé-Pareta et al. (Eds.): QofIS 2004, LNCS 3266, pp. 134–143, 2004.
© Springer-Verlag Berlin Heidelberg 2004

newly developed MPLS standard have scarcely been addressed by the IP community. A trace-based analysis of the complexity of a hybrid IGP+MPLS network was presented in [2]. This analysis uses a trigger-based mechanism to (1) differentiate flows based on their measured bandwidth during the last minute and (2) route these flows differently according to their bandwidth characteristics. Other papers [3,4,5,6] have addressed the problem of routing IP flows using offline IGP+MPLS approaches where the network topology and traffic matrix are known a priori.

This paper presents a TE model which uses a hybrid IGP+MPLS routing approach to achieve efficient routing of flows in IP networks. These flows can be high bandwidth demanding flows (HBD) such as real-time streaming protocol flows or low bandwidth demanding flows (LBD) such as best-effort FTP flows. We assume an online network design process where an optimal network configuration is built around optimality, reliability and simplicity. We adopt a path selection model where traffic flows are classified into LBD and HBD traffic classes at the ingress of the network and handled differently in the core where LBD flows are routed using traditional IGP routing while HBD flows are carried over MPLS tunnels. The main contributions of this paper are
- **Reliability-related optimality.** Reliability and optimality are the two key drivers for traffic engineering. These two TE objectives have been addressed separately by the IP community though neither objective considered alone is a true measure of the efficiency of a network. We present a routing approach referred to as *Hybrid* which combines reliability and optimality into a mixed routing metric minimizing both the number of flows to be re-routed under link failure (reliability) and the magnitude of flows rejected under heavy load conditions (optimality).
- **Routing simplicity.** *Hybrid* uses a simple design process where no changes to the traditional routing algorithms are required besides designing the link cost metric to integrate reliability and optimality. By moving complexity into the design of the link cost, *Hybrid* leads to simplicity in terms of time/logical complexity and easy implementation by ISPs. *Hybrid* implements a path selection model where a small number of LSPs are set up into an IGP network to overcome the network complexity resulting from the signaling operations required to setup and tear down LSPs in a pure MPLS network implementation.
- **Routing performance improvements.** Multiple metric routing using different TE objectives has been suggested to be used at best as an indicator in path selection [7] since (1) it may result in unknown composition rules for the different TE objectives and (2) it does not guarantee that each of the TE objectives is respected. An application to a 50-node network reveals that, using multiple metric routing, *Hybrid* achieves an optimal network configuration outperforming IGP routing in terms of network optimality and reliability and MPLS routing in terms of the network simplicity.

The reminder of this paper is organized as follows. Section 2 presents the IGP+MPLS routing approach. An application of the IGP+MPLS routing approach to compute paths for the flows offered to a 50-node network is presented in section 3. Our conclusions are presented in section 4.

2   The IGP+MPLS Routing Approach

Consider a network represented by a directed graph $(\mathcal{N}, \mathcal{L})$ where $\mathcal{N}$ is a set of nodes, $\mathcal{L}$ is a set of links, $N = |\mathcal{N}|$ is the number of nodes and $L = |\mathcal{L}|$ represent the number of links of the network. Assume that the network carries IP flows that belong to a set of service classes $\mathcal{S} = \{LBD, HBD\}$ where LBD and HBD define the class of low bandwidth demanding (LBD) and high bandwidth demanding (HBD) flows respectively. Let C_ℓ denote the capacity of link ℓ and let $\mathcal{P}_{i,e}$ denote the set of paths connecting the ingress-egress pair (i, e). Assume a flow-differentiated services where a request to route a service class $s \in \mathcal{S}$ flow of $d_{i,e}$ bandwidth units between an ingress-egress pair (i, e) is received and that future demands concerning IP flow routing requests are not known.

For each path p let $L_p = \sum_{\ell \in p} L_\ell(n_\ell, r_\ell, \alpha(s), \beta(s))$ denote the path cost where $L_\ell(n_\ell, r_\ell, \alpha(s), \beta(s))$ is the cost of link ℓ when carrying n_ℓ flows, r_ℓ is the total bandwidth reserved by the IP flows traversing link ℓ and $(\alpha(s), \beta(s))$ is a pair of network calibration parameters depending on the flow service class s. The flow routing problem consists of finding the best feasible path $p_s \in \mathcal{P}_{i,e}$ where

$$L_{p_s} = \min_{p \in \mathcal{P}_{i,e}} L_p \tag{1}$$

$$d_{i,e} < \min_{\ell \in p_s}(C_\ell - r_\ell). \tag{2}$$

Equations (1) and (2) express respectively the optimality of the routing process and the feasibility of the flows.

We consider a flow routing algorithm using a route optimization model based on a mixed cost model and a path selection model based on flow differentiation.

2.1   The Route Optimization Model

We consider a new cost metric which combines optimality and reliability with the expectation of routing the IP flows so that fewer flows are rejected under heavy load conditions and fewer flows are re-routed under link failure.

- **Reliability: the link loss.** The main reliability objective of our routing approach is to minimize the damage to the network transport layer under failure. This damage is expressed by the number of re-routed flows under failure. Let $\mathcal{F}$ be a set of possible failure patterns, w_f the probability of the failure pattern $f \in \mathcal{F}$ and n_f the number of re-routed flows under failure f. The expected number of re-routed flows under the set of failure patterns $\mathcal{F}$ is defined by

$$W = \sum_{f \in \mathcal{F}} W_f = \sum_{f \in \mathcal{F}} w_f n_f \tag{3}$$

where $W_f = w_f n_f$ expresses the damage to the network transport layer under failure event f.

Assuming that a fiber cut is the most likely failure event in optical networks, we consider the set of failure events $\mathcal{F} = \mathcal{L}$ and define a measure of reliability expressing the link loss by

$$W_\ell = w_\ell \sum_{r \in \mathcal{R}} \delta_{\ell,r} \tag{4}$$

where w_ℓ is the probability for the IP flows to traverse link $\ell \in \mathcal{L}$ referred to as the *link loss probability*, $\mathcal{R} = \cup_{i,e} \mathcal{R}_{i,e}$ is the set of flows carried by the network, $\mathcal{R}_{i,e}$ is the subset of flows from node i to node e, $n_\ell = \sum_{r \in \mathcal{R}} \delta_{\ell,r}$ is the total number of flows carried by link ℓ referred to as its *interference* and

$$\delta_{\ell,r} = \begin{cases} 1 & \text{flow } r \text{ traverses link } \ell \\ 0 & \text{otherwise} \end{cases} \tag{5}$$

- **Optimality: the congestion distance.** Most routing algorithms which maximize bandwidth (optimality) assume a fair bandwidth sharing process where the different flows receive the same service on a link ℓ. This service is expressed by the link residual bandwidth $C_\ell - r_\ell$. We consider a measure of optimality referred to as the *link congestion distance* defined by

$$D_\ell(r_\ell, \beta(s)) = C_\ell - \beta(s)(r_\ell + d_{i,e}) \tag{6}$$

where $\beta(s)$ is a calibration parameter expressing the subscription for bandwidth to the network link. Note that, in contrast to the residual bandwidth $C_\ell - r_\ell$ which is independent of the demand, the *link congestion distance* includes in its definition the bandwidth demand $d_{i,e}$ which can be low for LBD flows and high HBD flows. It is expected that by introducing unfairness among flows, our measure of optimality will lead to a link sharing model which improves the overall network performance by allowing the different flows to meet their service needs.

- **A mixed cost metric.** The routing optimization adopted in this paper is based on the assumption that a link metric minimizing the link loss (or equivalently the number of flows on a link) and maximizing the link congestion distance (or equivalently its inverse) can balance the number and magnitude of flows over the network to reduce rejection under heavy load conditions and re-routing under link failure. *Hybrid* achieves this objective by multiplying the link loss probability by power values of the link interference and the link congestion distance to form the mixed metric expressed by

$$L_\ell(n_\ell, r_\ell, \alpha(s), \beta(s)) = w n_\ell^{\alpha(s)} / (C_\ell - \beta(s)(r_\ell + d_{i,e}))^{1-\alpha(s)} \tag{7}$$

where $0 \le \alpha(s) \le 1$ is a calibration parameter expressing the trade-off between reliability and optimality.

2.2 The Path Selection Model

The basic idea behind our path selection model is to differentiate flows into classes based on their bandwidth requirements and route these flows using different cost metrics according to their service needs.

- **Flow differentiation.** In *Hybrid* flows are classified into LBD and HBD traffic classes depending on their bandwidth requirements ($d_{i,e}$). The two traffic classes are defined by

$$LBD = \{\text{class of flows requesting } d_{i,e} \text{ bandwidth units} \mid d_{i,e} < \tau\} \tag{8}$$

$$HBD = \{\text{class of flows requesting } d_{i,e} \text{ bandwidth units} \mid d_{i,e} \ge \tau\} \tag{9}$$

where each flow bandwidth demand $d_{i,e}$ is uniformly distributed in the range $[0, M]$ and $0 \le \tau \le M$ is a cut-off parameter defining the limit between LBD and HBD flows.

- **Routing metrics.** In *Hybrid* the IP flows are routed using different routing metrics expressing their service needs: the paths followed by LBD flows are found using the IGP-based OSPF model while HBD flows are routed over MPLS Label Switched Paths (LSPs). This is achieved using the link cost (7) which can lead to different routing metrics depending on the values of the link loss probability w_ℓ and the set of parameters $(\alpha(s), \beta(s))$. The routing metrics leading to IGP-based OSPF and MPLS routing are obtained by setting (1) the link loss probability to a constant value expressing an *equal probability* assumption where each link ℓ has the same probability $w_\ell = 1/L$ to carry any traffic flow and (2) the set of parameters $(\alpha(s), \beta(s))$ to $(\alpha, 1)$ for MPLS routing and $(0, 0)$ for IGP-based OSPF routing. Note that the link loss probability w_ℓ can be assigned appropriate weights to move the traffic away from bottleneck links such as proposed in the OSPF weight optimization TE model [8]. However the optimal weight setting may require an offline computation process which is beyond the scope of this paper.

- **Routing algorithm.** Consider a request to route a class s flow of $d_{i,e}$ bandwidth units between two nodes i and e. The algorithm proposed (hereafter referred to as HYBR) executes the following steps to route this flow

1. **Network calibration.**
   set $(\alpha(s), \beta(s)) = (0, 0)$ if $s \in LBD$, or
   set $(\alpha(s), \beta(s)) = (\alpha, 1)$ if $s \in HBD$.
2. **Path selection.**
   a) **Traffic aggregation.** if $s \in HBD$
      - find an existing LSP which can accommodate the new HBD request,
      - if found then (a) set $p_s := p$ where p is the path carrying the existing LSP and (b) goto step 3.
   b) **Prune the network.** Set $L_\ell(n_\ell, r_\ell, \alpha(s), \beta(s)) = \infty$ for each link ℓ whose link slack $C_\ell - r_\ell \leq d_{i,e}$.
   c) **Find a new least cost path.** Apply Dijkstra's algorithm using the link cost (7) to find a new least cost path $p_s \in \mathcal{P}_{i,e}$. This path is used to setup a new MPLS LSP for HBD flows when $s \in HBD$.
3. **Route the request.**
   - Assign the traffic demand $d_{i,e}$ to path p_s.
   - Update the link occupancy and interference. For each link $\ell \in p_s$
     set $r_\ell := r_\ell + d_{i,e}$ and $n_\ell := n_\ell + 1$.

Note that the path selection algorithm has the same complexity as Dijkstra's algorithm: $O(N^2)$.

3   An Implementation

This section presents simulation experiments conducted using a 50-node test network to compare the performance of (1) IGP routing using the Open Shortest Path First (OSPF) model [9] (2) MPLS routing using the Least Interference Optimization Algorithm (LIOA) [10] and (3) hybrid IGP+MPLS routing using the newly proposed HYBR algorithm. The 50-node network used in our experiments includes 2450 ingress-egress

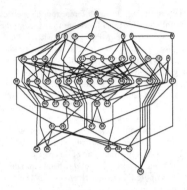

	Simulation expriments			
	Exp 1	Exp 2	Exp 3	Exp 4
α	0.5	–	0.5	0.5
λ	2	2	2	2
$1/\mu$	1	1	1	1
τ	150	150	100K	0.33M
M	500	500	500	250+50K
T	50	50	50	50
R	50000	50000	50000	50000

(a) The 50-node network   (b) The parameter values

Fig. 1. The 50-node network and simulation parameters

pairs and 202 links capacitated with 38,519,241 units of bandwidth. Each node is a potentially edge node. Figure 1 presents a graphical representation of the 50-node network and
the simulation parameter values. The parameter values for the simulation experiments
are presented in terms of the offered flow routing requests, the value of the calibration
parameter α, the flow request rates λ, the flow holding time $1/\mu$, the cut-off parameter τ,
the maximum bandwidth demand M, the number of simulation trials T and the number
of flow requests per trial R. The flow arrival and services processes are Poisson. We considered short-lived flows only since initial experiments revealed the same performance
patterns for short-lived ($1/\mu = 1$) and long-lived flows ($1/\mu = \infty$).

3.1 Performance Parameters

The relevant performance parameters used in the simulation experiments are (1) the
quality of the paths expressed by the *path length*, the *path multiplicity* and the *preferred
path usage* (2) the network optimality expressed by the percentage flow acceptance
ACC and the average link utilization *UTIL* (3) the network reliability expressed by
the average link interference *AV* and the maximum link interference *MAX* and (4) the
network simplicity expressed by the network gain *GAIN*. The *path length* determines the
average path length in number of hops. It gives an indication on resource consumption:
a longer path ties up more resources than a shorter path. The *path multiplicity* expresses
the average number of paths used by a source-destination pair. It gives an indication
of the load balancing capability of an algorithm: a higher value of this parameter is an
indication of a more balanced network. The *preferred path usage* expresses how often
a path finding algorithm routes flows on the preferred path connecting an I-E pair (the
path most used by an I-E pair is defined as the preferred path). *ACC* is the percentage of

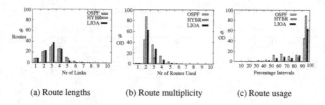

(a) Route lengths   (b) Route multiplicity   (c) Route usage

Fig. 2. Experiment 1. The quality of the paths

flows which have been successfully routed by the network. *UTIL* defines the average link load. It expresses how far a link is from its congestion region. *AV* is the average number of flows carried by the network links and *MAX* is the number of flows carried by the most interfering link. *AV* and *MAX* express the number of flows which must be re-routed upon failure: an algorithm which achieves lower interference is more reliable since it leads to re-routing fewer flows upon failure. *GAIN* determines the reduction in number of signaling operations (LSP setup and tear down) resulting from the implementation of *Hybrid* instead of a pure MPLS implementation.

3.2   Simulation Experiments

Four simulation experiments were conducted to analyze (1) the quality of paths used by the LBD flows, the dominant flows in the network (2) the impact of the calibration parameter α on the network efficiency (3) the impact of the cut-off parameter τ on the network efficiency and (4) the impact of the demand range $[0, M]$ on the network efficiency. The results of the experiments are presented in Figures 2, 3, 4 and Table 1. Each entry in Table 1 presents the average of each of the performance parameter described above (*ACC*, *UTIL*, *AV*, *MAX*, and *GAIN*). These averages (considered for different values of the α parameter) are computed at 95% confidence interval within 0.1% of the point estimates. Our choice of the best performance values is based on a trade-trade-off between the different values achieved by the different performance parameters. The best performance is indicated in bold.

Experiment 1. *The quality of paths for LBD flows.* The quality of the paths carrying LBD flows illustrated by Figure 2 shows that approximatively 70% of the routes used by the three routing algorithms are at most 3 hops long (Figure 2 (a)). The three algorithms thus perform equally well in terms of resource consumption. HYBR achieves the best route multiplicity and route usage. OSPF performs worse in terms of path multiplicity (Figure 2 (b)) and path usage (Figure 2 (c)). These results show that HYBR achieves the best stability in terms of path selection and balances the flows over the network better than the LIOA and OSPF models.

Experiment 2. *The impact of the calibration parameter α.* The results presented in Table 1 show that IP routing using OSPF achieves the highest values of the link interference (*AV* and *MAX*) while still exhibiting the lowest average link utilization *UTIL*. This is in agreement with the routing in the Internet where OSPF routing can lead to low link

Table 1. Experiment 2. The calibration parameter α

α	OSPF model					CSPF model				
	ACC(α)	UTIL(α)	AV(α)	MAX(α)	GAIN(α)	ACC(α)	UTIL(α)	AV(α)	MAX(α)	GAIN(α)
–	82	55	450	1801	-	90	55	435	1578	-
α	LIOA model					HYBR model				
	ACC(α)	UTIL(α)	AV(α)	MAX(α)	GAIN(α)	ACC(α)	UTIL(α)	AV(α)	MAX(α)	GAIN(α)
0.0	90	55	435	1578	-	90	55	438	1715	30
0.2	92	56	428	1424	-	91	56	433	1621	30
0.4	92	57	426	1333	-	91	58	432	1539	30
0.6	**92**	**59**	**426**	**1161**	**-**	**91**	**60**	**432**	**1374**	**30**
0.8	89	61	426	995	-	88	62	434	1239	30
1.0	86	62	428	864	-	84	64	440	1068	30

utilizations (typically 30%) on some links while other links are carrying the majority of the traffic flows. MPLS routing using LIOA achieves (1) the lowest values of the *MAX* and *AV* parameters and (2) higher acceptance *ACC* compared to OSPF routing. For the cut-off parameter setting ($\tau = 150$), HYBR reduces the network complexity by setting up 30% fewer LSPs than a pure MPLS routing using the LIOA model. The rest of this section will show that this percentage can be increased by choosing a suitable value of the cut-off parameter τ but at the price of reducing the flow acceptance and increasing the link interference. For the type of traffic considered, mapping new HBD flows to existing LSPs did not increase significantly the network gain compared to the case where HBD flows were not aggregated. This results from the fact that under the short-time holding time assumption the LSPs created on an I-E pair are torn down before the arrival of new HBD requests on the same I-E pair.

Experiment 3. *The impact of the cut-off parameter τ.* The results presented in Figure 3 show that in IGP+MPLS routing, the network gain (simplicity) is increased at the price of the degradation of the network optimality (Figure 3 (a)) and the network reliability (Figure 3 (b)). Finding a network operational point which balances reliability, optimality and simplicity is therefore an important aspect of the IGP+MPLS routing model. Consider the two functions $A(x)$ and $M(x)$ representing respectively the flow acceptance $ACC(\tau)$ and scaled values of the maximum interference $MAX(\tau)$ where the values $M(x)$ are defined by $M(x) = 100 * MAX(x)/\max_{k \in \{\tau\}} MAX(k)$. Figure 3 (c) depicts two curves obtained by approximating the data (values of the functions $M(x)$ and $A(x)$) with a Bezier curve of degree $n = 11$ (the number of data points) that connects the endpoints. This figure shows that for the network model under consideration, a network operational point which balances optimality, reliability and simplicity may be found at the intersection of the two curves $M(x)$ and $A(x)$ located around $\tau = 300$; a cut-off value corresponding to approximately 60% network gain, 89% flow acceptance and the reliability values $AV = 437$ and $MAX = 1643$ where AV values are in the average interference range $[426, 451]$ and *MAX* values are in the maximum interference range $[1267, 1739]$.

Experiment 4. *The impact of the demand range $[0, M]$.* The curves depicted by Figure 4 (a), (b), (c) represent the values of the flow acceptance, the average link utilization and the link interference respectively for different demand ranges. These curves reveal

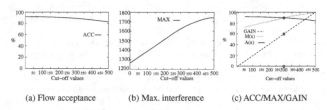

(a) Flow acceptance   (b) Max. interference   (c) ACC/MAX/GAIN

Fig. 3. Experiment 3. The cut-off parameter τ

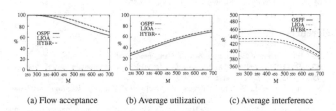

(a) Flow acceptance   (b) Average utilization   (c) Average interference

Fig. 4. Experiment 4. The demand range $[0, M]$

that for the three algorithms the flow acceptance ACC decreases with increasing range while the link utilization increases. Figure 4 (a) shows that (1) LIOA and HYBR achieve the same and higher flow acceptance compared to OSPF (2) the flow acceptance is the same for all the three algorithms for small values of M ($M \leq 300$) and (3) for higher values of M ($M > 300$), HYBR and LIOA perform better than *OSPF*. HYBR , LIOA and OSPF routing lead the network into a congestion region for a value of $M = Thr$ where on average the network rejects more than 20% of its offered flows. We refer to this value as *the network congestion threshold*. Simulation reveals that this value is lower ($Thr = 525$) for OSPF than for HYBR and LIOA ($Thr = 600$). This finding reveals that under varying traffic conditions a network implementing OSPF routing enters into its congestion region quicker than a network implementing HYBR or LIOA routing. Figures 4 (b) and (c) reveal the same performance pattern as Table 1 where OSPF achieves the lowest link utilization but higher interference compared to the HYBR and LIOA models.

4   Conclusion

This paper presents *Hybrid* , an IGP+MPLS routing approach which combines route optimization and efficient path selection to achieve efficient routing of flows in IP networks. The route optimization model adopted by *Hybrid* combines reliability and optimality to re-route fewer flows under link failure and reject fewer flows under heavy load conditions. Preliminary simulation results using a 50-node network reveal that the hybrid IGP+MPLS routing approach performs as good as an MPLS solution in terms of optimality. IGP+MPLS routing outperforms IGP routing in terms of network reliability and network optimality. Finally the IGP+MPLS routing approach performs better than MPLS routing in terms of network simplicity and finds better paths than both MPLS routing and IGP routing.

In *Hybrid* IGP routing is achieved by setting the OSPF link metric inversely proportional to the link capacity. It is expected that hybrid routing approaches combining MPLS routing and OSPF weight optimization should lead to further performance improvements. The focus of our ongoing work in the hybrid IGP+MPLS framework lies on these improvements. The performance of the IGP+MPLS approach when routing a mix of short- and long-lived flows is also under investigation.

References

1. The MPLS Resource Center, *http://www.mplsrc.com/*.
2. S. Uhlig, O. Bonaventure, "On the cost of using MPLS for interdomain traffic",*Proceeding of the QofIS2000, Berlin*, September 2000.
3. W. Ben-Ameur et al.,"Routing Strategies for IP-Networks", *Telekntronikk Magazine*, 2 March 2001.
4. E. Mulyana, U. Killat, "An Offline Hybrid IGP/MPLS Traffic Engineering Approach under LSP constraints", *Proceedings of the 1st International Network Optimization Conference INOC 2003, Evry/Paris France*, October 2003.
5. S. Koehler, A. Binzenhoefer, "MPLS Traffic Engineering in OSPF Networks- A Combined Approach", *18th ITC Specialist Seminar on Internet Traffic Engineering and Traffic Management*, August-September 2003.
6. A. Riedl, "Optimized Routing Adaptation in IP Networks Utilizing OSPF and MPLS", *Proceedings of the ICC2003 Conference*, May 2004.
7. Z. Wang, W. Crowcroft, "Quality-of-Service Routing for Supporting Multimedia Applications", *IEEE Transactions on Communications*, Vol.14, pp 1228-1234, September 1996.
8. B. Fortz, M. Thorup, "Internet Traffic Engineering by Optimizing OSPF Weights", *Proceedings of IEEE INFOCOM*, March 2000.
9. J. Moy, "OSPF Version 2", *Request For Comments, http://www.ietf.org/rfc/rfc1583.txt*, March 1994.
10. A. Bagula, M. Botha, and A.E Krzesinski, "Online Traffic Engineering: The Least Interference Optimization Algorithm", *Proceedings of the ICC2004 Conference, Paris*, June 2004.

A Fast Heuristic for Genetic Algorithms in Link Weight Optimization*

Christoph Reichert and Thomas Magedanz

Fraunhofer FOKUS, Kaiserin-Augusta-Allee 31, 10589 Berlin, Germany
{reichert,magedanz}@fokus.fraunhofer.de

Abstract. Genetic algorithms are a useful tool for link weight optimization in intra-domain traffic engineering where the maximum link load is to be minimized. As a local heuristic, the weight of the maximum loaded link is increased to speed up the search for a near-optimal solution. We show that implementing this heuristic as *directed mutation* outperforms an implementation as an inner loop in both quality of the result and number of calls to the objective function when used together with caching. Optimal mutation rates result in surprisingly high cache hit ratios.

1 Introduction

One goal in intra-domain traffic engineering is to distribute the load of the network so that no link is congested. For routing protocols based on link weights and shortest paths like OSPF [1] and IS-IS [2], this goal can be achieved through selecting the link weights which determine the shortest paths in such a way that the maximum over all link loads is minimal. The problem of computing optimal link weights for a given network topology with link capacities and traffic matrix is NP hard [3][4].

Besides other optimization schemes like tabu search [4][5][6][7][8], genetic algorithms have been used to solve this optimization problem [9][10][11][12][13]. Genetic algorithms resemble the evolutionary adaption of a biological population to its environment (given by an objective function) through selection ("survival of the fittest"), crossover (exchange of genetic substrings between chromosomes) and mutation [14]. Beyond the plain genetic algorithm, a local heuristic can be applied for link weight optimization, which increments the weight of the most loaded link in the network.

The local heuristic is typically implemented as a loop which increments the weight of the most loaded link until the overall maximum of all loads increases. In this paper, this strategy is analyzed and a different approach called *directed mutation* is proposed and evaluated. We also use caching to avoid calls to the objective function for chromosomes which have been evaluated already during earlier generations. The results show that the new strategy outperforms the

---

* This work was partially funded by the Bundesministerium für Bildung und Forschung (ministry for education and research) of the Federal Republic of Germany under contract 01AK045. The authors alone are responsible for the content of the paper.

J. Solé-Pareta et al. (Eds.): QofIS 2004, LNCS 3266, pp. 144–153, 2004.
© Springer-Verlag Berlin Heidelberg 2004

existing one in terms of both quality of the result and number of calls to the
(expensive) objective function. The latter is the case due to high cache hit ratios.

The fact that an optimal mutation rate yields relatively high cache hit ratios
is surprising because a high cache hit ratio clearly indicates that many points
in the search space are visited repeatedly. Increasing the mutation rate should
therefore lead to better exploration of the search space and improve the result.
We found this assumption to be wrong and present a strange effect: the best
strategy investigated re-creates on average each potential solution twice.

The paper is organized as follows: In Sec. 2 we resume the objective function,
genetic algorithms and the current implementation strategy of the local heuristic. In Sec. 3, we analyze this approach, propose a new strategy called *directed
mutation*, and caching. The new scheme is evaluated in Sec. 4, and discussed in
Sec. 5. Sec. 6 concludes the paper with further possible improvements. Related
work is discussed throughout the paper.

2   Background

An instance of the problem to be solved is given as a tuple (G, κ, μ, ϕ) where

- $G = (V, E)$ is the network *graph* where V is set of nodes and $E \subset V \times V$ is
  the set of directed links without self loops,
- $\kappa(u, v) > 0$ is the *capacity* for each link $(u, v) \in E$,
- $\mu(u, v) > 0$ is a *metric* i.e., the link weights for all $(u, v) \in E$, and
- $\phi(s, t) \geq 0$ is the bit rate of a *flow* from source $s \in V$ to target $t \in V$.

With ECMP [1], the shortest paths are fully determined by graph G and link
weights μ. The traffic matrix ϕ and the link capacities κ furthermore fully determine the relative load $\rho(u, v) \geq 0$, which is the ratio of (absolute) load and
capacity for each link. If G, ϕ and κ are fixed, the link load ρ and, in particular,
the maximum link load ρ^* depend only on the link weights μ. This defines the
objective function Ω we are interested in as

$$\rho^* = \Omega(\mu),$$

which returns the maximum load for a given set of link weights. The goal of the
optimization problem is to find link weights μ_0 so that $\rho^* = \Omega(\mu_0)$ is minimal.

Other objective functions or additional constraints are also possible. For example, link failures can be taken into account to minimize the maximum load for
the worst case failure [8][15][16], or to find new link weights so that only a minimal subset of weights has to be configured in routers after a failure [5]. Instead
of minimizing the maximum link load, a function thereof can be minimized, like
the piece-wise linear function in [4][5][6][15][16].

Throughout this paper, we consider only the base case described above and
mention only that this work is not restricted to this.

2.1 Genetic Algorithms

Genetic Algorithms (GAs) belong to the class of evolution strategies used in optimization. They resemble the process of biological evolution, where each individual of a population with a given size is described by its genetic code, called a chromosome. The chromosomes here are the sets of link weights μ, ordered as a sequence. The initial population is made up of random sequences. Each chromosome has a chance to be selected for reproduction, where the probability to be selected increases with the fitness of the chromosome. In our case, the fitness is the reciprocal value of the maximum link load $1/\rho^*$, thus, lower values for ρ^* increase the probability for selection.

During reproduction, a crossover operator is applied to a pair of selected chromosomes. With N-point crossover, N randomly selected positions in both parents yield N+1 subsequences, which are exchanged between the parents, so that two merged offsprings are formed. Whether crossover is applied is determined by a usually high probability (0.9 - 1.0). The idea behind crossover is that good properties of both parents are combined in a child, creating an even better chromosome. More complex crossover operators are also possible, for example the one described in [10].

After crossover, the resulting chromosomes are mutated by replacing randomly selected weights with a new random value. Mutation asserts variability in the search space by introducing new genetic information. The mutation probability has normally a low value (about 0.05 per link weight).

Finally, the best chromosome of the current population is always copied to the new generation, a strategy called *elitism* [14]. The entire process is repeated until a termination condition is met, usually just a given number of generations. The result is the best chromosome ever evaluated.

2.2 The Local Heuristic as an "Inner Loop"

GAs as described above are applicable to any optimization problem as long as a representation as chromosomes with a crossover operator and an objective function are defined. For example, Riedl et al. [17] uses a GA to optimize entire network topologies.

For the specific problem of link weight optimization, an additional local heuristic can be applied. After evaluation of a chromosome, let (u, v) be the maximum loaded link with $\rho(u, v) = \rho^*$. It is then reasonable to increase the weight $\mu(u, v)$, since this "lengthens" all shortest paths containing (u, v) and tends to shift off load, so one can expect that the maximum loaded link (u, v) carries less load than before.

In [10][11], this local heuristic is implemented in form of an inner loop, basically as follows: When a chromosome is evaluated, the objective function is called to determine the maximum load and the associated link weight. Then a loop is entered which increments this weight by one, and the objective function is called again, resulting in a new maximum load and link. This is repeated until

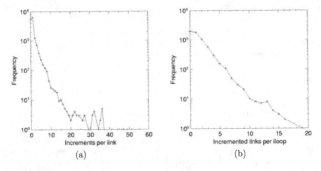

Fig. 1. Frequencies of weight increments (a) and number of incremented weights per invocation of the inner loop (b).

the new maximum load becomes greater than the previous one. After each iteration, it is possible that the maximum loaded link changes, so that the weights of several links might have been incremented when the loop is left. A single link weight could have been incremented also more than once. In other words, the inner loop finds a nearby local minimum.

The objective function needs to compute the shortest path sink trees for all nodes and to propagate the traffic demands along the computed paths. Since the inner loop changes only one link weight per iteration, incremental algorithms which reuse state from previous computations can be used to increase efficiency [10].

A GA with a local heuristic is also called "memetic algorithm" [9].

3   Analysis and New Approach

The GA runs about 340 generations for a population of 21 chromosomes, making 31521 calls of the objective function, for a network with 20 nodes and 106 uni-directional links (cf. Sec. 4 on further parameters). This means that on average $31521/340 \approx 92.7$ calls to the objective function per generation are made, resulting in an average of $92.7/21 \approx 4.4$ calls per chromosome.

Fig. 1(a) shows the frequencies of weight increments, and Fig. 1(b) shows the frequencies of the number of incremented weights per invocation of the inner loop. The inner loop increments a single weight up to 51 times, and up to 19 weights per invocation. But most often, a weight is incremented only once or not at all, and most often, no weight is successfully incremented. Repeating the run with different seeds of the random number generator yields similar results.

These results do not imply the inner loop is useless; the GA converges faster with the local heuristic than without it. But the inner loop is still expensive.

For the case that one weight is incremented just by one, the objective function is called *three* times: The first time to get the initial maximum load and most loaded link, the second time after the first increment, and the third time just to find that the second increment worsened the result and therefore needs to be undone.

Furthermore, the cases where many weights are incremented by large amounts occur during the early generations of a run, where the population still contains much random information from the initial, randomly created chromosomes. Such chromosomes are still far from global minima. For example, it is not unlikely for randomly generated weights to violate the triangle in-equation[1]. In other words, the inner loop is most efficient in finding local minima which are unlikely to be also global minima.

3.1   Directed Mutation

Instead of implementing the local heuristic as an inner loop which always "falls" into a nearby local minimum, we implemented it as *directed mutation*, as follows: A chromosome is evaluated normally by calling the objective function only once, but the most loaded link for this chromosome is remembered. Later, if the chromosome is selected for the next generation, not only random mutation is applied, but also the weight of the remembered link is incremented by one. This can be seen as a non-random mutation in the sense that it is a modifying operation on a single chromosome (while crossover operates on at least two). This approach makes the local heuristic only a *trend* towards a nearby local minimum, instead of a strict rule which always explores the minimum, and no additional calls of the objective function are required.

The drawback is that directed mutation interferes with crossover, which is normally performed before mutation. Crossover changes the chromosome, so that the remembered information about the most loaded link is invalidated. Therefore, directed mutation can be applied only to chromosomes which are not changed by crossover. To deal with this problem, we simply decrease the probability for crossover to leave more unchanged chromosomes[2].

Directed mutation is similar, but less complex than the approach described in [13], where *all* link weights are increased if their loads exceed certain thresholds. Additionally, *all* link weights are decreased if there loads are less than other certain thresholds. We found that decreasing weights of little loaded links does not improve the convergence of the GA.

---

[1] The triangle in-equation is violated if a link weight is greater than the weight of a detour connecting the endpoints of the link. With shortest path routing, such links remain unused for every destination.

[2] The other approach, doing directed mutation *before* crossover, leads to worse results.

3.2   Caching

In addition, we added an evaluation cache to increase efficiency. The cache stores each chromosome which has not yet been evaluated together with maximum load and associated link. There are three cases where a cache hit can occur:

- The best chromosome is always copied unchanged to the next generation (elitism), where it will be evaluated again.
- Sometimes the same chromosome is selected as both parents for crossover. Crossover will have no effect in such cases and simply return two identical clones of the identical parent chromosomes. Since the mutation rate is low, both children may also pass the mutation step and eventually reach the new population without having been modified.
- A chromosome is simply re-created by chance. Such cases are explicitly dealt with in e.g., tabu search, but not in GAs.

The original motivation for the cache was just to save some CPU time in the cases described above, because the objective function is by far the most time consuming function of the optimizer, especially for large networks[3]. We did not expect a high cache hit ratio, since this would imply that many points in the search space are visited repeatedly, indicating badly chosen parameters like e.g., the probability for random mutation. The surprising result is that this assumption was incorrect, as we will show in the following section.

4   Evaluation

We implemented both strategies for the local heuristic, inner loop and directed mutation, in C++. Rank selection [14] is used as selection scheme: the n chromosomes are sorted according to their fitness and the selection probability p_i depends on rank i, as follows:

$$p_i = \frac{2n - i}{\sum_{k=0}^{n-1} 2n - k},$$

which makes the probability for the worst chromosome half the probability of the best chromosome. N-point crossover is used, where the number of split points N is half the number of directed links.

All runs are performed with a population of 21 chromosomes. The initial population contains 20 random chromosomes with weights uniformly distributed in the range $[1-30]$, and one chromosome with all weights equal to 15. Although the upper bound (30) is considerably less than the maximum weight allowed in e.g., OSPF ($2^{16} - 1$), it significantly shrinks the search space without restricting it too much to exclude good solutions. The inner loop and directed mutation operations, however, may exceed this limit if required.

---

[3] The complexity for the (non-incremental) objective function is $O(V^2 \log V + VE)$ with a Fibonacci heap in Dijkstra's shortest path algorithm.

Table 1. Investigated Networks

Network	Nodes	Link Weights	Generations	Evaluations	Capacities
Cost	11	52	700	14700	inhomogeneous
Labnet	20	106	1500	31521	homogeneous
GT40	40	344	2000	42021	homogeneous

The crossover probability differs for both strategies. For the inner loop, it is set to 1.0, which means that crossover is applied always. For directed mutation, the crossover probability is set to the considerably smaller value 0.3, because we want increase the possibility for directed mutation, as described in Sec. 3.1. Both values have been determined by repeated experiments with different networks, so each strategy operates on its favorable working point. Both strategies use the evaluation cache.

The mutation probability for random mutation in both strategies is one of the parameters we want to study and will be in the range $[0.01 - 0.1]$. Directed mutation in the second strategy will be applied always when possible, i.e., when crossover is not performed, so the probability for directed mutation can be given as $1 - 0.3 = 0.7$.

We want to compare the performance of both strategies in terms of resulting maximum load and number of evaluations, i.e., calls to the objective function including cache hits. To give both strategies the same budget of evaluations, a run is terminated after a given number of evaluations has been performed (instead of generations). The last population is always allowed to complete, which constitutes a negligible bias in favor of the inner loop strategy.

Information about the investigated networks is given in Tab. 1. The generation numbers (quotient of evaluations and population size) correspond to the number of evaluations made for the directed mutation strategy. Networks with equal capacities for all links are called homogeneous.

The results are depicted in Fig. 2, Fig. 3 and Fig. 4. For each network, the left diagram (a) shows the maximum load as a function of the mutation rate, the middle diagram (b) shows the cache hit ratio in percent, and the right diagram (c) shows the running times in CPU seconds on a Pentium III Linux PC, 800 MHz, 256 MB. Each diagram shows the curve for the inner loop (IL) and for directed mutation (DM). Each data point represents the mean of 50 repeated experiments[4] with the given mutation rate, but different random seeds; the error bars give the 95% confidence interval.

Except for the 20-node network, directed mutation always yields lower maximum load than the inner loop strategy. For the 20-node network, the difference is not significant for mutation rates above 0.04. For mutation rates greater than 0.02, directed mutation always leads to much higher cache hit ratios and requires therefore much less CPU time. Directed mutation not only yields better results, but is also much more efficient in terms of calls to the objective function. The latter holds for all three test networks.

---

[4] For the 20-node network 200 samples were taken.

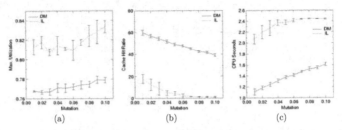

Fig. 2. 11 Node Network with 14700 evaluations.

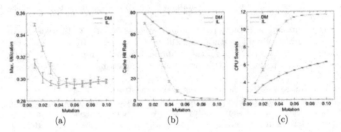

Fig. 3. 20 Node Network with 31521 evaluations.

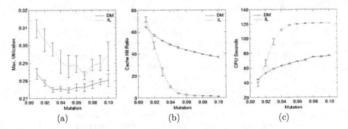

Fig. 4. 40 Node Network with 42021 evaluations.

If the incremental algorithms for the objective function are used, different calls to the objective function require different amounts of CPU time. We did not further analyze this difference, but mention that directed mutation would also benefit from incremental algorithms, since 70% of all chromosomes are subject only to random and directed mutation. The small number of weight changes in such cases make incremental algorithms attractive also for the directed mutation strategy.

5   Discussion

The fact that caching with high hit ratios improves CPU performance is fine but not very surprising. But cache hit ratios tell us also something about the search process. High cache hit ratios clearly indicate that the GA visits many points in the search space more than once. But the results for directed mutation are never worse, and often significantly better, than other strategies with lower cache hit ratios. This implies that simply exploring more potential solutions does not increase the quality of the result!

For the inner loop, we showed in Sec. 3 that many evaluations are wasted by exploring nearby local minima.

Greater random diversification through higher mutation rates also wastes evaluations. The reason behind this phenomenon seems to be the *error threshold*, as in biological populations. For asexual (no crossover), biological populations the error threshold for the mutation rate is well known [18][19]. Mutation is essentially an error occurring during the "transmission" of a chromosome to the next generation. Simply stated, if the mutation rate for a population is too low, the population evolves slowly. If the mutation rate is above the error threshold, the population dies out. The optimal value for the mutation rate is just below the error threshold, to evolve as quickly as possible. Although the population cannot die out in genetic algorithms by design since the population size is fixed, the error threshold remains effective in the sense that there is a mutation rate above which results become worse.

On the other hand, we find it strange that the best of the investigated strategies creates each chromosome about two times, which also wastes half the evaluations. While caching alleviates this effect and leads to better CPU performance, we find it hard to believe that a strategy which creates each potential solution twice should be optimal.

6   Conclusion

This work resumed genetic algorithms in link weight optimization and a local heuristic which increments the weight of the most loaded link. We analyzed the current implementation strategy of the heuristic and found that it is still expensive in terms of calls to the objective function. Based on the analysis, an implementation strategy called directed mutation has been proposed and evaluated together with caching. The new approach outperform the current strategy both in terms of the quality of the result and in terms of calls to the objective function, i.e., CPU usage.

There is still room for further improvements of this approach. For example, using the cache as a tabu list and not only to return the known result from the cache, but to use the tabu list to really create new chromosomes might even further improve performance for optimizations in this area.

References

1. Moy, J.: OSPF version 2. RFC 2328 (1998)
2. Callon, R.: Use of OSI IS-IS for routing in TCP/IP and dual environments. RFC 1195 (1990)
3. Lorenz, D.H., Orda, A., Raz, D., Shavitt, Y.: How good can IP routing be? Technical Report DIMACS TR: 2001-17, Center for Discrete Mathematics and Theoretical Computer Science (2001)
4. Fortz, B., Thorup, M.: Increasing internet capacity using local search. Technical Report IS-MG 2000/21, Universit Libre de Bruxelles (2000)
5. Fortz, B., Thorup, M.: Internet traffic engineering by optimizing OSPF weights. In: INFOCOM. (2000) 519–528
6. Fortz, B., Thorup, M.: Optimizing OSPF/IS-IS weights in a changing world. IEEE J. Select. Areas Commun. **20** (2002) 756–767
7. Fortz, B., Rexford, J., Thorup, M.: Traffic engineering with traditional IP protocols. IEEE Commun. Mag. **40** (2002) 118–124
8. Nucci, A., Schroeder, B., Bhattacharyya, S., Taft, N., Diot, C.: IGP link weight assignment for transient link failures. In: ITC 18. (2003)
9. Buriol, L.S., Resende, M.G.C., Ribeiro, C.C., Thorup, M.: A memetic algorithm for OSPF routing. In: INFORMS. (2002) 187–188
10. Buriol, L.S., Resende, M.G.C., Ribeiro, C.C., Thorup, M.: A hybrid genetic algorithm for the weight setting problem in OSPF/IS-IS routing (2003)
11. Riedl, A.: A hybrid genetic algorithm for routing optimization in IP networks utilizing bandwidth and delay metrics. In: IEEE IPOM. (2002)
12. Ericsson, M., Resende, M.G.C., Pardalos, P.M.: A genetic algorithm for the weight setting problem in OSPF routing. Technical report, ATT Shannon Laboratory (2001)
13. Mulyana, E., Killat, U.: An alternative genetic algorithm to optimize OSPF weights. In: 15-th ITC Specialist Seminar. (2002)
14. Goldberg, D.E.: Genetic Algorithms in Search, Optimization and Machine Learning. Addison-Wesley (1989)
15. Fortz, B., Thorup, M.: Robust optimization of OSPF/IS-IS weights. In: INOC. (2003) 225–230
16. Yuan, D.: A bi-criteria optimization approach for robust OSPF routing. In: IEEE IPOM. (2003)
17. Riedl, A., Bauschert, T., Frings, J.: A framework for multi-service IP network planning. In: 10th International Telecommunication Network Strategy and Planning Symposium (Networks 2002). (2002)
18. Eigen, M.: Selforganization of matter and the evolution of biological macromolecules. Naturwissenschaften **58** (1971) 465–523
19. Nowak, M., Schuster, P.: Error thresholds of replication in finite populations mutation frequencies and the onset of muller's ratchet. J. theor. Biol. **137** (1989) 375–395

A New Prediction-Based Routing and Wavelength Assignment Mechanism for Optical Transport Networks[1]

Eva Marín-Tordera, Xavier Masip-Bruin, Sergio Sánchez-López,
Josep Solé-Pareta, and Jordi Domingo-Pascual

Departament d'Arquitectura de Computadors, Universitat Politècnica de Catalunya
Avgda. Víctor Balaguer, s/n– 08800 Barcelona, Spain
{eva, xmasip, sergio, pareta, jordid}@ac.upc.es

Abstract. In optical transport networks algorithms dealing with the lightpath selection process select routes and assign wavelengths based on the routing information obtained from the network state databases. Unfortunately, due to some factors, in large dynamic networks this routing information may be non-accurate enough to provide successful routing decisions. In this paper we suggest a new prediction-based routing mechanism where lightpaths are selected based on prediction decisions. Consequently, the routing information is not required at all, so updating this information is neither required. In short, the signaling overhead produced by the updating process is practically removed.

Keywords: Optical routing, routing inaccuracy, prediction-based routing.

1 Introduction

Internet traffic demands are extensively growing in the last years due to the real time applications such as video, multimedia conferences or virtual reality. Optical wavelength-division multiplexing (*WDM*) networks are able to provide great bandwidth to support this growing traffic demands. Unlike traditional IP networks where the routing process only involves a physical path selection, in *WDM* networks the routing process involves both a physical path selection and a wavelength assignment, i.e., the routing and wavelength assignment (*RWA*) problem. The *RWA* problem is often tackled by being divided into two different sub-problems, the routing sub-problem and the wavelength assignment sub-problem. The first approach to the routing sub-problem in a *WDM* network focuses on always selecting the same route between each source-destination node pair, known as static routing. This route is calculated for example in the Fixed-shortest path, by means of the Dijkstra's [1] algorithm or the Bellman-Ford's algorithm. However, since the performance of the fixed-shortest path algorithm is limited, the Fixed-Alternate routing is proposed [2]. According to this, more than one fixed route is calculated for every source-destination node pair. For each new connection request the routing algorithm tries to send the traffic through the

---

[1] This work was partially funded by the MCyT (Spanish Ministry of Science and Technology) under contract FEDER-TIC2002-04344-C02-02 and the CIRIT (Catalan Research Council) under contract 2001-SGR00226.

J. Solé-Pareta et al. (Eds.): QofIS 2004, LNCS 3266, pp. 154–163, 2004.
© Springer-Verlag Berlin Heidelberg 2004

calculated fixed routes in sequence. This solution substantially reduces the number of connection blocked respect the fixed-shortest path.

The main problem of the static routing is that it does not consider the current network state. Hence, the second approach for the routing sub-problem in *WDM* networks is the dynamic (or adaptive) routing, which selects routes based on the current network state. There are different approaches for this scenario, such as the adaptive shortest-cost-path routing and the Least-Congested Path algorithm, *LCP* [3]. In short, in spite of the fact that *LCP* performs better than Fixed-Alternate routing, it is worth noting that in adaptive routing source nodes require of continuously receiving update messages about the changes in the network state.

The static wavelength assignment sub-problem consists in given a set of established routes for a set of lightpaths, to assign the wavelength to each route. In this paper we focus on Wavelength Selective (*WS*) networks, that is, networks without wavelength conversion capabilities. The main restriction in WS networks is that routes sharing the same link (or links) must have different wavelengths, i.e., the same wavelength must be assigned to the lightpath on all the links in its route.

If connection requests arrive by an incremental or dynamic traffic, heuristic methods are used to assign wavelength to the lightpaths. In this case the number of available wavelength is supposed to be fixed. A large number of different heuristic algorithms have been proposed in the literature as shown in [4], such as Random, First-Fit, Least-Used, Most-Used, Min-Product, Least-Loaded, Max-Sum, Relative Capacity Loss, Protecting Threshold, and Distributed Capacity Loss

Most *RWA* solutions proposed in the recent literature use distributed mechanisms based on source-routing. In this scenario the routing inaccuracy problem (*RIP*) comes up. The *RIP* describes the impact on global network performance because of taking *RWA* decisions according to inaccurate routing information. In highly dynamic networks, inaccuracy is mainly due to the restriction of aggregating routing information in the update messages, the frequency of updating the network state databases and the latency associated with the flooding process. It is worth noting that two factors are negative collateral effects of their inclusion to reduce the signaling overhead produced by the large amount of update messages required to keep accurate routing information. It has been clearly shown [5] that the routing inaccuracy problem, that is, to select a path based on outdated network state information, may significantly impact on global network performance significantly increasing the number of blocked connection requests.

In this paper we propose the prediction-based routing as a mechanism that does not only address the *RWA* problem but also the *RIP*, achieving a drastic reduction in the signalling overhead. In short, the prediction-based routing mechanism selects routes not based on the 'old' or inaccurate network state information but based on some kind of 'predicted' information. Hence, since routing information from network state databases is not required, we may eliminate the need of flooding update messages (except those required for connectivity).

The remainder of this paper is organized as follows. Section 2 reviews main significant contributions existing in the recent literature about the Routing Inaccuracy Problem. Then, Section 3 proposes the Predictive Routing Algorithm, Section 4 evaluates our proposal and finally, Section 5 concludes the paper.

2   Handling the Routing Inaccuracy Problem

Most of the Dynamic *RWA* algorithms assume that the network state databases (named Traffic Engineering Databases, TEDs when including QoS attributes) contain accurate information of the current network state. Unfortunately, when this information is not perfectly updated routing decisions can be wrongly performed at the source nodes producing a significant connection blocking increment (i.e., the routing inaccuracy problem). Most recent related work is summarized in the following paragraphs.

In [5] the effects produced in the blocking probability because of having inaccurate routing information when selecting lightpaths are shown by simulation. The authors indeed verify over a fixed topology that the blocking ratio increases when routing is done under inaccurate routing information. The routing uncertainty is introduced by adding an update interval of 10 seconds. Some other simulations are performed to show the effects on the blocking ratio due to changing the number of fibers on all the links. Finally, the authors argue that new Routing and Wavelength Assignment *(RWA)* algorithms that can tolerate imprecise global network state information must be developed for dynamic connection management in *WDM* networks.

In [6] the routing inaccuracy problem is addressed by modifying the lightpath control mechanism, and a new distributed lightpath control based on destination routing is suggested. The mechanism is based on both selecting the physical route and wavelength on the destination node, and adding rerouting capabilities to the intermediate nodes to avoid blocking a connection when the selected wavelength is no longer available at set-up time in any intermediate node along the lightpath. There are two main weaknesses of this mechanism. Firstly, since the rerouting is performed in real time in the set-up process, wavelength usage deterioration is directly proportional to the number of intermediate nodes that must reroute the traffic. Secondly, the signaling overhead is not reduced, since the *RWA* decision is based on the global network state information maintained on the destination node, which must be perfectly updated.

Another contribution on this topic can be found in [7] where authors propose a mechanism whose goal is to control the amount of signaling messages flooded throughout the network. Assuming that update messages are sent according to a hold-down timer regardless of frequency of network state changes, authors propose a dynamic distributed bucket-based Shared Path Protection scheme (an extension of the *Shared Path Protection, SPP* scheme). Therefore, the amount of signaling overhead is limited by both fixing a constant hold-down timer which effectively limits the number of update messages flooded throughout the network and using buckets which effectively limits the amount of information stored on the source node, i.e. the amount of information to be flooded by nodes. The effects of the introduced inaccuracy are handled by computing alternative disjoint lightpaths which will act as a protection lightpaths when resources in the working path are not enough to cope with those required by the incoming connection. Authors show by simulation that inaccurate database information strongly impacts on the connection blocking. This increase in the connection blocking may be limited by properly introducing the suitable frequency of update messages. According to the authors, simulation results obtained when applying the proposed scheme along with a modified version of the *OSPF* protocol, may help network operators to determine that frequency of update messages which better maintains a trade-off between the connection blocking and the signaling overhead.

In [8] authors propose a new adaptive source routing mechanism named BYPASS Based Optical Routing (*BBOR*), aiming to reduce the routing inaccuracy effects, i.e., blocking probability increment and non-optimal path selection, in WS networks. In [9] authors extend the mechanism to be applied to networks with conversion capabilities. The *BBOR* mechanism is based on bypassing those links which cannot forward the setup message because of lacking the selected wavelength. The bypass is achieved by forwarding the setup messages through a previously precomputed alternative path (bypass-path).

3   New Proposal of Prediction-Based Routing

The main idea of the Prediction-based Routing (*PBR*) mechanism is based on extending the concepts of branch prediction used in computer architecture [10]. In this field, there are several methods to predict the direction of the branch instructions. The prediction of branch instructions is not made knowing exactly the state of the processor but knowing the previous branch instructions behavior. The prediction can be either wrong or correct but the goal is to maximize the number of correct predictions. Considering this idea, the *PBR* mechanism is based on predicting the route and wavelength assignment between two nodes according to the routing information obtained in previous connections set-up. Thus, the *PBR* avoids the use of inaccurate network state information obtained from the Traffic Engineering databases, therefore removing the need of frequent updating. It is necessary to mention that a minimal updating is required to ensure connectivity just reporting about link/node availability.

The main objective is to optimize the routing algorithm decision, considering the state 'history' for each path, that is, every source node must keep previous information about both wavelength and route allocated to this path established between itself and a destination node This history is repeated all through the time and is stored in a history register, which will be used as a pattern of behavior, which is used to train a new table, named Prediction Table (*PT*).

It must be noticed that in order to generate the history, every source node must keep not only the last information but also previous information of the wavelength and routes used. With all this information it creates an index which is then used to index the *PT*. This *PT*, has different entries, each keeping information about a different pattern by means of a counter. The prediction is obtained reading the counter value from the table. These counters are updated (increased or decreased) in order to learn [10].

3.1   Wavelength History Registers

Before defining a prediction algorithm it is necessary to introduce the parameter used to decide when the history registers may be modified. We define indeed a cycle as the basic unit of time where the history state is susceptible to be modified.

As it is mentioned above every source node must know the history state information, and for this reason the history state is kept in history registers. There are one of such registers for every wavelength on every path to every destination node. We name these registers as wavelength registers (WR).

We propose a method to register the history of the network state in every source node based on assuming that for each cycle, each WR is updated with a 0 value when this wavelength on this path is used on that cycle. Otherwise, the register of an unused wavelength on a path is updated with a 1. It must be noticed that the expression "a path is used" means that it has been selected by the prediction algorithm and actually the decision is right since the path is available. On the other hand, "a path is unused" when no incoming connection is assigned to this path.

3.2   Prediction Tables

The prediction tables are the base to be able to predict a wavelength and a path. In the source nodes one prediction table, PT, is needed for every feasible circuit between any source-destination node pair. The prediction table for a wavelength on a path is accessed by an index obtained from the corresponding WR. For example, a source node sends traffic towards two different destination nodes and every source-destination node pair has two different paths (two shortest-paths). Moreover, if we assume the existence of 6 wavelengths then 24 PTs are needed on the source node, one for every path and wavelength. In every source node there is the same number of wavelength registers than of prediction tables.

Every entry in the prediction tables has a counter, which is read when accessing the table. This value is compared to a threshold value. If the value from the table is lower, the prediction result is to accept the request through the wavelength on this path. Otherwise, the path is predicted to be not available. The counters are two-bit saturating counter, where 0 and 1 account for the availability and 2 and 3 accounts for path unavailability [10]. The use of two values to account for the availability or the unavailability has been well studied in the area of branch prediction. As it is presented in [10] a two bit counter gives better accuracy than a one bit counter. The use of a one bit counter means that it predicts what happened last time. If last time the traffic request was blocked and the counter has only one bit, the next time that the history is repeated the prediction will be that there will not be availability, or if the traffic was accepted last time the prediction will be that there will be availability. On the other hand if the counter has two bits it is necessary that the traffic request has been blocked (or accepted) two times for the same history to change the direction of the prediction. It is also exposed in [10] that going to counters larger than two bits does not necessarily give better results. This is due to the "inertia" that can be built up with a large counter. In that case more than two changes in the same direction are necessary to change the prediction. Saturating counter means that when counter has a value of 0 and it is decreased its new value is also 0, and when its value is 3 and it is increased its value remains at 3.

As explained above, in the source nodes there is one prediction table, PT, for every wavelength on every path and for every destination. The tables have to be updated with the same index used on the prediction. When a new connection request is set up the table of the selected wavelength and path is updated, decreasing the counter. On the other hand, when the connection request has been blocked the counter is increased. The rest of the tables of the unused paths are not updated. Note that when a connection request is set up only the prediction table of the wavelength and path used is updated, but all the wavelength registers corresponding to that destination are updated, of the used with 0 and of the unused wavelengths with 1.

It is worth noting that the updating of prediction tables in the source nodes is done immediately the prediction is done and it is known if the connection request is set up or blocked. For this reason it is not necessary to flood update message through the network to update the network state databases.

3.3   RWA Prediction Algorithm

We define a new *RWA* prediction algorithm, Route and Wavelength Prediction, *RWP*, inferred from the *PBR* mechanism, which utilizes the information contained in the prediction tables to decide about which path and wavelength will be selected. When a new request arrives at the source node demanding a connection to one destination node, all the prediction tables of the corresponding destination are accessed. It must be noticed that one prediction table, *PT*, and one wavelength register, *WR*, exist for every wavelength on every path to every destination. We assume that two shortest paths are computed for every source-destination node pair, SP_1 and SP_2. Prediction tables are accessed by one index per table which is built with the wavelength histories contained in the *WR*. As a consequence of reading the prediction tables, the 2-bit counters are obtained. As an example, Fig. 1 shows the accesses to existing PTs for the shortest path (either SP_1 or SP_2). In Fig. 2 we can see the *RWP* flow chart, supposing W wavlengths in every link. The *RWP* algorithm always starts considering the value of the counter of the *PT* of the first wavelength on the shortest path, for instance SP_1. If the counter is less than 2 (0,1), and this wavelength is free in the node's outgoing link towards SP_1, the prediction algorithm decides to use this wavelength on this path. Otherwise (counter=2, 3 or outgoing link not available) this wavelength is not used. In this last case, the value of the counter of the next PT is examined. Notice that next *PT* corresponds to the second wavelength on SP_1. When the counters of the *PTs* of all the wavelengths of SP_1 have been examined, that is, either the counters always are greater than 2 or all wavelengths on the outgoing link towards SP_1 are not avail-

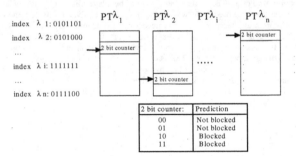

Fig. 1. Example of Prediction Table access and values of the 2-bit counters

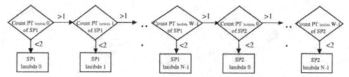

Fig. 2. *RWP* flow chart

able, the prediction algorithm checks the PTs of the next path, SP₂, and so on. When the prediction algorithm, after checking all *PTs*, decides that all the feasible wavelengths on the two paths are blocked, then it tries to forward the connection request through the first available wavelength on the outgoing link towards one of the two shortest path either SP₁ or SP₂. The information about of the outgoing links of the source node is always known by the source node.

Wavelength registers (*WR*) are updated depending on which wavelength is used and whether the request is blocked or not. Also the prediction table, *PT*, of the used wavelength and path is updated by either increasing (means connection blocked) or decreasing (means connection not blocked) the counter of the corresponding entry in the *PT*.

It is worth noting that counters of every wavelength on all the feasible paths between a source-destination node pair can be read, so allowing the prediction to be made, before a new connection request reaches the source node. It is a very significant factor which significantly reduces the cost involved with the *PBR* mechanism. In fact, even though several tables must be accessed to make the prediction, these accesses can be done offline. For every possible new request, the decision of which path to use is already done.

4   Performance Evaluation

We have developed a tool to check the Prediction-Based Routing performance. Simulations are obtained by applying the PBR to a topology test composed by 15 nodes and 27 links, with 2 source nodes and 2 destination nodes. All these nodes are connected by one fiber-links and the number of lambdas is a variable in the range of 2 and 5. Connection arrivals are modeled by a Poisson distribution and each arrival connection requires a full wavelength on each link it traverses. Each *WR* keeps information about the last 5 cycles, 5 bits, so there are 32 entries of 2 bits in each *PT*. In order to show the capacity overhead in terms of bits because of applying the PBR we propose as an example the following: we assume that 2 shortest paths, SP1 and SP2, are computed with 5 lambdas each, therefore will be 20 *PTs* in every source node. Such a scenario represents a total capacity of 1280 bits, which can perfectly be considered as negligible.

The initial goal is to verify that the RWP can know the network behavior, in terms of routing and wavelength assignment, using the prediction tables. We compare the performance of both the *RWP* and First-Fit algorithm. When applying the First-Fit algorithm we vary the updating frequency and the number of available wavelengths on every fiber. As a nomenclature, we define a cycle as the basic unit of time. Fig. 3

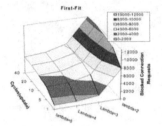

Fig. 3. Blocked Connection Requests for the First-Fit Algorithm

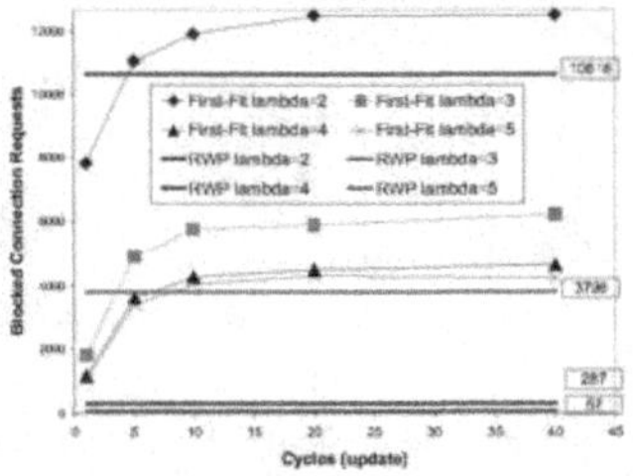

Fig. 4. First-Fit versus *RWP* Algorithm

shows the blocking obtained by the First-Fit algorithm, assuming a total number of 62000 connection requests, when varying the update interval from 1,5,10, 20 and 40 cycles. The Y-axis in Fig.3 depicts the number of blocked requests, consisting in both those requests rejected at any intermediate node and those requests blocked because of lacking resource enough in the path selection process. Fig. 3 also shows the effects of varying the number of available wavelengths. We can see that a minimum number of blocked requests (1054, that is a 1.7%) is obtained when N=1 (update messages every cycle) and the number of lambdas is 5. Fig. 4 shows a comparison between RWP and First-Fit algorithm for several lambda values. Analyzing the results, we demonstrate that from lambda=4 the RWP behaves better than the First-Fit. Therefore, for lambda=4 the result for RWP is of 287 blocked requests and for First-Fit is of 1066, and for lambda=5 the results are 56 blocked requests for RWP and 1054 for First-Fit.

There are two origins of blocked requests. The first is produced when there is no available path for a connection request. The second occurs when the algorithm fails in the route assignment, so that the set-up connection is blocked in an intermediate node.

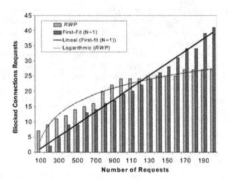

Fig. 5. Evolution of Blocked Connections Requests for the First-Fit and *RWP*.

The *RWP* achieves a number of blocked requests less than the First-Fit (e.g. for lambda=4) due to the fact that the First-Fit fails more in the route assignment. even when the update messages reaches the source node every cycle (N=1). This case occurs when two connections are requested at two nodes at the same time, one node assigns route before the other. Thus, the second node assigns route utilizing network information out of date. This case does not happen in the *RWP* because it has more capability of learn which route is the best for each request. In Fig. 5 we present the evolution of blocked requests (for lambda=4) every 100 new request since the total number of request is 0 to 2000, for both the First-Fit and *RWP*. Initially the prediction algorithm fails more (7 and 0 blocked requests for the first 100 requests for the *RWP* and First-Fit algorithm respectively), then when the number of requests is 1400 the number of blocked requests is equal for both algorithms, and for 2000 requests the results are 27 and 41 for *RWP* and First-Fit respectively. We can conclude that the prediction algorithm learns about its fails and the slope of rising decreases (logarithmic approximation), but the First-Fit algorithm has a constant rising in the number of blocked requests (lineal approximation).

It is worth noting that we have compared the *RWP* with the First-Fit algorithm assuming N=1. However, it is well known that the signaling overhead involved by this updating frequency is non-affordable. Hence, when simulations take into account more realistic values, for instance N=40, *RWP* is still much better than the First-Fit algorithm.

5   Conclusions

In this paper authors propose the Prediction-Based Routing (*PBR*) mechanism to tackle the *RWA* problem in *WDM* networks. The main skill of *PBR* is to provide source nodes with the capability of taking routing decisions without using the traditional routing information, that is the network state information contained in their Traffic Engineering databases (TEDs). Two immediate benefits may be inferred from the PBR mechanism. The former, the *PBR* removes the update messages required to update the TEDs (only connectivity messages are required), so significantly reducing the signaling overhead. The latter, in highly dynamic networks the *PBR* can efficiently change the routing decisions after a training period. Simulation results show that the *PBR* mechanism behaves better than the First-Fit algorithm even when an update frequency of 1 cycle is set for the First-Fit.

References

1. Lawler E., "Combinational Optimization: Networks and Matroids", Holt, Rinehart and Winston, 1976.
2. Harari H., Massayuki M., Miyahara H., "Performance of Alternate Routing Methods in All-Optical Networks". IEEE INFOCOM'96, 1996.
3. Chan K.M., Yun T.S.P:, "Analysis of Least Congested Path Routing Methods in All-Optical Switching Networks" Proc. Of INFOCOM'97, 1997.

4. Zang H., Jue J.P., Mukherjee B., "A Review of Routing and Wavelength Assignment Approaches for Wavelength-Routed Optical WDM Networks" Optical Network Magazine, January 2000.
5. Thou J., Yuan X., "A Study of Dynamic Routing and Wavelength Assignment with Imprecise Network State Information", ICPP Workshop on Optical Networks, August 2002.
6. Zheng J., Yuan X., "Distributed lightpath control based on destination routing in wavelength-routed WDM networks", Optical Networks Magazine, July/August 2002, vol.3, n°:4, pp.38-46.
7. Darisala S., Fumagalli A., Kothandaraman P. Tacca M., Valcarenghi L., "On the Convergence of the Link-State Advertisement Protocol in Survivable WDM Mesh Networks", in Proc. of ONDM'03, pp. 433-447, Budapest, Hungary, February 2003.
8. Masip-Bruin X., Sánchez-López S., Solé-Pareta J., Domingo-Pascual J., "A QoS Routing Mechanism for Reducing the Routing Inaccuracy Effects", QoS-IP, Milan, Italia, February 2003.
9. Masip-Bruin X., Sánchez-López S., Solé-Pareta J., Domingo-Pascual J., Colle D. "Routing and Wavelength Assignment under Inaccurate Routing Information in Networks with Sparse and Limited Wavelength Conversion", IEEE GLOBECOM, 2003, San Francisco, USA, December 2003.
10. Smith J.E., "A study of branch prediction strategies", In Proc. of 8th International Symposium in Computer Architecture, Minneapolis 1981.

On the Construction of QoS Enabled Overlay Networks

Bart De Vleeschauwer, Filip De Turck, Bart Dhoedt, and Piet Demeester

Sint-Pietersnieuwstraat 41, B-9000 Gent, Belgium.
Tel.: +32 9 264 99 91, Fax: +32 9 264 99 60
bart.devleeschauwer@intec.ugent.be

Abstract. The lack of QoS support in the Internet makes it difficult for service providers to give guarantees regarding the timely delivery and quality of their services. For multimedia services like video on demand and video conferencing however, the delay should be minimal. One of the main causes of the absence of QoS is the inter- and intradomain routing scheme in the Internet, minimizing the hop count instead of optimizing the QoS. In this paper we will discuss the construction of an overlay network that is able to meet the requirements of time-critical services. An optimal algorithm, for the placement of overlay servers in the Internet, will be described together with a number of heuristics. These server placement algorithms will be evaluated by comparing the resulting overlay network to the standard Internet in terms of number of connections accepted and number of overlay servers required.

Keywords: QoS, Overlay Network, Server Placement, Integer Linear Programming.

1 Introduction

The usage of the Internet is evolving from html and file traffic towards advanced multimedia service delivery. The current infrastructure lacks the ability to provide the QoS required by these services. Applications like video on demand and video conferencing are characterized by both a significant need for bandwidth coupled with minimal delay. These parameters are not taken into account by the standard Internet routing algorithms, which select the shortest path in terms of hop count. As a result, the Internet doesn't offer any guarantees regarding the delay/bandwidth on a path. Other reasons for the lack of QoS support are the inability of the Internet to route around congested links and the fact that autonomous systems (ASs) often eject packets destined for other ASs early to minimize the load on their own network, regardless of the effect this might have on the delay [1]. We propose the use of overlay network technology to overcome these problems. An overlay network is a network built on top of an existing network. Overlay networks facilitate the introduction of new network functionality whilst keeping the underlying network unchanged. Examples of the use of overlay networks include MBone [2] and 6Bone [3]. There has already been research

J. Solé-Pareta et al. (Eds.): QofIS 2004, LNCS 3266, pp. 164–173, 2004.
© Springer-Verlag Berlin Heidelberg 2004

on the possibilities of allowing alternative route selection via an overlay network. In [4] the authors discuss an overlay network that is able to dynamically react to path outages and [5] discusses QoS aware routing algorithms for overlay networks.

An overlay network can be used to route traffic by sending the data between overlay servers. Monitoring the condition of the overlay links allows an overlay network to get information regarding the state of the Internet and to dynamically react to link congestion or quality degradation by sending its traffic via routes that still fulfill the QoS demands of a connection. Essential to the functioning of such an overlay network is the location of the different overlay servers. Good server placement algorithms will greatly increase the quality of the overlay networks. In [6] the server placement problem for overlay networks is studied by determining locations of servers such that the distance of every client to its nearest overlay server is minimized. Here however, we argue that an important aspect of QoS degradation is situated in the core network. We will thus study algorithms that place the overlay servers on a best effort network in such a way that connections will have end-to-end paths that fulfill both their bandwidth and their delay requirements.

Fig. 1 illustrates the use of an overlay network to provide a route selection infrastructure. We see two clients in a small IP network, consisting of five routers. The delay on every link is shown and if standard IP routing is used, the delay between the routers 1 and 2 will be 200 (situation (a)). However, by deploying an overlay server near router 4 the traffic can be redirected. Thus the route that is chosen for the traffic between the clients will have a much lower delay of 30 (situation (b)).

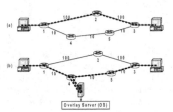

Fig. 1. Routing via an overlay network

This paper is organized as follows: section 2 gives a full description of the problem we want to solve and discusses an optimal server placement algorithm and a number of heuristics. In section 3 we present the evaluation of the different algorithms. Conclusions are drawn in section 4 and future work is addressed in section 5. As multimedia services are often characterized by a multicast nature and overlay networks are also able to provide multicast functionality [7], we have evaluated the algorithms for both the unicast and the multicast case.

2   Algorithms

The goal of the algorithms is finding the location of a number of overlay servers, such that the resulting overlay network supports as many connections as possible from a given set of requested connections, guaranteeing connection bandwidth and a bounded end-to-end delay. By formulating the problem as an Integer Linear Programming (ILP) problem, we can find an optimal solution using standard techniques [8]. For completeness we developed a formulation for the multicast problem, where all connections can have multiple destinations. The unicast formulation can be easily derived from the multicast case. The solution will not only determine the location of the servers, but will also determine the routes that are followed by the different connections. These routes will thus load balance the network, as they are chosen in such a way that as much connections as possible can be supported. We also designed three heuristics for determining a server placement.

2.1   Multicast ILP Solution

Input. Following notations were introduced: $G_{IP} = (V, A_{IP})$ denotes the directed graph of the IP network with set of vertices V and set of arcs A_{IP}. $G_{OV} = (V, A_{OV})$ is used to denote the full mesh overlay network constructed on top of the IP network. The nodes in the IP network represent routers and the arcs IP links connecting two routers. The arcs in the overlay graph are mapped to routes in the IP network, for the delay of an overlay arc, we took the sum of the delays of the IP arcs on the corresponding route. The delay on an IP arc was assumed to be constant. The delay of an overlay arc a_{OV} is denoted by DELAY (a_{OV}). For every arc a_{IP} in the IP graph, there is a set U (a_{IP}), containing all the overlay arcs using a_{IP}. Every IP arc a_{IP} has an available bandwidth, denoted by BW (a_{IP}). For every node v of the overlay network there is a set $in (v)$ with all the incoming arcs and a set $out (v)$ with all the outgoing arcs for that node.

The multicast connections that we want to be able to support are bundled in a set C. A multicast connection $c \in C$ has a source vertex $S (c)$ and a set of target vertices $T (c)$. It also has a required bandwidth BW (c) and a maximal delay DELAY (c). A multicast connection can be seen as originating from a client located behind the source router and destined to clients located behind the destination routers. We assumed that an overlay server can be connected to every router in the IP-network. Furthermore, we assumed that the traffic bottleneck is in the actual IP-network and therefore do not take into account the delay/bandwidth restrictions for the links connecting the overlay servers and clients to the routers.

We also included the possibility to limit the total number of servers used, this number should not exceed the value of n, a parameter of the algorithm.

ILP Problem formulation. Following decision variables have been introduced:

- a_c: an a_c variable equals 1 if a connection c is supported by the overlay network.
- x_v: an x_v variable equals 1 if there is an overlay server connected to router v.
- $y_{a,c}$: these binary variables describe the multicast tree for a connection c. If $y_{a,c}$ equals 1, the overlay link a is used to connect the source node of c to a target node of c. If a connection c is not supported by the overlay, the values of $y_{a,c}$ equal 0 for every overlay link a.
- $z_{a,c,t}$: these variables determine the end-to-end path for a target node t of a connection c, so if an overlay arc a is used for connecting the source node to a target node t of a connection c, the $z_{a,c,t}$ variable equals 1.

The ILP formulation contains a set of constraints specifying the requirements on the solution and an objective function, describing the goal. A first set of constraints are the flow conservation constraints, they determine all the paths connecting the source of a connection to every target node of that connection, for all the connections. Constraints (1) and (2) make sure that a source node can only send traffic if the connection is supported and that a source node never has incoming links in an end-to-end path. Constraints (3) and (4) enforce analogous constraints on the target nodes of the connection. (5) states that an intermediary node only forwards traffic if it receives traffic.

$$\forall c \in C, \forall t \in T(c) : \sum_{a \in out(S(c))} z_{a,c,t} = a_c \qquad (1)$$

$$\forall c \in C, \forall t \in T(c) : \sum_{a \in in(S(c))} z_{a,c,t} = 0 \qquad (2)$$

$$\forall c \in C, \forall t \in T(c) : \sum_{a \in in(t)} z_{a,c,t} = a_c \qquad (3)$$

$$\forall c \in C, \forall t \in T(c) : \sum_{a \in out(t)} z_{a,c,t} = 0 \qquad (4)$$

$$\forall c \in C, \forall t \in T(c), \forall v \in V \setminus (t, S(c)) : \sum_{a \in in(v)} z_{a,c,t} - \sum_{a \in out(v)} z_{a,c,t} = 0 \qquad (5)$$

A next set of constraints bundles the variables describing the end-to-end paths (z variables) in the variables describing a multicast tree for every connection (y variables). This is done by letting an arc be present in the tree if there is a source-target path that crosses that arc (6) and making sure that no arcs that are not used in the end-to-end paths are present in the tree (7) . Constraint (8) states that there is only one incoming arc for every node in the overlay, this is done to enforce the tree property on the y variables.

$$\forall c \in C, \forall t \in T(c), \forall a \in A_{OV} : z_{a,c,t} \le y_{a,c} \qquad (6)$$

$$\forall c \in C, \forall a \in A_{OV} : \sum_{t \in T(c)} z_{a,c,t} - y_{a,c} \geq 0 \qquad (7)$$

$$\forall c \in C, \forall v \in V \setminus (S(c)) : \sum_{a \in in(v)} y_{a,c} \leq 1 \qquad (8)$$

The following constraints are used to enforce the QoS requirements of the connections and to take into account the capacity limitations of the arcs. Constraint (9) makes sure that on every IP link there is no more bandwidth consumed than is available and (10) states that the end-to-end delay from the source to every destination must be lower than the required maximal delay.

$$\forall a_{IP} \in A_{IP} : \sum_{c \in C} \sum_{a_{OV} \in U(a_{IP})} (y_{a,c} \times \mathrm{BW}(c)) \leq \mathrm{BW}(a_{IP}) \qquad (9)$$

$$\forall c \in C, \forall t \in T(c) : \sum_{a \in A_{OV}} (z_{a,c,t} \times \mathrm{DELAY}(a)) \leq \mathrm{DELAY}(c) \qquad (10)$$

A last set of constraints are the server presence constraints. They determine where the servers for enabling the multicast trees have to be placed and capture that the number of servers that are placed is limited to n. Constraint (11) makes sure that a node is present in the overlay if it has to forward traffic in the multicast tree of a connection.

$$\forall c \in C, \forall v \in V \setminus (S(c)), \forall a \in out(v) : x_v \geq y_{a,c} \qquad (11)$$

$$\sum_{v \in V} x_v \leq n \qquad (12)$$

The objective function will describe the goal of the ILP problem. As we want to maximize the number of connections that is supported by the overlay network, the model will minimize the following function:

$$\alpha \times \sum_{v \in V} x_v - \beta \times \sum_{c \in C} a_c \qquad (13)$$

The α and β parameters will allow to adapt the relative importance of the number of servers placed in comparison to the number of connections that are supported.

Solution. To solve the ILP problem, we made use of the dual simplex method as described in [8].

2.2   Heuristics

Random Heuristic. This heuristic determines an arbitrary server placement. To do this n nodes are selected at random from the collection of nodes and overlay servers are placed at those nodes. This heuristic is only used for reference purposes.

Best Path Heuristic (BP). First this heuristic calculates the end-to-end minimal delay path between every pair of nodes in the network. All the nodes of the network are then ranked according to their occurence in these paths. The heuristic then places n overlay servers in the top n nodes of the network. This heuristic chooses the nodes that are on the most minimal delay paths.

Minimal Servers Per Path Heuristic (MSPP). Using the optimal algorithm, this heuristic determines a path for every pair of nodes in the network. The requirements, in terms of delay and bandwidth, associated with each of those connections are chosen in such a way that they show the same behavior as the connections we expect the overlay network to support. The resulting paths have a bounded delay and use a minimal number of overlay servers. The nodes of the network are then ranked according to their presence as overlay servers in these paths and the n nodes that occur the most are selected as locations for overlay servers.

3   Evaluation

We have evaluated the algorithms for both the unicast and the multicast case. We used a 16-node network with 46 directed arcs, this network is shown in fig. 2. The bandwidth on the arcs was uniformly distributed in $[10, 40]$ and the delay on the arcs was uniformly distributed in $[10, 80]$. We generated sets of connections and used the algorithms to determine the servers needed for those sets of connections with the different algorithms. The results shown are averaged over a number of iterations. As a benchmark, we also calculated the maximal number of connections that can be supported with both standard IP and IP Multicast [9]. To evaluate the different algorithms, we used an ILP formulation and the dual simplex method to determine the maximal number of connections that can be set up, given a certain server placement for the algorithms and given the IP topology for the IP cases. In the tests it was assumed that standard routing in the Internet follows the shortest paths between the two end points and that the multicast trees setup in IP multicast follow the shortest paths for every connection. In every test, background traffic was not taken into account. The values of the α and β parameters were chosen in such a way that the number of connections supported was more important than the number of servers needed.

3.1   Unicast

To generate a unicast connection, we chose 2 nodes of the network, a source node and a target node, all nodes had equal probability to be chosen. The bandwidth required by every connection was 5 Mbit/s and for the delay bound we calculated the minimal delay between the end nodes and added 15 percent to that value.

In a first test we executed the optimal algorithm for increasing numbers of connections. The maximal number of servers that could be placed was 4. For

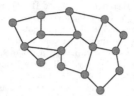

Fig. 2. The network on which the tests were executed

every number of connections, 250 sets were generated. For every set of connections, we also determined the maximal number of connections that could be supported simultaneously with standard IP routing. In fig. 3, the acceptance rate for the overlay is compared to the acceptance rate of a standard best effort network and the number of servers needed to form the overlay is shown. The results clearly show the advantage of using an overlay network over the standard Internet. As the size of the set of connections gets larger, we see a decrease in the acceptance rate for both the overlay network and standard IP. This is a result of congestion. As more connections have to be routed, some links in the network will get congested, resulting in a lower acceptance rate. We also see that the number of servers that is needed to form the overlay increases as the size of the set increases. The reason for this is twofold, as more connections have to be supported, some links get congested and we need overlay servers to route around those congested links. Another reason is that the overlay network needs to cover the whole network as more connections are supported.

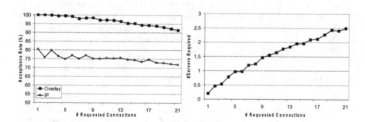

Fig. 3. Average acceptance rate and average number of servers needed for optimal overlay network

In a second test we evaluated the heuristic algorithms. The number of iterations for this test was 50. The heuristics were used to determine the locations of 5 overlay servers. Fig. 4 illustrates the performance of the resulting overlays by comparing them to each other and to standard IP. We see that all the heuristics result in an overlay network with a far greater acceptance rate. The gain relative to the standard IP approach is also significant. As the number of requested connections is increased, the gain will decrease up until a point where it seems to stagnate. This behavior can be explained as follows: as the number of connections increases, the overlay network will congest the links that are good in terms of delay. The IP network however will spread its load more equally over the total collection of links. However, from a certain point onwards, both the IP and the overlay approach will have a lot of congested links. Then the overlay has still got a higher acceptance rate by routing around these links. The random heuristic is clearly outperformed by the more intelligent heuristics, proving that determining the location of the overlay servers is not trivial.

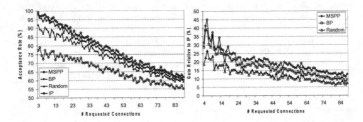

Fig. 4. Average acceptance rate and average achieved gain relative to IP of heuristics

In a last figure we will show the effect of an increased number of servers on the acceptance rate of the requested connections. The heuristic we used was the MSPP heuristic and the number of iterations was again 50. Fig. 5 clearly shows that an increased number of servers will result in a higher number of accepted connections. It is also important to observe that even one overlay server will result in a much higher acceptance rate.

3.2 Multicast

To generate a multicast connection with k client nodes, a node is chosen at random from the network, this is the source node. From the remaining nodes, k nodes are chosen, these are the target nodes of the multicast connection. All the multicast connections required a bandwidth of 5 Mbit/s and for the delay we calculated the minimal delay between the source node and every target node and we added 15 percent to the maximum of those delays. A multicast connection is accepted by the network if all the destinations are reached within the delay bound.

Fig. 6 shows the acceptance rate of the different heuristics and IP. The number of clients of a connection was put to 3 and 5 servers were placed by the

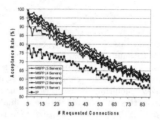

Fig. 5. Average acceptance rate for different numbers of servers with MSPP heuristic

algorithms. We have also indicated the expected behavior of the ILP algorithm. Due to the limited scalability of the ILP approach, it was not possible to perform tests with large number of connections for that algorithm. We see that the performance of the overlay network will be far higher than that of the standard Internet. This is of course a result of the ability of the overlay network to offer alternative routes to the connections. When comparing the unicast case to the multicast case, we see that the acceptance rate of IP unicast is a bit higher than that of IP multicast. This is because IP unicast will congest more links. As the number of connections that is asked to the network increases, the acceptance rate will decrease as a result of the congestion of the network. This decrease of the acceptance rate is also faster than with the unicast case, this is explained by the fact that a multicast connection will have 3 clients instead of 1. Although the overlay network can make use of its multicast functionality, the load on the network will still be higher than in the unicast case, resulting in a faster consumption of the available bandwidth. We also point out that the intelligent heuristics again outperform the random placement. The influence of the number of servers used is also shown in fig. 6, here we can draw the same conclusions as in the unicast case.

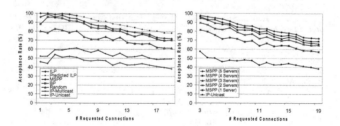

Fig. 6. Average acceptance rate in function of the number of requested connections for the different heuristics and for different numbers of overlay servers.

4   Conclusion

We have developed and evaluated several server placement algorithms for overlay networks. The results in this paper clearly prove the possibilities of overlay networks to deliver QoS for multimedia applications. For both the unicast and the multicast case, the achievable gain by using overlay networks is significant. Overlay networks can thus be used to provide an infrastructure that will offer guaranteed bounded delay to multimedia connections. Our results also show that intelligently placing the overlay servers will increase the functionality of the overlay network, the BP and MSPP heuristics outperform a random placement. It was also shown that the number of overlay servers deployed increases the overlay network performance.

5   Future Work

In next papers we will discuss mechanisms that make the ILP-model more scalable by intelligently pruning overlay edges and overlay nodes on a per connection basis. This will allow us to test the performance of overlay networks for larger topologies. We will also look at routing algorithms for overlay networks that give us the ability to dynamically react to QoS degradation in the Internet.

References

1. S. Savage, T. Anderson and et al., "*Detour: a case for informed internet routing and transport.*", IEEE Micro, 19(1):50–59, January 1999.
2. H. Erikson, "*MBone: The multicast backbone*", Communications of the ACM, vol. 37, no.8, pp.54-60, August 1994.
3. www.6bone.net
4. D. G. Andersen, H. Balakrishnan, M. Kaashoek, and R. Morris, "*Resilient overlay networks.*", In Proc. 18th ACM SOSP, Banff, Canada, October 2001.
5. Z. Li and P. Mohapatra, "*QRON: QoS-aware routing in overlay networks.*", IEEE JSAC, 2003.
6. S. Shi and J. Turner, "*Placing Servers in Overlay Networks.*", In International Symposium on Performance Evaluation of Computer and Telecommunication Systems (SPETS), July 2002
7. Y. Chu, S. G. Rao, S. Seshan and H. Zhang, "*Enabling Conferencing Applications on the Internet Using an Overlay Multicast Architecture.*", ACM SIGCOMM 2001, , San Diago, CA, August 2001.
8. G.L. Nemhauser, L.A. Wolsey, "*Integer and Combinatorial Optimization.*", John Wiley & Sons, 1988.
9. S. Deering, "*Multicast Routing in Internetworks and Extended Lans.*", In Proceedings of ACM SIGCOMM, August 1988.
10. L. Qiu, Y. R. Yang, Y. Zhang and S. Shenker, "*On Selfish Routing in Internet-Like Environments.*", Proceedings of the ACM SIGCOMM, pages 151-162, 2003.
11. S. Blake, et al., "*An architecture for differentiated services.*", RFC 2475, 1998
12. Z. Duan, Z. Zhang and Y. T. Hou, "*Service overlay networks: SLAs, QoS, and bandwidth provisioning.*", IEEE/ACM Transactions on Networking (TON), Volume 11 , Issue 6 , pages 870-883, 2003

MPLS Protection Switching Versus OSPF Rerouting
A Simulative Comparison

Sandrine Pasqualini[1], Andreas Iselt[1], Andreas Kirstädter[1], and Antoine Frot[2]

[1] Siemens AG, Corporate Technology, Information & Communication, Munich,
Germany
{sandrine.pasqualini|andreas.iselt|andreas.kirstaedter}@siemens.com
[2] École Nationale Supérieure des Télécommunications de Bretagne, Brest, France
antoine.frot@aitb.org

Abstract. Resilience is becoming a key design issue for future IP-based
networks having a growing commercial importance. In the case of
element failures the networks have to reconfigure in the order of a
few hundred milliseconds, i.e. much faster than provided by the slow
rerouting of current implementations. Several multi-path extensions
to IP and timer modifcations have been recently proposed providing
interesting alternatives to the usage of of MPLS below IP. In this paper
these approaches are first described in a common context and then
compared by simulations using very detailed simulation models. As
one of the main results it can be shown that an accelerated update
of the internal forwarding tables in the nodes together with fast
hardware-based failure detection are the most promising measures for
reaching the required reconfiguration time orders.

Keywords: Resilience, OSPF, MPLS, Simulation.

1 Introduction

The current situation of the Internet is marked by the development and introduction of new real-time connection-oriented services like streaming technologies
and mission-critical transaction-oriented services. Therefore, the Internet is gaining more and more importance for the economic success of single companies as
well as of whole countries and network resilience is becoming a key issue in the
design of IP based networks.

Originally, IP routing had been designed to be robust, i.e. to be able to re-
establish connectivity after almost any failure of network elements. However,
the applications mentioned only allow service interruptions on the order of a few
hundred milliseconds - a time frame that cannot be reached by today's robust
routing protocols. Therefore, several extensions and modifications have been proposed recently for speeding up IP protection performance: e.g. a simple reduction
of the most important routing timer values or the large-scale introduction of IP

J. Solé-Pareta et al. (Eds.): QofIS 2004, LNCS 3266, pp. 174–183, 2004.
© Springer-Verlag Berlin Heidelberg 2004

multi-path operation with a fast local reaction to network element failures. Increasingly, network operators also deploy a designated MPLS layer below the IP layer having its own rather fast recovery mechanisms and providing failure-proof virtual links to the IP layer.

The most important aspect in the comparison of all these approaches is the resulting recovery speed. In order to thoroughly investigate the time-oriented behaviour of the alternatives we developed very detailed simulation models of the corresponding router/switch nodes. We implemented the single state machines and timing constants as extensions to the basic MPLS and OSPF models of the well-known Internet protocol simulation tool NS-2 [1]. The resulting simulator then was integrated into a very comfortable tool chain that allows the flexible selection of network topologies, traffic demands and protection mechanisms.

The rest of this paper is organized as follows: section 2 first describes MPLS and OSPF starting with MPLS basics and the two most interesting MPLS recovery mechanisms. This is followed by the description of the basic mechanisms of OSPF, the main time constants that were considered in the simulator, and the proposed extensions for faster reaction. In section 3 we describe the simulation framework, the enhancements implemented in the common public domain simulator NS-2 and the resulting tool chain. Section 4 details on the measurements we ran on the selected network topology and discusses the results obtained. Conclusions and recommendations for future hardware and protocol generations are given in section 5.

2 Resilience Mechanisms

2.1 Multiprotocol Label Switching (MPLS)

Label Switching. The routing in IP networks is destination-based: routers take their forwarding decisions only according to the destination address of a packet. Therefore, routing tables are huge and the rerouting process takes a correspondig amount of time. With Multiprotocol Label Switching (MPLS) ingress routers add labels to packets. These labels are interpreted by transient routers known as Label Switching Routers (LSR) as connection identifiers and form the basis for their forwarding decision. Each LSR re-labels and switches incoming packets according to its forwarding table. Label Switching speeds up the packet forwarding, and offers new efficient and quick resilience mechanisms. The setup of a MPLS path consists in the establishment of a sequence of labels, called Label Switched Path (LSP) that the packet will follow through the network. This can be simply done using conventional routing algorithms. But the main advantage of Label Switching appears when the forwarding decision takes the Quality of Service or links reservation into consideration. Then more complicated routing algorithms have to be used in order to offer the most efficient usage of the network.

MPLS Recovery. MPLS Recovery methods provide alternative LSPs to which the traffic can be switched in case of a failure. We must distinguish two types of recovery mechanisms: protection Switching and Restoration. The former includes recovery methods where a protection LSP is pre-calculated, just needing a switching of all traffic from the working LSP to the backup LSP after the failure detection. In the latter case, the backup LSP is calculated dynamically after the detection. Another way to classify these recovery mechanisms depends on which router along the LSP takes the rerouting decision: it can be done locally, the node detecting a failure immediately switching the traffic from the working to the backup LSP, or globally when the failure is notified to upstream and downstream LSRs that reroute the traffic. This paper will focus on Protection Switching schemes. Hereby Link Protection, similar to Cisco's Fast Reroute, and the mechanism introduced by Haskin [2] are considered further.

Link Protection provides a shortest backup path for each link of the primary LSP. When a failure occurs on a protected link, the backup path replaces the failed link in the LSP: the upstream router redirects incoming traffic onto the backup path and as soon as traffic arrives on the router downstream of the failed link it will use the primary LSP again. The Haskin scheme uses a global backup path for the LSP from ingress to egress router. When a failure occurs on a protected link the upstream router redirects incoming traffic back to the ingress router, which will be advertised that a failure has occurred. Then these packets are forwarded on the backup path and reach the egress router.

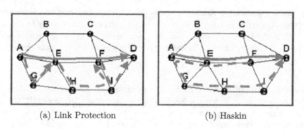

(a) Link Protection   (b) Haskin

Fig. 1. MPLS recovery mechanisms

Routes distribution. There are several possible algorithms to distribute labels through the network such as the Label Distribution Protocol (LDP), extended for Constraint-based Routing (CR-LDP). Another way is to distribute labels by piggybacking them onto other protocols, in particular the Reservation Protocol (RSVP) and its Traffic Engineering extension (RSVP-TE [3]).

2.2   OSPF

Today, one of the most common intra-domain routing protocols in IP networks is OSPF. This section shortly describes the OSPF mechanisms relevant for an understanding of the general behaviour and the various processing times and timers.

Basic OSPF mechanisms. The Hello protocol is used for the detection of topology changes. Each router periodically emits Hello packets on all its outgoing interfaces. If a router has not received Hello packets from an adjacent router within the "Router Dead Interval", the link between the two routers is considered down. When a topology change is detected, the information is broadcasted to neighbours via Link State Advertisements (LSA).

Each router maintains a complete view of the OSPF area, stored as a LSA Database. Each LSA represents one link of the network, and adjacent routers exchange bundles of LSAs to synchronise their databases. When a new LSA is received the database is updated and the information is broadcasted on outgoing interfaces.

Routes calculation: configurable cost values are associated to each link. Each router then calculates a complete shortest path tree[1]. However, only the next hop is used for the forwarding process.

The Forwarding Information Base (FIB) of a router determines which interface has to be used to forward a packet. After each computation of routes, the FIB must be reconfigured.

Main time constants. Considering the previous mechanisms, the convergence behaviour of OSPF in case of a failure can be divided into steps as follows : detection of the failure[2], then flooding of LSAs and - at the same time - scheduling of a SPF calculation, and launching a FIB update. Table 1 lists these times along with their typical values.

Proposed extensions to OSPF. Considering the standardized values, the OSPF protocol needs at least a few seconds to converge. To accelerate the convergence time, it is proposed to investigate the following two options: reduce delays, and associate multipath routing with local failure reaction. In the last years, there were several proposals [8, 9] to accelerate OSPF convergence time by reducing the main timers : $T_{spfDelay}$ and $T_{spfHold}$ set to 0, and sub-second T_{Hello} or hardware failure detection. These accelerated variants of OSPF will be refered to in the following sections as $OSPF_{hello}^{acc.}$ when only sub-second hellos are used, and $OSPF_{hard}^{acc.}$ when hardware detection is enabled in addition. A new approach, proposed in [10] is to associate multipath routing with local failure reaction. This would allow to reduce the impact of a link failure by continuing to send traffic on the remaining paths. The OSPF standard [11] already allows to use paths with equal costs[3] simultaneously. In practice it is not straightforward to find link cost assignments yielding equal cost for several paths [12]. [10] presents

---

[1] Shortest Path First (SPF) calculation
[2] by expiration of the Router Dead Interval or by reception of a new LSA
[3] Equal Cost Multi-Path (ECMP)

Table 1. Main time constants in OSPF

Name	Typically	Short Description
T_{Hello}	10s [4]	Interval between successive Hello packets
T_{Dead}	$4 \times T_{Hello}$ [5,6]	Router Dead Interval
T_{spf}	$\mathcal{O}(n.logn)$ $\mathcal{O}(n^2)^{(a)}$	SPF calculation
$T_{spfDelay}$	5s [5,6]	Minimum time between LSA reception and start of SPF computation
$T_{spfHold}$	10s [5,6]	Minimum time between consecutive SPF computations
T_{lsa}	0.6-1.1ms [7]	Process LSA : check if LSA is new and update LSA database
$T_{lsaFlood}$	33ms [7]	LSA flooding time : process LSA, bundle LSAs and pacing timer
T_{fib}	100-300ms [7]	Update the FIB : from end of LSA processing to end of new routes installation

[(a)] $2.53 \times 10^{-6} n^2 - 1.25 \times 10^{-5} n + 0.0012$, where n is the number of routers in the area, for details see [7]

a new routing scheme which provides each node in the network with two or more outgoing links towards every destination. Two or more possible next hops are then used at each router towards any destination instead of OSPF's single next hop. In [10] such paths are called *hammocks*, due to their general structure where the multiple outgoing paths at one node may recombine at other nodes. The routing algorithms for calculating the hammocks where designed in order to fulfill the following criteria:

1. The algorithm must propose at least two outgoing links for every node,
2. if the topology is such as it is impossible to fulfill the first requirement, the algorithm should minimize the number of excpetions,
3. the algorithm must provide loop-free routing,
4. and no "single point of failure"[4],
5. it should minimize the maximum path length.

A router detecting a link or port failure can then react locally, immediately rerouting the affected traffic over the remaining next hops. This local mechanism avoids the time-consuming SPF calculation and flooding of LSAs in the entire area in the case of a single link failure. However, if multiple link failures occur and there is no remaining alternative link at a router, the local reaction will trigger a standard OSPF reaction. This multipath variant of OSPF will be refered to in the following sections as $OSPF_{hello}^{hammock}$ and $OSPF_{hard}^{hammock}$, depending on which detection mechanism is used.

---

[4] Such a node would prevent at least one other node from reaching a destination if it fails

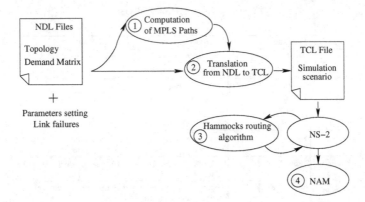

Fig. 2. Tool chain

3   Simulation Framework

In order to investigate the recovery performances of OSPF and MPLS, a simulation tool has been implemented. Based on the simulator NS-2 [1], it uses extensions such as the MPLS module MNS [13], the rtProtoLS module [14] and other protocol implementations, e.g. RSVP-TE Hellos. The OSPF implementation derives from rtProtoLS, to which a Hello protocol and timers have been added [15]. And the OSPF extensions were built from this implementation by changing the way the routes are calculated and the reactions to a failure are handled. The simulation scenario is specified in topology and traffic demand files, in NDL format (Network Description Language), an extension of GML [16]. NAM [1] is also used for the visualisation of the network activity. All tools are integrated into a comprehensive simulation framework, easily customizable through a simple GUI. This simulator automates the creation of OSPF or MPLS simulations for NS-2. Figure 2 shows how the different tools are articulated within the simulation framework. Given a topology, the MPLS Paths computation module① builds MPLS working and backup paths, using Dijkstra's algorithm, and exports them in NDL format. Supported recovery schemes are Link Protection, similar to Cisco's Fast Reroute, and the method of Haskin [2]. After giving some parameters, such as triggering link failures, a tool translates all NDL sources into one NS-2 simulation file②. For the OSPF simulations, the NS-2 simulator has been extended to allow local external routing algorithms③. This allows to use existing routing tools and to develop routing independently from NS-2. The results are visualized in NAM ④.

4   Measurements and Results

The focus of the investigations was on the speed of the traffic restoration after
a failure. As a main sample network, the Pan-European optical network from
the COST 239 project [17] was chosen because of its widespread use for network
investigations. This network, shown in Fig.3 contains 11 nodes and 26 links with
capacities of 20 Gbit/s. A full-mesh of equal flows between all nodes has been
used as demand pattern. To save simulation time, the link bandwidths are scaled
down by a factor of 1000. The sources send packet flows with 800kbit/s constant
bit rate (CBR) (packets of 500 bytes sent every 5 ms). This allows more than 20
simultaneous flows on one link without any packet loss. The simulation starts
with the establishment of the network configuration. For MPLS this includes
the set up of paths and backup paths. For OSPF this means the convergence of
the OSPF routing protocol. After starting the sources, a link failure is simulated
triggering failure detection, dynamic route calculation, if necessary, and switching to alternative routes. To get rid of synchronisation effects of hello timers with
failure times, the simulations are repeated with different periods of time between
the simulation start and the failure time. The simulation is also repeated for all
possible link failures, to average over the effect of different failure locations. To
characterise the effect of the failure, the sum of the rates of all traffic received
at sinks in the network is considered over the time. Fig. 4 shows the affected
traffic and the times for restoration for different MPLS protection switching and
IP rerouting approaches both with different timer values for the RSVP refresh
messages or for the OSPF hello protocol. Each curve in Fig. 4 shows the sum
of all traffic flows in the network. After the occurrence of a failure the sum rate
decreases since the traffic that is expected to be carried over the failed link is
lost. Just after the link is repaired, shortest routes are used again while packets
are still on the alternate routes, which results in more packets reaching their
destination during a few milliseconds. The four curves represent the cases:

- MPLS Link Protection[5] with RSVP-TE standard failure detection intervals
  of 5ms ⓐ and 100ms ⓒ.
- $OSPF_{hello}^{acc.}$ with modified hello intervals of 100ms ⓓ and $OSPF_{hard}^{acc.}$ with
  hardware failure detection of 5ms ⓑ.

It can be noticed that standard MPLS protection switching, ⓐ, is much faster
than both OSPF mechanisms. Even MPLS ⓒ, with the same T_{Hello} and T_{Dead}
timers as $OSPF_{hello}^{acc.}$ is still faster, in the order of 100ms. This results from the
computational effort, the signalling delay and mostly from the update of the
FIBs, which is more time consuming for the larger tables of OSPF - compared
to MPLS. Of course, this is a very implementation dependent parameter and
may be addressed in future router developments. The effect of hardware failure
detection is shown in Fig. 5. Obviously the hardware failure detection[6] speeds
up the OSPF recovery considerably. This figure also shows a difference between

---

[5] the Haskin cases give similar results regarding reconfiguration time

[6] this timer is set to 5ms, which is realistic regarding current physical possibilities

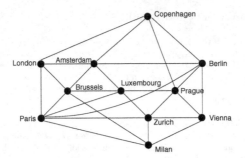

Fig. 3. COST239 network

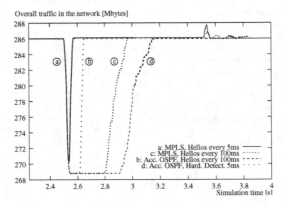

Fig. 4. Comparison between MPLS and accelerated OSPF recovery

shortest path routing ⓕⓖ, and multi-path routing ⓔⓗ, as it is described in [10]. With multi-path routing the traffic is distributed over a fan of paths, including paths longer than the shortest paths. Therefore the probability for such a path to be hit by a single link failure is higher. This results in the increased impact represented by the lower throughput in the case of a failure. Fig. 6 depicts the different times involved in the extended OSPF implementation, with the values used for the simulations. The predominant times here are the detection of failures and the updating of the forwarding tables. For larger networks, the LSA processing times also have to be considered. This indicates clearly where future improvements in OSPF and router technology are necessary: failure detection

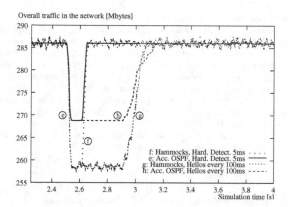

Fig. 5. Comparison between Hammocks and OSPF routing

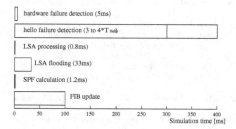

Fig. 6. Relative size of the various times involved in OSPF implementation.

and FIB update. To reduce the failure detection time, hardware failure detection already gives major relief. Moreover, where hardware failure detection does not help, short hello intervals will also allow faster failure detection. In [18] a protocol is proposed allowing the use of short hello intervals independent of the routing protocol. The other major time that has to be improved is the FIB update time. As already mentioned above, this requires changes in the router implementation.

5   Conclusion

The current Internet routing protocol OSPF as it is implemented and used today has major deficiencies with respect to network resilience. The simulative comparison with MPLS-enhanced networks shows the superior time behavior of

MPLS resilience. We have outlined that there are several proposed extensions to improve the resilience of routed networks. These proposals include optimization of timers and the use of multi-path routing with local failure reaction. At investigating the extensions by simulation it turned out that they are the first steps in the right direction. From the investigations it can be concluded that there are two major points to be addressed in order to improve the restoration speed of OSPF re-routing: speed-up of failure detection and acceleration of forwarding information base (FIB) update. For the former some very promising approaches, like hardware failure detection and fast hello protocols (e.g. BFD [18]) are already evolving. For the acceleration of the FIB updates the internal router architectures have to be improved. With these extensions OPSF routed networks will be able to reach sub-second restoration speeds in the future.

References

1. UCB/LBNL/VINT. [Online]. Available: http://www.isi.edu/nsnam/
2. D. Haskin and R. Krishnan, "A method for setting an alternative label switched paths to handle fast reroute," IETF," Internet Draft.
3. D. Awduche et al., "RSVP-TE: Extensions to RSVP for LSP tunnels," NWG, RFC 3209.
4. Cisco Corp. [Online]. Available: http://www.cisco.com
5. C. Huitema, *Routing in the Internet*, 2nd ed.   Prentice Hall PTR, 2000.
6. J. T. Moy, *OSPF: Anatomy of an Internet Routing Protocol*, nov 2000.
7. A. Shaikh and A. Greenberg, "Experience in black-box ospf measurement," in *ACM SIGCOMM Internet Measurement Workshop (IMW)*, nov 2001.
8. C. Alaettinoglu et al., "Toward millisecond IGP convergence," IETF," Internet Draft.
9. A. Basu and J. Riecke, "Stability issues in OSPF routing."   ACM SIGCOMM, aug 2001.
10. G. Schollmeier et al., "Improving the resilience in IP networks," in *HPSR 2003*, jun 2003.
11. J. Moy, "OSPF version 2," NWG, RFC 2328.
12. A. Sridharan et al., "Achieving near-optimal traffic engineering solutions for current OSPF/IS-IS networks."   INFOCOM 2003.
13. G. Ahn. Mns. [Online]. Available: http://flower.ce.cnu.ac.kr/~fog1/mns/
14. M. Sun. [Online]. Available: http://networks.ecse.rpi.edu/~sunmin/rtProtoLS/
15. C. Harrer, "Verhalten von IP-routingprotokollen bei ausfall von netzelementen," Master's thesis, Technische Universität München, LKN, apr 2001.
16. M. Himsolt. Gml: A portable graph file format. Universität Passau. [Online]. Available: http://www.infosun.fmi.uni-passau.de/Graphlet/GML/
17. P. Batchelor et al., "Ultra high capacity optical transmission networks," COST 239," Final report of Action, jan 1999.
18. D. Katz and D. Ward, "Bfd for ipv4 and ipv6 (single hop)," IETF," Internet Draft.

A Distributed-Request-Based DiffServ CAC for Seamless Fast-Handoff in Mobile Internet

Kyeong Hur[1], Hyung-Kun Park[2], Yeonwoo Lee[3], Seon Wook Kim[1], and Doo Seop Eom[1]

[1] Dept. of Electronics and Computer Engineering, Korea University, Seoul Korea
{hkyeong, seon, eomds}@korea.ac.kr
[2] School of Information and Technology, Korea University of Technology and Education, Chunan-Si Korea
hkpark@kut.ac.kr
[3] School of Engineering and Electronics, The University of Edinburgh, Edinburgh UK
yl@ee.ed.ac.uk

Abstract. The next-generation wireless networks are evolving toward an IP-based network that can provide various multimedia services seamlessly. To establish such a wireless mobile Internet, the registration domain supporting fast handoff is integrated with the DiffServ mechanism. In this paper, a Distributed-request-based CDMA DiffServ call admission control (CAC) scheme is proposed for the evolving mobile Internet. Numerical examples show that the forced termination ratio of handoff calls is guaranteed to be much less than the blocking ratio of new calls for a seamless fast-handoff while proposed scheme provides quality of service guarantee for each service class efficiently.

1 Introduction

Provision of various realtime multimedia services to mobile users is the main objective of the next-generation wireless networks, which will be IP-based and should internetwork with the Internet backbone seamlessly [1]. The establishment of such wireless mobile Internet is technically very challenging. Two major tasks are the support of seamless fast-handoff and the provision of quality of service (QoS) guarantee over IP-based wireless access networks. For realtime traffic, the handoff call processing should be fast enough to avoid high loss of delay-sensitive packets. Note that handoff call dropping never occurs in seamless wired networks. The forced termination probability of handoff calls in progress must be at least less than the new call blocking probability for GoS (Grade of Service) guarantees to users and a seamless networking with wired networks [2].

To achieve fast handoff requires both a fast location/mobility update scheme and a fast call admission control (CAC) scheme. The popular scheme for fast location update is a registration-domain-based architecture. The radio cells in a geographic area are organized into a registration domain (e.g., a foreign network in the TeleMIP architecture [3]), and the domain connects to the Internet through a foreign mobile agent (FA) [3], [4], [5]. When a mobile host (MH)

J. Solé-Pareta et al. (Eds.): QofIS 2004, LNCS 3266, pp. 184–193, 2004.
© Springer-Verlag Berlin Heidelberg 2004

moves into a registration domain for the first time, it will register the new care-of-address (the address of the FA) to its home agent. While it migrates within the domain, the mobility update message will only be sent to the FA, without registration with the home agent which is often located far away. This registration process significantly reduces mobility management signaling.

The differentiated services (DiffServ) architecture has been proposed as a scalable way to provide quality of service (QoS) in IP networks [6]. In this paper, to provide the DiffServ-based QoS over wireless networks, the registration domain is modeled as a DiffServ administrative domain, with the FA router as the edge router connecting to the Internet backbone and the base stations as the edge routers providing Internet access to mobile hosts (MHs). A bandwidth broker will manage the resource allocation over the DiffServ registration domain. We consider wireless links as bottleneck links in the domain, and the service level agreement (SLA) is negotiated mainly based on the wireless resource availability. In the DiffServ architecture, there are three differentiated services of premium service, assured service, and best-effort service. The premium service is ideal for real time services such as IP telephony, video conference and the like [7]. The assured service was proposed to ensure an expected capacity and a low delay for avoidance of an excessive delay of non-realtime applications [8]. In our scheme, voice and videophone services are considered as the premium services.

In this paper, we propose a distributed-request-based CDMA DiffServ call admission control (CAC) scheme over such a DiffServ registration domain, to achieve a seamless fast-handoff, QoS guarantee for each service class, and high utilization of the scarce wireless frequency spectrum. The time frame consists of an access slot and a transmission slot for a distributed-request-based code assignment. An access permission probability is adaptively given to each service user and we give higher access permission probability to the handoff calls than the new calls. For premium data services of voice and videophone services, the code reservation is allowed to guarantee throughputs and the forced termination probability of handoff calls is guaranteed to be less than the new call blocking probability. For the assured date service, proposed scheme can make the average transmission delay be much lower than that of the best-effort data service through provision of reserved codes. Thus, a higher data throughput than the best-effort data service can be guaranteed to the assured data service.

Numerical examples using an EPA (Equilibrium Point Analysis) method [9] show that the proposed CAC scheme can determine capacities for the differentiated services in a cell satisfying the requirements in next-generation wireless networks (e.g., a minimized forced termination ratio for a seamless fast-handoff, avoidance of an excessive delay of non-realtime applications in providing multimedia services), which capacities are required to determine the resource requirement in the SLA negotiation between a DiffServ registration domain and the Internet service provider.

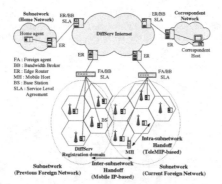

Fig. 1. A DiffServ registration-domain-based wireless network architecture

2   Proposed DiffServ CAC Scheme

As illustrated in Fig. 1, the system under consideration is a DiffServ registration-domain-based wireless networks architecture where the TeleMIP is used to manage mobility and support fast handoff. The FA router is the interface connecting to the DiffServ Internet backbone, where an SLA is negotiated to specify the resource allocated by the Internet service provider to serve the aggregate traffic flowing from/into the FA router. The FA router conditions the aggregate traffic for each service class according to the SLA resource commitments. The base stations provide MHs with access points to the Internet, and perform per-flow traffic conditioning and marking when data flow in the uplink direction. We focus on a CAC scheme for seamless QoS provisions in the following.

2.1   System Description

We have considered a multi-rate transmission CDMA system that is required to support the multi-class services such as voice and videophone premium services, assured and best-effort data services. In this paper, an uplink CAC scheme is proposed for the system considered. The time scale is organized in frames containing the access slot and the transmission slot as shown in Fig. 2. When each terminal has new data ready to transmit, it first sends an access packet spread by a randomly chosen access code through the access slot according to a given access permission probability, in order to reserve a code for data transmission. If any other contending terminal does not send an access packet with the same access code, no collision with its access packet occurs, and the base station can identify the terminal that sent the access packet. The base station then returns an acknowledgment with the assignment information of an available code for data transmission, over the downlink. The terminals succeeding in code reservation send traffic packets spread by assigned transmission codes through the transmission slot [10]. Therefore, the terminals can send data without collision after code reservation.

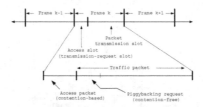

Fig. 2. Uplink frame structure

If packet collision occurs in the access slot and the base station sends no response, the terminal will retry for code reservation in the next frame. If the access packet of a terminal is successfully received by the base station but there is no assignment of a transmission code to the terminal due to lack of transmission codes, the terminal also retries for code reservation in the next frame. This is because the base station does not have the request table for recoding such terminals in the proposed protocol. By using this distributed-request-based protocol, the mobile host can perform a request of call set-up in both uplink and down link data transmissions, which makes proposed scheme perform a faster admission control than a scheme where a sender-initiated-request of call set-up is done even in wireless networks after a large location/mobility-update-delay of Mobile IP protocol passes [3], [5].

2.2   Distribution of Codes and Access Permission Probabilities

To reduce the packet collisions in the access slot, the transmission code reservation through the piggybacking request and the access permission probability are presented. All the voice and the videophone premium service users except the data service users can demand further reservation of the acquired transmission code for the next frame during their call duration, while sending the traffic packets having the piggybacking request bit indicating an additional transmission request or no more transmission. According to the piggybacking request informations from the users, the base station keeps the reservation of the assigned transmission codes for the next frame during their call duration, to guarantee throughputs for premium service users.

As shown in Fig. 3, K_a spreading codes are used for sending the access packets, and a separate set of K_t spreading codes are used for sending the traffic packets. However, only the $K_a - K_v$ access codes are actually used among all the service users. The remaining K_v access codes are fixedly assigned to those voice users that succeed in the reservation of a transmission code using one of the $K_a - K_v$ access codes. On the other hand, K_v codes among the K_t transmission codes are used for voice users, and then K_{vi} and K_{as} transmission codes are used for the videophone and the high-rate assured data service users respectively. Note that there are no reserved transmission codes for the low-rate best-effort service users. Instead, they use the transmission codes that remain after assigning the codes among K_v transmission codes to voice users. Also, they can use the transmission codes that are already assigned to voice users but are not used during silent periods of voice calls, to utilize transmission codes efficiently. For the voice users, the assigned transmission codes are released during the silent periods of voice calls through the piggybacking request, because traffic packets are not transmitted during that time. However, when the voice user enters the talk-spurt state from the silent state, the user transmits the access packet denoted as A in Fig. 3 to demand re-reservation of the acquired transmission code. Since the access packet A is spread by the fixedly assigned access code (i.e., one of the K_v access codes) during its entire call duration, there are no voice packets dropped. When the voice call is over, the voice user transmits the access packet denoted as T to indicate call termination with the fixedly assigned access code.

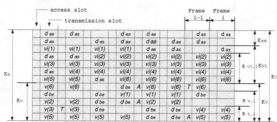

Fig. 3. Assignment scheme of transmission codes

Next we will describe the access permission probability. Based on the number of available transmission codes for each service and the estimated number of contending users for each service, it is calculated every frame by the base station, and then it is broadcast to all users of each service in a cell over the downlink. As shown in Fig. 3, $R_{v,i}$ and $R_{vi,i}$ denote the numbers of reserved transmission codes for the voice and the videophone connected users at the i th frame, respectively. The $R_{t,i}$ denotes the number of the voice users in talk-spurt state at the i th frame. Then, from the Fig. 3, the numbers of the available transmission codes for the voice, the videophone, the high-rate assured data, and the low-rate best-effort data services at the i th frame are determined as $K_v - R_{v,i}$, $K_{vi} - R_{vi,i}$, $K_{as} + K_{vi} - R_{vi,i}$, and $K_v - R_{t,i}$, respectively. Since the available transmission codes are assigned firstly to the voice users, the low-rate best-effort data service users whose access packets are received successfully use the remaining $K_v - R_{v,i}$ codes not assigned to the voice users and the $R_{v,i} - R_{t,i}$ codes in the silent periods of voice calls. On the other hand, we denote the estimated numbers of contending users of the new call and the handoff call for the voice service as C_{v_n} , C_{v_h} , those of the new call and the handoff call for the videophone service as C_{vi_n} , C_{vih} , and those for the high-rate assured and the low-rate best-effort data services as C_{as} , C_{be} , respectively. Then, the access permission probabilities of the new calls and the handoff calls for the voice and the videophone services P_{v_n} , P_{v_h} , P_{vi_n} , P_{vi_h} , , and those for the assured and the best-effort data services P_{as} , P_{be} are given as in Eq. (1). The priority is given to the handoff calls by permitting P_{v_h} , P_{vi_h} as 1. Also, we set the access permission probabilities for the forcedly terminated handoff calls as 1.

$$P_{v_n} = min\{1, \frac{K_v - R_{v,i}}{C_{v_n}}\}, P_{vi_n} = min\{1, \frac{K_{vi} - R_{vi,i}}{C_{vi_n}}\}, P_{v_h} = P_{vi_h} = 1$$

$$P_{as} = min\{1, \frac{K_{as} + K_{vi} - R_{vi,i}}{C_{as}}\}, P_{be} = min\{1, \frac{(K_v - R_{v,i}) + (R_{v,i} - R_{t,i})}{C_{be}}\}$$

$$(1)$$

3   Numerical Examples and Discussions

In this section, we discuss the steady state performances of the proposed CAC scheme through numerical examples obtained using the EPA method. We chose the multi-processing gain CDMA technique for the multi-rate transmission CDMA system where using random code sequences and the BPSK modulation scheme in the B bandwidth [11]. The performance of the proposed CAC scheme is influenced by the parameters of K_v, K_{vi}, K_{as}, and K_a whenever using any kind of codes. We assume that all data are transmitted with the same bit energy to noise ratio 20dB E_b/N_0 and there is no ICI (Inter Cell Interference). Then, by using the multi-processing gain CDMA technique, the CDMA system can provide voice and low-rate best-effort data services of 24kbps bit rate with 10^{-4} BER, and the videophone and the high-rate assured data services of 72kbps bit rate with 10^{-5} BER, by setting K_v , K_{vi} , and K_{as} as 8, 5, and 3, respectively. Also, K_a is set as 43 to guarantee a 3×10^{-4} BER for the access packet of 24kbps data rate [11].

We assume that the call processing and the transmission code assignment are completed simultaneously and the data service users have infinite buffer size. The wireless bandwidth B is set 4.096MHz as in UTRA W-CDMA, and the frame rate of videophone service using the H.263 coding technique is set 50 frames/sec from the frame length of 20msec [12]. The forced termination ratio of handoff calls is set equal to the blocking ratio for handoff calls because the delay constraint on the handoff call is set zero frame. In the voice service subsystem model, the mean talk-spurt duration and the mean silent duration are set 1sec and 1.35sec. The voice activity ratio is set 0.43, and the mean call holding times of the voice and the videophone services are set 3min. Also, the activity ratio of data traffic is set 0.05 by assuming that the mean active time and the mean idle time are 19/57 sec and 360/1080 sec for the low-rate best-effort/the high-rate assured data services [9]. For voice and videophone service users, λ_{v_n}, λ_{v_h}, λ_{vi_n} and λ_{vi_h} denote arrival rates of new calls and handoff calls for each service. λ_{v_n} is set as 152 calls/hour and λ_{vi_n} is set as 95 calls/hour. Also, We set $\lambda_{v_h}/\lambda_{v_n}$ and $\lambda_{vi_h}/\lambda_{vi_n}$ ratios as 0.5 which reflect the mobility of the users. New call generation rates for the services are set as 0.2 times of call termination rates.

Figure 4 and 5 respectively show how the blocking ratios for voice and videophone services vary under given $\lambda_{v_h}/\lambda_{v_n}$ and $\lambda_{vi_h}/\lambda_{vi_n}$ ratios when changing the total number of users for each service in a cell (i.e., M_v and M_{vi}). We can see that the blocking ratio for each service becomes larger as the total number of users increases. It is because the number of contending users for each service is proportional to the total number of its users in a cell. On the other hand, we can see that the blocking ratio of the new call increases as the arrival rate of the handoff call with higher access permission probability increases. Also, the forced termination ratio of the handoff call increases due to the increase in the number of contending users as the arrival rate ratio of the handoff call to the new call increases. In both figures, we can observe that there are turning points where the blocking ratios increase radically. If M_v or M_{vi} is larger than the turning point, the blocking ratio becomes very large. On the whole, the forced termination ratio of the handoff call is smaller than the blocking ratio of the new call, and the difference between the ratios becomes more significant over the turning point. It is because we give priority to the handoff call over the new call through the access permission probability and the access permission probability of the new call becomes much smaller than that of the handoff call as the number of contending users increases (See Eq. (1)). If the delay constraint on the handoff call is set as a larger non-zero frames, this effect for GoS guarantees to users and a seamless fast-handoff also becomes larger while guaranteeing 72kbps and 24kbps throughputs for the premium services, respectively.

In the proposed CAC scheme, the low-rate best-effort data users who have accessed successfully can use the remaining transmission codes of $K_v - R_{v,i}$ not assigned to the voice users, and the high-rate assured data users can use the remaining codes of $K_{vi} - R_{vi,i}$ not assigned to the videophone users in i th frame. Therefore, we investigate the impact of the numbers of contending users for the voice and the videophone services upon the average delay performances of both data services, respectively. We have considered an average packet transmission delay to achieve a reference 24 kbps throughput, to compare the performance. Figure 6 shows how the average delay for the low-rate best-effort data service varies under given numbers of contending users for the voice service when changing the total number of users for the low-rate best-effort data service in a cell M_{be} . We can see that the average delay increases rapidly as the number of

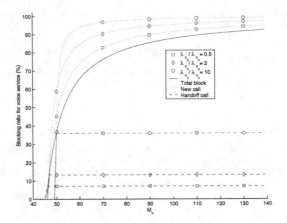

Fig. 4. New call blocking and forced termination ratios for voice service

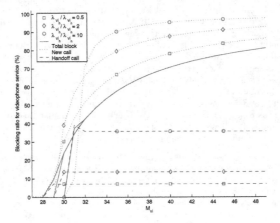

Fig. 5. New call blocking and forced termination ratios for videophone service

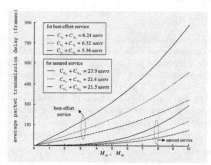

Fig. 6. Average delay comparison for the data services

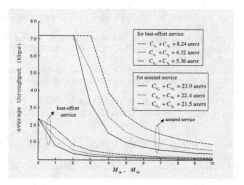

Fig. 7. Average throughput comparison for the data services

contending users for the voice service or the Mbe becomes larger. We can obtain similar results for the high-rate assured data service. However, for assured data service, the average packet transmission delay is much lower than that of best-effort service through the provision of Kas codes resource reservation. Thus, a higher data throughput than the best-effort data service is guaranteed to the assured data service as shown in Fig. 7. However, three assured data service users can have the 72 kbps data throughput if the number of contending users for the videophone service is not larger than 21.5. This is because more contending users can cause more packet collisions in the access slot, so that the data throughput is decreased.

4   Conclusion

In this paper, for seamless fast-handoff in wireless mobile Internet, we have proposed a distributive-request-based CDMA DiffServ CAC scheme. Through the presented code assignment and access permission probability schemes, we have shown that it can guarantee a constant GOS to handoff calls. Thus, their forced termination probability is guaranteed to be much less than the new call blocking probability for a seamless fast-handoff. Furthermore, proposed CAC scheme can provide QoS guarantee for each service class efficiently in multi-rate transmission cellular CDMA systems.

Acknowledgements. This research was supported by University IT Research Center Project.

References

1. Bos, L., Leory, S.: Toward an All-IP-Based UMTS System Architecture. IEEE Network. Vol. 15 (2001) 36–45
2. Eom, D. S., Sugano, M., Murata, M., Miyahara, H.: Call Admission Control for QoS Provisioning in Multimedia Wireless ATM Networks. IEICE Transactions on Communications. Vol. E82-B (1999) 14–23
3. Das, S., Misra, A., Agrawal, P.: TeleMIP: Telecommunications-Enhanced Mobile IP Architecture for Fast Intradomain Mobility. IEEE Personal Communications. Vol. 7 (2000) 50–58
4. Perkins, C.E.: IP Mobility support. IETF, RFC 2002, (1996)
5. Eom, D. S., Lee, H. S, Sugano, M., Murata, M., Miyahara, H.: Improving TCP handoff performance in Mobile IP based networks. Computer Communications. Vol. 25 (2002) 635–646
6. Blake, S., Black, D., Carlson, M., Davies, E., Wang, Z., Weiss, W. : An Architecture for Differentiated Services. IETF, RFC 2475, (1998)
7. Jacobson, V., Nichols, K., Poduri, K.: An Expedited Forwarding PHB. IETF, RFC 2598, (1999)
8. Heinanen, J., Baker, F., Weiss, W., Wroclawski, J.: Assured Forwarding PHB Group. IETF, RFC 2597, (1999)
9. Tan, L., Zhang, Q. T.: A Reservation Random-Access Protocol for Voice/Data Integrated Spread-Spectrum Multiple-Access Systems. IEEE Journal of Selected Areas on Communications. Vol. 14 (1996) 1717–1727
10. Choi, S., Shin, K. G.: An Uplink CDMA System Architecture with Diverse QoS Guarantees for Heterogeneous Traffic. IEEE/ACM Transactions on Networking. Vol. 7(1999) 616–628
11. Ottosson, T., Svensson, A.: Multi-Rate Schemes in DS/CDMA Systems. In Proceedings of IEEE Vehicular Technology Conference. (1995) 1006–1010
12. Rijkse, K.: H.263 : Video Coding for Low-Bit-rate Communication. IEEE Communications Magazine. Vol. 7(1996) 42–45

Cross-Layer Analytical Modeling of Wireless Channels for Accurate Performance Evaluation

Dmitri Moltchanov, Yevgeni Koucheryavy, and Jarmo Harju

Institute of Communication Engineering,
Tampere University of Technology,
P.O.Box 553, Tampere, Finland
{moltchan,yk,harju}@cs.tut.fi

Abstract. An intention to adopt IP protocol for future mobile communication and subsequent extension of Internet services to the air interface calls for advanced performance modeling approaches. To provide a tool for accurate performance evaluation of IP-based applications running over the wireless channels we propose a novel cross-layer wireless channel modeling approach. We extend the small-scale propagation model representing the received signal strength to IP layer using the cross-layer mappings. Proposed model is represented by the IP packet error process and retains memory properties of initial signal strength process. Contrarily to those approaches developed to date, our model requires less restrictive assumptions regarding behavior of the small-scale propagation model at layers above physical. We compare results obtained using our model with those, published to date, and show that our approach allows to get more accurate estimators of IP packet error probabilities.

1 Introduction

While next generation (NG) mobile systems are not completely defined, there is a common agreement that they will rely on IP protocol as a consistent end-to-end transport technology. The motivation is to introduce a unified service platform for future 'mobile Internet' known as 'NG All-IP' mobile systems.

To date only a few studies devoted to IP layer performance evaluation at the air interface have been published. Survey of literature has shown that most studies were devoted to analysis of the data-link layer protocols [1,2]. Additionally, approaches developed to date, adopt quite restrictive assumptions regarding the performance of wireless channels at layers above physical. As a result, they may lead to incorrect estimation of IP layer performance parameters.

In this paper we propose a novel cross-layer wireless channels modeling approach. We extend the small-scale propagation model representing the received signal strength to IP layer using the cross-layer mappings. The proposed model is represented by the IP packet error process, retains memory properties of initial signal strength process and captures specific peculiarities of protocols at layers below IP. We show that our approach allows to get more accurate estimators of IP packet error probabilities compared to those approaches used to date.

J. Solé-Pareta et al. (Eds.): QofIS 2004, LNCS 3266, pp. 194–203, 2004.
© Springer-Verlag Berlin Heidelberg 2004

Our paper is organized as follows. In Section 2 we overview propagation characteristics of wireless channels and models used to capture them. In section 3 we propose our extension to IP layer and define model that provides IP packet error probabilities. In Section 4 we provide numerical comparison of our approach with that one widely used in literature. Conclusions are drawn in the last section.

2   Propagation Characteristics of Wireless Channels

The propagation path between the transmitter and a receiver may vary from simple line-of-sight (LOS) to very complex ones due to diffraction, reflection and scattering. To represent performance of wireless channels propagation models are used. We distinguish between large-scale and small-scale propagation models [3].

When a mobile user moves away from the transmitter over large distances the local average received signal strength gradually decreases. This signal strength is predicted using the large-scale propagation models. However, such models [4, 5,6] do not take into account rapid variations of the received signal strength. As a result, they cannot be effectively used in performance evaluation studies. Indeed, when a mobile user moves over short distances the instantaneous signal strength varies rapidly. The reason is that the received signal is a sum of many components coming from different directions. Propagation models characterizing rapid fluctuations of the received signal strength over short time duration are called small-scale propagation models. In the presence of dominant non-fading component the small-scale propagation distribution is Rician. As the dominant component fades away Rician distribution degenerates to Rayleigh one.

The small-scale propagation models capture characteristics of wireless channel on a finer granularity than large-scale ones. Additionally, these models implicitly take into account movements of users over short travel distances [7,8]. In what follows, we restrict our attention to the small-scale propagation models.

2.1   Model of Small-Scale Propagation Characteristics

Assume a discrete-time environment, i.e. time axis is slotted, the slot duration is constant and given by $\Delta t = (t_{i+1} - t_i)$, $i = 0, 1, \ldots$. We choose Δt such that it equals to the time to transmit a single symbol at the wireless channel. Hence, the choice of Δt depends on properties of the physical layer.

Small-scale propagation characteristics are often represented by the stochastic process $\{L(n), n = 0, 1, \ldots\}$ modulated by the discrete-time Markov chain $\{S_L(n), n = 0, 1, \ldots\}$, $S_L(n) \in \{1, 2, \ldots, M\}$ each state of which is associated with conditional probability distribution function of the received signal strength [9,10]. The underlying modulation allows to take into account autocorrelation properties of the signal strength process. Since it is allowed for the Markov process $\{S_L(n), n = 0, 1, \ldots\}$ to change state in every time slot, every bit may experience different received signal strengths.

An illustration of such a model is shown in the Fig. 1 where states are associated with conditional distribution functions $F_L(k\Delta f|i)(\Delta f) = Pr\{L(n) =$

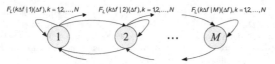

Fig. 1. An illustration of the Markov model for small-scale propagation characteristics.

$k\Delta f|S_L(n) = i\}$, $k = 1, 2, \ldots, N$, $i = 1, 2, \ldots, M$, where N is the number of bins to which the signal strength is partitioned and Δf is the discretization interval. Let D_L and $\boldsymbol{\pi_L} = (\pi_1, \pi_2, \ldots, \pi_M)$ be the one-step transition probability matrix and the stationary probability vector of $\{S_L(n), n = 0, 1, \ldots\}$ respectively. Parameters M, D_L, $F_L(k\Delta f|i)(\Delta f)$, must be estimated from statistical data [9, 10]. For the ease of notation we will use $F_L(k|i)$ instead of $F_L(k\Delta f|i)(\Delta f)$.

3 Wireless Channel Model at IP Layer

The small-scale propagation model of the received signal strength defined in the previous section cannot be directly used in performance evaluation studies and must be properly extended to IP layer at which QoS usually is defined. To do so we have to take into account specific peculiarities of layers below IP including modulation schemes at the physical layer, data-link error concealment techniques and possible segmentation procedures between different layers. As a result, the IP layer wireless channel model must be a complex cross-layer function of underlying layers and propagation characteristics.

In the following subsections we define models of incorrect reception of the protocol data units (PDU) at different layers. For this purpose we implicitly assume that these PDUs are consecutively transmitted at corresponding layers.

3.1 Bit Error Process

Consider a certain state i of the Markov chain $\{S_L(n), n = 0, 1, \ldots\}$ associated with the conditional probability distribution function $F_L(k|i)$, $k = 1, 2, \ldots, N$, of the received signal strength. Since the probability of a single bit error is the deterministic function of the received signal strength [3], all values of $F_L(k|i)$ that are less or equal to a computed value of the so-called bit error threshold B_T cause bit error. Those values which are greater than B_T do not cause bit error. So that each state i, $i = 1, 2, \ldots, M$ of the Markov process $\{S_L(n), n = 0, 1, \ldots\}$ can be now associated with the following bit error probability $p_{E,i}$:

$$p_{E,i} = Pr\{E(n) = 1|S_E(n) = i\} = \sum_{k=1}^{B_T} Pr\{L(n) = k|S_L(n) = i\}, \qquad (1)$$

where $\{E(n), n = 0, 1, \ldots\}$, $E(n) \in \{0, 1\}$ is the bit error process for which 1 denotes an incorrectly received bit, 0 denotes a correctly received bit, $\{S_E(n), n = 0, 1, \ldots\}$ is the underlying Markov chain of $\{E(n), n = 0, 1, \ldots\}$. Note that

$\{S_L(n), n = 0, 1, \dots\}$ and $\{S_E(n), n = 0, 1, \dots\}$ are actually the same and $\pi_E = \pi_L$, $D_E = D_L$, where D_E and π_E are one-step transition probability matrix and stationary distribution vector of $\{S_E(n), n = 0, 1, \dots\}$ respectively. B_T must be estimated based on a modulation scheme and other specific features of physical layer utilized at a given wireless channel [3].

Let us denote by $d_{E,ij}(k) = Pr\{E(n) = k, S_E(n) = j | S_E(n-1) = i\}$, $k = 0, 1$, the transition probability from the state i to state j with correct ($k = 0$) and incorrect ($k = 1$) bit reception respectively. These probabilities can be represented in a compact form using matrices $D_E(1)$ and $D_E(0)$ such that $D_E(1) + D_E(0) = D_E$. In our case the state from which the transition occurs completely determines the bit error probability. The state to which transition occurs is used for convenience of matrix notation useful in the following.

3.2   Frame Error Process Without FEC

Assume that the length of the frame is constant and equals to m bits. The sequence of consecutively transmitted bits, denoted by gray rectangles, is shown in the Fig. 2, where $(l-1)$, l, $(l+1)$ denote time intervals whose length equals to the time to transmit a single frame; k, i, j, denote the state of the Markov chain $\{S_E(n), n = 0, 1, \dots\}$ in the beginning of these intervals.

Consider the stochastic process $\{N(l), l = 0, 1, \dots\}$, $N(l) \in \{0, 1, \dots, m\}$, describing the number of incorrectly received bits in consecutive bit patterns of the length m. This process is doubly stochastic one modulated by the underlying Markov chain $\{S_N(l), l = 0, 1, \dots\}$. $\{N(l), l = 0, 1, \dots\}$ and can be completely defined via parameters of the bit error process.

Let us denote the probability of going from the state i to the state j for the Markov chain $\{S_N(l), l = 0, 1, \dots\}$ with exactly k, $k = 0, 1, \dots, m$ incorrectly received bits in a bit pattern of the length m by $d_{N,ij}(k) = Pr\{N(l) = k, S_N(l) = j | S_N(l-1) = i\}$. These transition probabilities can be found using $D_E(k)$, $k = 0, 1$ and π_E:

$$d_{N,ij}(0) = \pi_E D_E^m(0)e,$$

$$d_{N,ij}(1) = \pi_E \sum_{k=m-1}^{0} D_E^{m-k-1}(0)D_E(1)D_E^k(0)e,$$

$$\dots$$

$$d_{N,ij}(m) = \pi_E D_E^m(1)e, \tag{2}$$

where e is the vector of ones of appropriate size.

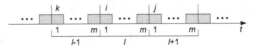

Fig. 2. Sequence of consecutively transmitted bits at the wireless channel.

Let us now introduce the frame error process $\{F(l), l = 0, 1, \dots\}$, $F(l) \in \{0, 1\}$, where 0 indicates the correct reception of the frame, 1 denotes incorrect frame reception. Process $\{F(l), l = 0, 1, \dots\}$ is modulated by the underlying Markov chain $\{S_F(l), n = 0, 1, \dots\}$. Note that $\{S_F(l), l = 0, 1, \dots\}$ and $\{S_N(l), l = 0, 1, \dots\}$ are the same. Let us denote the probability of going from the state i to the state j for the Markov chain $\{S_F(l), l = 0, 1, \dots\}$ with exactly k, $k = 0, 1$ incorrectly received frames by $d_{F,ij}(k)$. Process describing the number of bit errors in consecutive frames can be related to the frame error process $\{F(l), l = 0, 1, \dots\}$ using the so-called frame error threshold F_T:

$$d_{F,ij}(0) = \sum_{k=0}^{F_T-1} d_{N,ij}(k), \qquad d_{F,ij}(1) = \sum_{k=F_T}^{m} d_{N,ij}(k). \qquad (3)$$

Expressions (3) are interpreted as follows: if the number of incorrectly received bits in the frame is greater or equal to a computed value of the frame error threshold ($k \geq F_T$) the frame is incorrectly received and $F(l) = 1$, otherwise ($k < F_T$) the frame is correctly received and $F(l) = 0$.

Assume that FEC is not used at the data-link layer. It means that every time a frame contains at least one bit error, it is received incorrectly ($F_T = 1$). Thus, the probabilities (3) of the frame error process take the following form:

$$d_{F,ij}(0) = d_{N,ij}(0), \qquad d_{F,ij}(1) = \sum_{k=1}^{m} d_{N,ij}(k) = 1 - d_{F,ij}(0). \qquad (4)$$

The slot durations of $\{N(l), l = 0, 1, \dots\}$ and $\{F(l), l = 0, 1, \dots\}$ are the same $\Delta t'$ and related to the slot duration of the received signal strength process $\{L(n), n = 0, 1, \dots\}$ as $\Delta t' = nl\Delta t$, $n = 0, 1, \dots$.

3.3 Frame Error Process with FEC

The frame error threshold F_T depends on FEC correction capabilities. Assume that the number of bit errors that can be corrected by a FEC code is l. Then, $F_T = (l + 1)$ and the frame is incorrectly received when $k \geq (l + 1)$. Otherwise, it is correctly received. Thus, the transition probabilities (3) of the frame error process take the following form:

$$d_{F,ij}(0) = \sum_{k=0}^{l} d_{N,ij}(k), \qquad d_{F,ij}(1) = \sum_{k=l+1}^{m} d_{N,ij}(k). \qquad (5)$$

3.4 IP Packet Error Process

Assume that IP packet is segmented into z frames of equal size at the data-link layer.[1] The sequence of consecutively transmitted frames, denoted by gray

---

[1] Assumption of the constant frame size does not restrict the generality of the results as long as only one traffic source is allowed to be active at any instant of time for which only data-link error concealment techniques are possible (e.g., IP telephony).

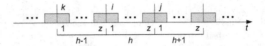

Fig. 3. Sequence of consecutively transmitted frames at the wireless channel.

rectangles, is shown in the Fig. 3, where $(h-1)$, h, $(h+1)$ denote time intervals whose length equals to the time to transmit a single packet; k, i, j, denote the state of the Markov chain $\{S_F(n), n = 0, 1, \dots\}$ in the beginning of intervals.

Consider the stochastic process $\{M(h), h = 1, 2, \dots\}$, $M(h) \in \{0, 1, \dots, z\}$, describing the number of incorrectly received frames in a consecutive frame patterns of the length z. This process is modulated by the Markov chain $\{S_M(h), h = 0, 1, \dots\}$ and can be defined via parameters of the frame error process.

Let us denote the probability of going from state i to state j for the Markov chain $\{S_M(h), h = 0, 1, \dots\}$ with exactly k, $k = 0, 1, \dots, z$ incorrectly received frames in a frame pattern of length z by $d_{M,ij}(k) = Pr\{M(h) = k, S_M(h) = j | S_M(h-1) = i\}$. These transition probabilities can be found using $D_F(k)$, $k = 0, 1$ and $\boldsymbol{\pi_F}$ of $\{F(l), l = 0, 1, \dots\}$ as follows:

$$d_{M,ij}(0) = \boldsymbol{\pi_F} D_F^z(0)\boldsymbol{e},$$

$$d_{M,ij}(1) = \boldsymbol{\pi_F} \sum_{k=z-1}^{0} D_F^{z-k-1}(0) D_F(1) D_F^k(0)\boldsymbol{e},$$

$$\dots$$

$$d_{M,ij}(z) = \boldsymbol{\pi_F} D_F^z(1)\boldsymbol{e}, \tag{6}$$

where $\boldsymbol{\pi_F}$ is the stationary distribution vector of $\{S_F(l), l = 0, 1, \dots\}$.

Let us now introduce the packet error process $\{P(h), h = 0, 1, \dots\}$, $P(h) \in \{0, 1\}$, where 0 indicates the correct reception of the packet, 1 denotes incorrect packet reception. Process $\{P(h), h = 0, 1, \dots\}$ is modulated by the underlying Markov chain $\{S_P(h), h = 0, 1, \dots\}$. Note that $\{S_P(h), h = 0, 1, \dots\}$ and $\{S_M(h), h = 0, 1, \dots\}$ are the same. Let us denote the probability of going from the state i to the state j for the Markov chain $\{S_P(h), h = 0, 1, \dots\}$ with exactly k, $k = 0, 1$ incorrectly received packets by $d_{P,ij}(k)$. Process $\{M(h), h = 0, 1, \dots\}$ describing the number of incorrectly received frames in consecutively transmitted packets can be related to the packet error process $\{P(h), h = 0, 1, \dots\}$ using the so-called packet error threshold P_T:

$$d_{P,ij}(0) = \sum_{k=0}^{P_T-1} d_{M,ij}(k), \qquad d_{P,ij}(1) = \sum_{k=P_T}^{z} d_{M,ij}(k) = 1 - d_{M,ij}. \tag{7}$$

Expressions (7) are interpreted as follows: if the number of incorrectly received frames in a packet is greater or equal to a computed value of the packet error threshold ($k \geq P_T$) the packet is incorrectly received and $P(h) = 1$. Otherwise, it is correctly received and $P(h) = 0$. Since no error correction procedures are defined for IP layer, $P_T = 1$ and only $d_{M,ij}(0)$ must be computed in (6). That is, every time a packet contains at least one incorrectly received frame, the whole packet is received incorrectly.

The slot durations of $\{P(h), h = 0, 1, \dots\}$ and $\{M(h), h = 0, 1, \dots\}$ are the same $\Delta t''$ and related to the slot duration of the received signal strength process $\{L(n), n = 0, 1, \dots\}$ as $\Delta t'' = nlh\Delta t,\ n = 0, 1, \dots$.

3.5 Illustration of the Proposed Extension

An illustration of proposed cross-layer mapping is shown in Fig. 4 where time diagrams of $\{L(n), n = 0, 1, \dots\}$, $\{E(n), n = 0, 1, \dots\}$, $\{N(l), l = 0, 1, \dots\}$, $\{F(l), l = 0, 1, \dots\}$, $\{M(h), h = 0, 1, \dots\}$, $\{P(h), h = 0, 1, \dots\}$ are shown. Error thresholds B_T, F_T and P_T must estimated as outlined previously and then used to compute transition probabilities of error processes at different layers.

To define models of the incorrect reception of PDUs at different layers we implicitly assumed that appropriate PDUs are consecutively transmitted at corresponding layers. Hence, the IP packet error process is conditioned on the event of consecutive transmission of packets.

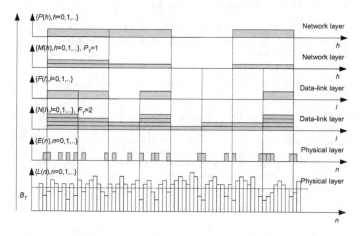

Fig. 4. Illustration of proposed cross-layer mapping.

4   Comparison of the Proposed Approach

Let us now compare the proposed approach with that one developed and used to date. We consider two cases: (1) the signal strength is assumed to be constant during the frame transmission time (2) the signal strength is mapped to the IP layer model in accordance with our approach. In what follows, we use subscripts 1 and 2 to denote performance estimators obtained using corresponding approach.

Assume that the wireless channel at the physical layer is represented by the Markov chain with $M = 4$, $p_{E,i} = 0$, $i = 1, 2, 3$, $p_{E,4} \neq 0$ and the following transition probability matrix:

$$P_E = \begin{pmatrix} 0.42 \; 0.18 \; 0.24 \; 0.16 \\ 0.18 \; 0.42 \; 0.04 \; 0.36 \\ 0.07 \; 0.03 \; 0.54 \; 0.36 \\ 0.03 \; 0.07 \; 0.09 \; 0.81 \end{pmatrix} . \tag{8}$$

We also assume that exactly one IP packet is mapped into one frame at the data-link layer. It can be easily shown that this assumption provides the best possible conditions at the IP layer.

4.1   Packet Error Processes Without FEC

Assume that the length of the frame is m bits and FEC is not used at the data-link layer. Then, the conditional mean of the incorrectly received packets[2] is given by the following expressions:

$$E_1[P] = \sum_{i=1}^{4} \pi_{E,i} \sum_{k=0}^{m-1} (1 - p_{E,i})^k p_{E,i}, \qquad E_2[P] = 1 - \boldsymbol{\pi_E} D_E^m(0)\boldsymbol{e}. \tag{9}$$

The estimated values of the conditional mean of the incorrectly received IP packets for different values of m and $p_{E,4}$ are shown in the Table 1. Comparing obtained results we note that the assumption of the same received signal strength during the frame transmission time significantly overestimates the actual performance of wireless channels when the channel coherence time is comparable with the time to transmit a single symbol.

4.2   Packet Error Processes with FEC

Consider now the effect of FEC. Assume that the FEC code may correct up to l bit errors i.e. $F_T = l + 1$. Then, the conditional mean of the incorrectly received

---

[2] This performance parameter can be interpreted as the mean number of the incorrectly received IP packets given that packets are generated according to Bernoulli process with probability of a single arrival set to 1.

Table 1. Conditional mean number of incorrectly received packets (no FEC)

m	$p_{E,4} = 0.3$		$p_{E,4} = 0.1$	
	$E_1[P]$	$E_2[P]$	$E_1[P]$	$E_2[P]$
100	0.622	1.000	0.622	0.998
500	0.622	1.000	0.622	1.000
2000	0.622	1.000	0.622	1.000

Table 2. Conditional mean number of incorrectly received packets (FEC)

(m, l)	$p_{E,4} = 0.3$		$p_{E,4} = 0.1$	
	$E_1[P]$	$E_2[P]$	$E_1[P]$	$E_2[P]$
(100,5)	0.622	1.000	0.622	0.587
(100,10)	0.622	0.976	0.622	0.054
(100,20)	0.622	0.328	0.622	8E-8
(500,20)	0.622	1.000	0.622	0.975
(500,40)	0.622	1.000	0.622	0.052
(500,100)	0.622	0.233	0.622	<10E-10
(2000,100)	0.622	1.000	0.622	0.986
(2000,200)	0.622	1.000	0.622	4E-6
(2000,300)	0.622	1.000	0.622	<10E-10

packets is given by the following expressions:

$$E_1[P] = \sum_{i=1}^{4} \pi_{E,i} \sum_{k=F_T}^{m} \binom{k}{m} (1 - p_{E,i})^k p_{E,i}^{m-k},$$

$$E_2[P] = \sum_{i=1}^{4} \pi_{E,i} \sum_{k=F_T}^{m} d_{N,ij}(k)e. \tag{10}$$

Results for different values of l, m and $p_{E,4}$ are shown in the Table 2. For illustrative purposes some non-realistic values of (m, l) are also included. Comparing obtained results we note that the assumption of the same received signal strength during the frame transmission time significantly underestimates or overestimates the actual performance of wireless channel depending on correction capabilities of FEC code. Our approach provides exact values of conditional mean of the incorrectly received packets when the channel coherence time is comparable with the time to transmit a single symbol at the wireless channel.

5   Conclusions and Future Work

We extended the small-scale propagation model representing the received signal strength of the wireless channel to IP layer using the cross-layer mappings. Our model is represented by the IP packet error process and retains memory properties of the initial signal strength process. We compare results obtained using our model with those presented in literature and show that our approach allows to get more accurate estimators of IP packet error probabilities when the channel coherence time is comparable with the time to transmit a single symbol.

The proposed model entirely relies on classic small-scale propagation model and does not take into account the signal strength attenuation caused by movements of the user over the large distances. The aim of our further work is to extend our approach to mobility-dependent propagation characteristics.

References

1. M. Krunz and J.-G. Kim. Fluid analysis of delay and packet discard performance for QoS support in wireless networks. *IEEE JSAC*, 19(2):384–395, February 2001.
2. M. Zorzi, R. Rao, and L. Milstein. ARQ error control for fading mobile radio channels. *IEEE Trans. on Veh. Tech.*, 46(2):445–455, May 1997.
3. T. Rappaport. *Wireless communications: principles and practice.* Communications engineering and emerging technologies. Prentice Hall, 2nd edition, 2002.
4. T. Okumura, E. Omori, and Fakuda K. Field strength and its variability in VHF and UHF land mobile service. *Review of electrical communication laboratory*, 16(9/10):825–873, September/October 1968.
5. J. Andersen, T. Rappaport, and S. Yoshida. Propagation measurements and models for wireless communications channels. *IEEE Comm. Mag.*, 33:42–49, Nov. 1995.
6. S. Seidel and T. Rappaport. Site-specific propagation prediction for wireless in-building personal communication system design. *IEEE Trans. on Veh. Tech.*, 43(4):879–891, November 1994.
7. A. Saleh and R. Valenzuela. A statistical model for indoor multipath propagation. *IEEE JSAC*, 5(2):128–137, February 1987.
8. D. Durgin and T. Rappaport. Theory of multipath shape factors for small-scale fading wireless channels. *IEEE Trans. on Ant. and Propag.*, 48:682–693, May 2000.
9. Q. Zhang and S. Kassam. Finite-state markov model for Rayleigh fading channels. *IEEE Trans. on Comm.*, 47(11):1688–1692, November 1999.
10. J. Swarts and H. Ferreira. On the evaluation and application of markov channel models in wireless communications. In *Proc. VTC'99*, pages 117–121, 1999.

Efficiency Issues in MPLS Transport for the UMTS Access Network

Enrique Vázquez, Manuel Álvarez-Campana, and Ana B. García

Dept. of Telematics, Technical University of Madrid,
ETS Ing. Telecomunicación, Ciudad Universitaria, 28040 Madrid, Spain
{enrique, mac, abgarcia}@dit.upm.es

Abstract. Multiprotocol Label Switching (MPLS) offers a simple and flexible transport solution for multiservice networks. Therefore, many UMTS operators are currently considering the use of a MPLS backbone. However, the efficient transport of short voice and data packets in the UMTS access network requires multiplexing and segmentation functions not provided by MPLS. This paper investigates the use of ATM Adaptation Layer 2 (AAL2) over MPLS to perform both functions. The efficiency of AAL2/MPLS is analyzed for different traffic types and compared with other transport options. The results indicate that significant capacity savings can be obtained with this solution.

1 Introduction

The 3G Universal Mobile Telecommunications System (UMTS) is a network designed to support multiple applications (telephony, video-conferencing, audio and video streaming, games, Web access, e-mail, etc.) with very different traffic patterns and quality of service (QoS) requirements. The initial UMTS architecture defined by the Third Generation Partnership Project (3GPP) comprises a Wideband CDMA radio interface, an access network based on Asynchronous Transfer Mode (ATM), and a core network evolved from 2G networks. The core has two domains. The circuit-switched domain, founded on GSM, handles voice and other circuit-mode traffic. The packet-switched domain, derived from GPRS, handles IP traffic. This structure, described in the 3GPP Release 99 specifications, is progressively changing towards a unified model based on IP (Releases 4, 5, and 6).

In contrast with the core network, the UMTS Terrestrial Radio Access Network (UTRAN) relies on an integrated transport infrastructure for all traffic types (voice, data, etc.). The UTRAN protocol architecture is divided into a Radio Network Layer (RNL), designed specifically for UMTS, and a Transport Network Layer (TNL) that reuses existing transport technologies. In Releases 99 and 4 the TNL consists of ATM connections with ATM Adaptation Layer type 2 (AAL2) or 5 (AAL5). See Table 1.

Release 5 specifications [1], [2] allow two TNL alternatives: ATM, as outlined above, or IP with UDP in the user plane, and SCTP (Stream Control Transmission Protocol) for signaling. IP version 6 is mandatory and IP version 4 is optional, although a dual stack is recommended. The default encapsulation for IP packets is PPP

J. Solé-Pareta et al. (Eds.): QofIS 2004, LNCS 3266, pp. 204–213, 2004.
© Springer-Verlag Berlin Heidelberg 2004

Table 1. UTRAN interfaces and ATM Adaptation Layers used. The air interface (Uu) does not use ATM

Interface	User Plane	Control Plane
Iub: Node B – Radio Network Controller (RNC)	AAL2	AAL5
Iur: Serving RNC – Drift RNC	AAL2	AAL5
Iu-CS: S-RNC – Mobile Switching Center (MSC)	AAL2	AAL5
Iu-PS: S-RNC – Serving GPRS Support Node (SGSN)	AAL5	AAL5

(Point-to-Point Protocol) with HDLC framing, but other layer 2 protocols are also possible.

A typical UTRAN topology will have many Node Bs, often connected by low-capacity links. Therefore, the efficiency of the protocol stack in the Iub interface between Node B and RNC becomes an important design issue. Two main problems must be solved. Firstly, UDP, IP, and layer 2 headers add a large overhead to voice packets, even if header compression procedures are used. This problem can be alleviated by multiplexing voice communications. In this way, several short voice packets are concatenated before transport, so the overhead of the headers added after the concatenation is shared by all the packets in the group. Secondly, long packets may introduce unacceptable delay variations when transmitted over a slow link. To avoid this, packets that exceed a certain maximum size are segmented before transmission. Note that segmentation reduces efficiency, but it may be necessary to meet QoS requirements (delay variation).

The structure of the rest of this paper is as follows. Sect. 2 reviews various solutions proposed to implement concatenation and segmentation in the Iub protocol stack. Sect. 3 presents a new proposal that relies on AAL2 [3] for packet concatenation/segmentation, and MPLS [4] for transport. Sect. 4 evaluates the performance of AAL2/MPLS for voice and data, and compares it with other options. Finally, Sect. 5 summarizes the contributions of the paper.

2 Alternatives for Transport in the UTRAN

ATM in the UTRAN is studied in [5] and [6], focusing on traffic modeling, QoS analysis, and simulation experiments with voice and data traffic. The 3GPP has investigated IP transport in the UTRAN in an ad hoc group created within the Technical Specification Group Radio Access Network, Working Group 3 (TSG RAN WG3), which is responsible for the overall UTRAN design. A technical report [7] presents the conclusions of this work and makes some general recommendations for Release 5 specifications.

Different Iub stacks based on IP are simulated and compared in [8]. The concatenation and segmentation functions mentioned in the previous section may be located above UDP/IP or below it. Protocols such as Composite IP (CIP) and Lightweight IP Encapsulation (LIPE) [9] concatenate and, optionally, segment packets above UDP/IP. Alternatively, both functions may be implemented at layer 2 by using

PPPmux [10] for concatenation, and the multi-class, multi-link extensions to PPP (MC/ML-PPP) [11] for segmentation. This is the option recommended by 3GPP in [7]. The application of the Differentiated Services model in the UTRAN is addressed in several studies. For example, in [12] voice calls are multiplexed with CIP and transferred with the Expedited Forwarding Per-Hop Behavior [13].

While these studies indicate that IP can be a viable alternative to ATM for the UTRAN, it is necessary to implement a complicated protocol stack to achieve the required performance. Firstly, UDP/IP headers have to be compressed to reduce the overhead added to short packets (e.g. voice). Many compression algorithms have been proposed, for example IP Header Compression, Compressed RTP, Enhanced Compressed RTP, Robust Header Compression, and others described in RFCs or Internet drafts. For references and a discussion of the advantages and drawbacks of each method see [7]. PPP encapsulation and framing overhead must also be minimized by using the simplified header formats foreseen in the standard.

Secondly, PPPmux and MC/ML-PPP must be implemented. When there are intermediate IP routers between Node B and RNC, PPPmux and MC/ML-PPP may be terminated at the edge router (ER) next to the Node B or at the RNC. In the former case, header compression, concatenation, and segmentation are applied only in the last hop between the Node B and the ER, where capacity is likely to be more limited. Packets are routed individually between the ER and the RNC. In the latter case, those functions are moved to the RNC, so the efficiency gains are maintained all the way up to the RNC. The disadvantage of this approach is that layer 2 frames must be tunneled between the Node B and the RNC. This implies that another IP header is added to the information transported across the network. See Fig. 1.

The Radio Network Layer (RNL) includes other protocols not shown in Fig. 1, namely Medium Access Control, Radio Link Control, and Packet Data Convergence Protocol. Non-access stratum protocols, which are transparent for the UTRAN, are located on top of the RNL, for example IP packets exchanged between the user equipment and the Gateway GPRS Support Node (GGSN) in the UMTS core network, plus the higher layer protocols required by the user applications (TCP, HTTP,

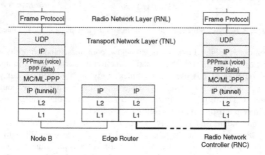

Fig. 1. UDP/IP transport stack in the user plane of the Iub interface, with end-to-end concatenation/segmentation between Node B and RNC

etc.). With the architecture of Fig. 1 these IP packets would be encapsulated with two additional IP headers, resulting in a poor efficiency.

The use of MPLS in the UTRAN has been considered mainly in terms of protocol functionality, for example: mapping of traffic flows to MPLS Label Switched Paths (LSPs) and signaling for LSP establishment [14], combination of MPLS traffic engineering (MPLS-TE) functions with the Differentiated Services model (DiffServ) [15], and mobility support [16], [17]. These references point out potential advantages of MPLS for transport in the UTRAN, but no numerical results are presented.

The MPLS/Frame Relay Alliance (MFA) defined an ad-hoc multiplexing protocol for efficient transport of voice calls over MPLS [18]. More recently, the MFA has proposed to use AAL protocols (without ATM cells) for voice trunking [19] and for TDM circuit emulation [20] over MPLS networks.

3   AAL2/MPLS Transport in the UTRAN

This section proposes a simpler TNL architecture with two main components: AAL2 and MPLS. See Fig. 2. Specifically, we propose to use the Service Specific Segmentation and Reassembly (SSSAR) sublayer [21], which is more adequate for integrated voice and data transport, instead of the convergence sublayer for trunking [22] included in previous proposals that focused on voice traffic only.

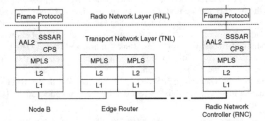

Fig. 2. AAL2/MPLS transport protocol stack in the user plane of the Iub interface

Our solution, explained in more detail below, combines standard protocols in such a way that each one performs the essential function it was designed for, and complements the other. AAL2 is used to multiplex many variable-bit-rate, delay-sensitive traffic flows, and MPLS provides a flexible tunnelling mechanism without the overhead of ATM. The result is a simple protocol architecture that offers significant advantages in comparison with other solutions (see Sect. 3.1.)

The Common Part Sublayer (CPS) [3] of AAL2 concatenates voice and data packets. Each CPS packet has a 3-byte header and a maximum payload length of 45 (default) or 64 bytes. See Fig. 3. The SSSAR sublayer accepts data units up to 65568 bytes and segments them up to the maximum length admitted by CPS. The last segment of each packet is marked in the UUI bits of the CPS header, so the SSSAR adds

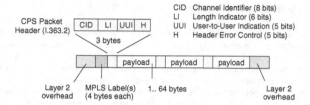

Fig. 3. Concatenated AAL2 CPS packets over MPLS

no extra overhead. CPS and SSSAR are used in ATM-based UTRANs as defined in Release 99 and Release 4 specifications. CPS packets are mapped to cells sent over ATM virtual connections at the Iub interface. The first byte of each cell payload, called Start Field, indicates the next CPS packet boundary. The time that a partially filled cell waits for the next CPS packet is limited by Timer_CU (Combined Use). If it expires, the cell is completed with padding bytes and transmitted.

In the proposed AAL2/MPLS solution, AAL2 performs the same functions as in the AAL2/ATM case, except that CPS packets are not mapped to cells. They are concatenated up to Timer_CU expiration or up to a given maximum length Lmax (e.g. set to comply with a maximum transmission time in the Node B - Edge Router link), and transmitted over an LSP preceded by the MPLS label stack. CPS payloads contain Frame Protocol data units, segmented by SSSAR if necessary.

3.1 Advantages of the Proposed Solution

AAL2/MPLS is conceptually very similar to AAL2/ATM. ATM virtual connections and related traffic management procedures are replaced by MPLS LSPs, possibly supporting traffic engineering and class-of-service differentiation (DiffServ-Aware MPLS Traffic Engineering or DS-TE) [23]. AAL2/MPLS is considerably more efficient, because the ATM cell header overhead is eliminated. See the results of Sect. 4.

The AAL2 protocol is implemented only at the end points (Node B and RNC), and intermediate ATM switches are not needed. Since AAL2 is still used, the interface offered by the Transport Network Layer to the Radio Network Layer is the same as in UMTS Releases 99 and 4. Moreover the standard signaling procedures used to establish and release AAL2 channels (Access Link Control Application Part, ALCAP) can be reused.

Compared with IP-based alternatives, AAL2/MPLS is simpler. IP tunnels, header compression, PPP multiplexing, and Multi-Class Multi-Link PPP are not required in the TNL (although IP is still used by applications in the non-access stratum, as mentioned in Sect. 2). In the control plane, new signaling procedures (provisionally named "IP-ALCAP" by 3GPP) are not needed, because the standard ALCAP is used. Regarding efficiency, AAL2/MPLS compares to the best IP based solutions, with values in the 90-95% range. See Sect. 4.

4   Performance Evaluation

The header formats and procedures of the different protocols mentioned in the preceding sections have been analyzed in order to evaluate the efficiency of alternative transport stacks for the UTRAN. The analysis considers a simple case where every concatenated packet has the same length. The results presented in Sect. 4.1 and 4.2 show the improvements that can be obtained with the AAL2/MPLS solution.

A higher transport efficiency should allow the network to carry more traffic maintaining the QoS levels or, alternatively, to reduce the capacity needed to carry a given amount of traffic with the required QoS. To verify this assumption, we have simulated in more detail each protocol stack, transporting various mixes of voice and Web data traffic in the Iub interface between base stations (Node Bs) and controllers (RNCs). Optimizing this interface is particularly important for the operators due to the high number of Node Bs that must be connected in a typical UMTS network, so we focused the simulation study on it. However, the solution proposed here can be used in other UMTS interfaces as well. Sect. 4.3 gives a sample of the simulation results obtained.

4.1   Analysis of Efficiency for Voice Traffic

In this case, we analyze the transport of voice payloads of constant size $P=40$ bytes, corresponding to 32 bytes generated the Adaptive Multi Rate codec used in UMTS, plus 8 bytes added by the RNL [5]. The background noise description packets sent by the AMR codec during silence periods are not considered in the analysis. The number of concatenated packets per group (N) is equal to the number of voice connections in the active state. Segmentation is not necessary.

The overhead per packet (H) and the overhead per group (Hg) take different values depending on the transport protocol stacks considered. For example, AAL2/MPLS over PPP with simplified HDLC framing (AAL2/MPLS/PPP/ HDLC) gives $H=3$ bytes and $Hg=9$ bytes. UDP/IP with headers compressed to 4 bytes and concatenation at layer 2 (cUDPIP/PPPmux/HDLC) gives $H=Hg=5$ bytes. The details of the model and the complete set of parameter values used can be found in [24].

Fig. 4 shows the efficiency (number of RNL bytes transported divided by the total number of bytes transmitted by the TNL) as a function of the number of active voice connections. The best alternative is AAL2/MPLS, with efficiency above 90% even for moderate values of N. The efficiency of UDP/IP with headers compressed to 4 bytes (cUDPIP) over PPPmux is not as good, because concatenation at layer 2 gives a higher overhead per packet (H). If the end-to-end configuration between Node B and RNC illustrated in Fig. 1 is used, the extra IP tunnel increases the overhead per group (Hg) and reduces the efficiency, especially when there are few packets per group.

The curve labeled MPLSx2/PPP/HDLC corresponds to the case where AAL2 is not used and each packet is transmitted with a stack two MPLS labels: the inner label is used as a channel identifier, and the outer one serves to route the packets to their destination. The two labels (4 bytes each) plus the PPP/HDLC header add a total of 13 bytes to each voice payload of 40 bytes. Therefore, the efficiency is $40/53=75.5\%$,

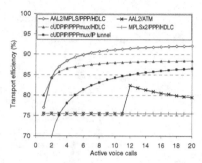

Fig. 4. IP and MPLS transport options vs. AAL2/ATM (voice traffic)

which happens to be the same value obtained with AAL2/ATM when N voice pay-loads (N·40 bytes) are transmitted in N ATM cells (N·53 bytes). This is true for N<12. When N=12, 12 voice payloads are carried in 12 concatenated CPS packets (12·43=516 bytes), which still fit in 11 cells (11·47=517 bytes), so the efficiency of AAL2/ATM increases to 12·40/11·53=82.3% as illustrated in Fig. 4.

Other transport options not shown in the graph give poor results. For example, the efficiency of a simple UDP/IP/PPP/HDLC stack without header compression and no multiplexing is only 55.6%. This value corresponds to IP version 4 headers (20 bytes). With IP version 6 headers (40 bytes) it would be even lower: 43.5%. If IP is sent over AAL5/ATM instead of PPP/HDLC, each voice packet is encapsulated in 2 cells and the efficiency drops further to 37.7%.

4.2 Analysis of Efficiency for Data Traffic

Fig. 5 shows the transport efficiency for data payloads of variable size P up to 1500 bytes. As explained in previous sections, short packets may be concatenated up to a maximum size Lmax, and packets longer than Lmax are segmented before transmis-sion. In Fig. 5 Lmax has been set to 1000 bytes. H is the overhead per packet or seg-ment and Hg is the overhead per group, with different values for each protocol stack as in the previous case.

With compressed UDP/IP (cUDPIP) and MPLS using a 2-label stack (MPLSx2), segmentation is done at layer 2 by MC/ML-PPP. Both exhibit a high efficiency for long data packets, although they were not as good for short voice packets (compare the corresponding curves in Fig. 4 and Fig. 5). In MPLSx2, the internal label identi-fies each data channel, and the external one is used for routing. AAL2/MPLS reaches approximately the same value, close to 95%, for both voice and data. In this case, MC/ML-PPP is not needed because AAL2 takes care of segmentation at the SSSAR sublayer, so AAL2/MPLS uses the default PPP/HDLC encapsulation.

Fig. 5 indicates that MPLSx2 outperforms AAL2/MPLS for packets longer than 250 bytes approximately. Therefore, in scenarios where the traffic mix includes sig-

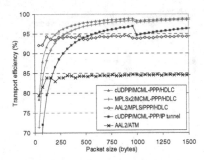

Fig. 5. IP and MPLS transport options vs. AAL2/ATM (data traffic)

nificant amounts of short voice packets (a few tens of bytes) and long data packets (a few hundreds of bytes or more), the overall efficiency can be optimized by separating the traffic of each Node B in two Label Switched Paths: LSP 1 for voice with AAL2, and LSP 2 for data without it. In this case, and assuming that voice packets are always smaller than 65 bytes, the SSSAR part of AAL2 can be removed in LSP 1. Packets sent via LSP 2 are segmented, if necessary, by Multi-Class Multi-Link PPP. Note that the network operator may prefer to use separate LSPs for voice and for data anyway, even if AAL2 is used in both.

4.3 Simulation Results

The Iub transport protocol stacks considered in the preceding sections have been simulated with voice and Web traffic, in order to compare the delay and loss performance of the different options. The simulator was developed in C language, and we are currently porting it to OPNET Modeler [25].

The traffic source modules generate Frame Protocol data units (see Figs. 1 and 2). The data unit sizes and inter arrival times are set taking into account the relevant characteristics of the UMTS radio access bearers used, as well as the overheads added by the RNL protocols. The transmission rate and the Transmission Timing Interval (TTI) are the most decisive parameters. In our simulations, voice is coded at 12.2 kbit/s with TTI=20 ms. Including the RNL overhead, this corresponds to one FP data unit of 40 bytes every 20 ms. During silence periods the data unit size is reduced to 13 bytes. Web pages are downloaded at 64 kbit/s with TTI=40 ms, which corresponds to one data unit of 331 bytes every 40 ms, also including RNL overhead. For AAL2/MPLS, the simulator can be configured to use the same LSP for all traffic, or separate LSPs for voice and for data. In these experiments we chose the latter option.

Fig. 6 is a sample of the results obtained. The curves show the Iub delay (0.95- or 0.99-quantile) vs. the number of active users. The capacity available for user traffic is set to 1.92 Mbit/s, and delays are measured in the RNC-to-Node B direction (Web traffic is higher in this direction). As anticipated by the previous analysis of effi-

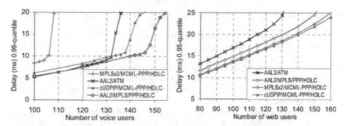

Fig. 6. Simulation of delay at the Iub Interface for different transport options

ciency, AAL2/MPLS gives the lowest delay for voice. MPLS without AAL2 is a good option for data traffic (right) but not for voice (left), because it gives a much higher delay than the other options. For a given maximum delay in the Iub interface, these curves may be used to estimate the number of users that can be served with the available capacity.

A detailed description of the simulated scenarios and additional results, which cannot be presented here due to lack of space, can be found in [5,24].

5 Conclusions

The UMTS transport layer is expected to migrate from ATM to packet-switched architectures that can provide the required quality of service at a lower cost. In this scenario, AAL2 (with SSSAR and CPS) over MPLS is a simple solution that offers a functionality similar to ATM, but in a more flexible and efficient way. This paper has discussed the main issues in transporting voice and data traffic from base stations connected with low capacity links, and has compared the performance of AAL2/MPLS with other proposals, both analytically and by simulation. Although the study focused on the access network, MPLS infrastructure may also be used in the UMTS core, and may even be shared by traffic from external IP networks.

AAL2/MPLS is more efficient than AAL2/ATM (typical differences are between 12% and 20%), so a larger fraction of the available capacity is dedicated to carry user traffic, and more customers can be served with the required QoS. Our simulator can be used to estimate the benefits of AAL2/MPLS in practical scenarios with different traffic loads and multiplexing strategies. AAL2/MPLS is particularly well suited for short packets, which are the most affected by protocol overhead.

References

1. 3GPP Technical Specification 25.426, v5.2.0: UTRAN Iur and Iub interface data transport & transport signalling for DCH data streams, Release 5 (Sep. 2002)
2. 3GPP Technical Specification 25.414, v5.4.0: UTRAN Iu interface data transport and transport signalling, Release 5 (Mar. 2003)

3. ITU-T Rec. I.363.2: B-ISDN ATM Adaptation Layer Specification: Type 2 AAL (Sep. 1997)
4. Rosen, E.: Multiprotocol Label Switching Architecture. RFC 3031 (Jan. 2001)
5. García Hernando, A.B.: Dimensioning and Quality of Service Support in the UMTS Access Network. Ph.D. Thesis. Technical University of Madrid (2002)
6. García Hernando, A.B., García, E., Álvarez-Campana, M., Berrocal, J., Vázquez, E.: A Simulation Tool for Dimensioning and Performance Evaluation of the UMTS Terrestrial Radio Access Network. Lecture Notes in Computer Science, Vol. 2515. Springer-Verlag, Berlin Heidelberg New York (2002)
7. 3GPP Technical Report 25.933, v5.3.0: IP transport in UTRAN (June 2003)
8. Gelonch, A. (ed.): Research results on UTRAN (RRM, QoS and packet data compression) Phase2. Project IST-1999-10699 Wine Glass, deliverable D11 (Dec. 2001)
9. Kempf, J. (ed.): IP in the RAN as a Transport Option in 3rd Generation Mobile Systems. Mobile Wireless Internet Forum, Technical Report MTR-006, release 2.0.0 (June 2001)
10. Pazhyannur, R., Ali, I., Fox, C.: PPP Multiplexing. IETF RFC 3153 (Aug. 2001)
11. Bormann, C. The Multi-Class Extension to Multi-Link PPP. IETF RFC 2686 (Sep. 1999)
12. Venken, K., de Vleeschauwer, D., de Vriendt, J.: Designing a diffserv-capable IP-backbone for the UTRAN. 2nd International Conference on 3G Mobile Communication Technologies. London, UK (Mar. 2001)
13. Davie, B.: An Expedited Forwarding PHB (Per-Hop Behavior). RFC 3246 (Mar. 2002)
14. Guo, Y., Antoniou, Z., Dixit, S.: IP transport in 3G radio access networks: an MPLS-based approach. IEEE Wireless Communications and Networking Conference, WCNC 2002. Orlando, Florida (Mar. 2002)
15. Feng, L.: QoS Support in IP/MPLS-based Radio Access Networks. Institute for Communications Research (ICR) Seminar (Oct. 2002)
16. Jabbari, B., Papneja, R., Dinan, E.: Label Switched Packet Transfer for Wireless Cellular Networks. IEEE Wireless Communications and Networking Conference, WCNC 2000 (Aug. 2000)
17. Chiussi, F., Khotimsky, D., Krishnan, S.: Mobility Management in Third-Generation All-IP Networks. IEEE Communications Magazine, Vol. 40, Num. 9 (Sep. 2002)
18. MPLS Forum Technical Committee: Voice over MPLS - Bearer Transport Implementation Agreement. MPLSF 1.0 (July 2001)
19. MPLS/Frame Relay Alliance Technical Committee: I.366.2 Voice Trunking Format over MPLS. MPLS/FR 5.0.0 (Aug. 2003)
20. MPLS/Frame Relay Alliance Technical Committee: TDM Transport over MPLS using AAL1. MPLS/FR 4.0 (June 2003)
21. ITU-T Recommendation I.366.1: Segmentation and Reassembly Service Specific Convergence Sublayer for the AAL type 2 (June 1998)
22. ITU-T Recommendation I.366.2: AAL type 2 Service Specific Convergence Sublayer for Trunking (Feb. 1999)
23. Fineberg, V.: QoS Support in MPLS Networks. MPLS/FR Alliance (May 2003)
24. Vázquez, E.: Analysis and Simulation of IP and MPLS Transport in the UTRAN. Technical University of Madrid (2003)
25. OPNET Modeler. web page: http://www.opnet.com/products/modeler/

QoS Constraints in Bluetooth-Based Wireless Sensor Networks

Veselin Rakocevic, Muttukrishnan Rajarajan, Kerry-Ann McCalla,
and Charbel Boumitri

Information and Biomedical Engineering Research Centre
School of Engineering and Mathematical Sciences, City University
Northampton Square, London E1CV 0HB, United Kingdom
{V.Rakocevic, R.Muttukrishnan}@city.ac.uk

Abstract. This paper investigates the case of a wireless sensor network deploying Bluetooth technology. The paper analyses main technological features of Bluetooth, such as error recovery, power consumption and connection establishment, and using some experimental results analyses their effect on the Quality of Service in the sensor network. The paper presents a case for a different view of QoS in sensor networks. We argue that the network availability, reliability and especially the originality (freshness) of data have emerged as crucial QoS issues in wireless sensor networking. The paper puts emphasis on master-side scheduling in Bluetooth and analyses using simulation the performance of three scheduling schemes in a sensor network model. We analyse the performance of the schemes in symmetric and asymmetric load environments and with different bit error probabilities. Two of the schemes are specially designed for sensor networks, and one of them, the Maximum Burst Delay First, shows very good results in the asymmetric load environment.

1 Introduction

Bluetooth [1][2] is a wireless communication technology, specially designed to replace wires in short-range applications. Bluetooth operates at 2.4GHz, using Frequency Hopping spread spectrum baseband technology to minimise interference. The Bluetooth specification defines a radio frequency interface and the set of communication protocols for device discovery, data exchange and error correction. The link speed, communication range (less than 10m), and transmit power level for Bluetooth were chosen to support low-cost, power-efficient, single-chip implementations.

Initially, Bluetooth was designed to replace wires in the PC systems, e.g. between mouse and PC or keyboard and PC. Recently, the concept of ubiquitous computing has emerged. Ubiquitous computing primarily refers to distributed systems of usually small computing devices able to communicate, exchange data and monitor each other's activities. Examples of ubiquitous systems are sensor networks and smart home networks. Sensor networks are network of small devices which have limited functionality and only basic communication abilities. The transmitting power of sensors must be minimal, in order to simplify the maintenance of sensor networks and

J. Solé-Pareta et al. (Eds.): QofIS 2004, LNCS 3266, pp. 214–223, 2004.
© Springer-Verlag Berlin Heidelberg 2004

minimise the cost. With respect to that, short-range low-consumption Bluetooth emerged as one of the most interesting technologies for communication in the sensor networks.

In a Bluetooth piconet (smallest Bluetooth network) one device must always be the master and the other devices are slaves. Each device can operate as either master or slave. One master can communicate to maximum 7 slaves. If more than 7 slaves are in the piconet they must not be active, i.e. they must be in the low-power mode.

In a typical Bluetooth application a master probes the environment searching for slaves. The discovery procedure is explained in more detail in section 3.2. Once the slaves have been found, the master polls them. The exact scheduling policy definition is omitted in the specification. Scheduling is one of the important research issues in Bluetooth. It is important to stress that the slaves can communicate only by replying to master's polling packet. This is often used as one of the main drawbacks of Bluetooth – if two slaves want to communicate, they have to do it through the master. Other disadvantages include connection establishment and piconet/scatternet establishment delay, and connection availability due to interference.

For an application to make some use of Bluetooth technology, it must be robust to delay, satisfied with the throughput provided, and find the range to be adequate. The use of Bluetooth in wireless sensor networks has already been investigated [3][4][5]. There is a general consensus that for Bluetooth to be used in sensor systems, a number of modifications need to be made. This paper contributes to the ongoing discussion about the use of Bluetooth in sensor systems by providing a parallel analysis of Bluetooth features that influence the Quality of Service in sensor systems. When it comes to sensor systems, the QoS analysis has to be done in a different way than with other communication systems. This paper presents a new approach to understanding QoS in wireless sensor systems, and proceeds to evaluate the performance of Bluetooth using both simulation and experimental testbed. We analyse three different scheduling schemes through simulation and derive conclusions on the optimal scheduling scheme for a sensor network.

2   Quality of Service in Wireless Sensor Networks

The traditional view of Quality of Service (QoS) in communication networks is concerned with end-to-end delay, packet loss, delay variation and throughput. Numerous architectures, scheduling techniques and admission control solutions have been presented in an effort to achieve guaranteed levels of network performance. Other performance-related features, such as network reliability, availability, communication security and robustness are often neglected in the QoS research. Matters are somewhat different in wireless personal and local area networks. 802.11 wireless LANs suffer greatly from problems in availability and reliability. The Quality of Service provided by these networks depends significantly on the level of interference in the environment, where the interference comes from other wireless and microwave devices, such as mobile phones, neon lights or microwave ovens. Situation with Bluetooth is slightly better, considering the short-range nature of Bluetooth communication. Still, we can say that the problems with overcoming interference and achieving satisfactory network availability present major challenge in design of local wireless network systems. With respect to local wireless networks, we can define network availability not only as the ability of user hosts to connect to the network, but also to gain at least some minimal communication rate. Network reliability can be defined as the number of times the network "is down" due to significant interference or just bad signal reception.

3 Bluetooth

3.1 Interference, Error Correction, and Transport Control

When it comes to network availability and reliability in wireless networks, they are primarily determined by the way the network reacts to packet losses. Bluetooth technology achieves reliability by retransmitting packets. Each Bluetooth packet carries a header with an acknowledgement bit in it. More specifically, in every packet header, there is an ARQN flag indicating the status of the previously received packet. ARQN=1 (ACK) means the packet has been received and correctly decoded. ARQN=0 (NAK) means the previous receive failed. A NAK occurs when the slave's response to a master transmission has been lost, or when HEC (Header Error Check) fails or when the CRC fails – this is the usual reason for a retransmit condition during connection. When a device receives an indication that the last packet was corrupted, it simply retransmits the packet. The slave will respond in the slave-to-master slot directly following the master-to-slave slot; the master will respond at the next event it will address the same slave. This retransmission carries on until it receives an acknowledgement that the packet got through correctly.

Bluetooth uses sequence numbers to inform the receiver whether the packet is being retransmitted or being sent for the first time.

In the case of the packet being lost, the sender keeps the timer and when the timeout expires without the acknowledgement from the receiver, the packet is retransmitted. This slows down the communication substantially.

To analyse the impact of the interference and bit errors on a end-to-end file transfer application, we have done a simple experiment. Our experimental testbed consisted of two notebook computers equipped with Bluetooth hardware and software. Fig. 1. shows the achieved goodput in transferring large files (380KB, 1.2MB and 3.6MB). We observe two environments – one in which there is no interference, and one in which an 802.11 WLAN Access Point is active in the close proximity of the notebooks. The Access Point is only transmitting control signals, with no data transfer happening in the WLAN, i.e. the level of interference is reasonably small. We can see how goodput decreases with the increase in the distance between the notebooks. Also, it is possible to quantify the impact of interference. We can see that, even with low-load interference, the impact on end-to-end file transfer time can be substantial. This result is interesting because raw application-layer file transfer time is observed.

In wireless sensor network, for example, we have a case of a chunk of data being fragmented into packets and sent over Bluetooth link. In the case of high interference and high packet loss, the packets will be repeatedly retransmitted, which will substantially increase the overall file transfer. In the sensor network, there is an issue of the maximum file transfer time. If the data that is being transmitted is some measurement data, there is likely to be certain time delay after which the measurement data is considered useless.

Further to this, the Bluetooth specification defines something called the flush timeout. The flush timeout gives the amount of time in milliseconds that a device will spend trying to transmit a packet segment before it gives up. If a packet segment does not get through before the flush timeout is exceeded, the segment can't get through and the whole packet is flushed. The flush timeout controls how many times the baseband can retransmit the packet

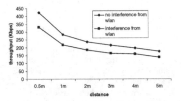

Fig. 1. Throughput measurements

3.2 Connection Establishment and Service Discovery

The complicated and slow process of connection establishment is one of the main disadvantages of the Bluetooth system. There are two delays in setting up a Bluetooth link. First, it takes time to discover devices in the neighbourhood. A second delay occurs in setting up the connection itself.

For two Bluetooth devices to start exchanging data, the nodes must discover each other first. The master sends small *inquiry* packets hopping frequencies at twice the rate in a normal connection. Slaves must be in an *inquiry scan* mode, changing the frequency very slowly. When a slave receives two consecutive inquiry packets, it waits for random time and then it responses. Before the two devices can establish a connection, they must be in *page* and *page scan* modes. The paging device initiates a connection, while the page scanning device responds.

The whole process is not short; Bluetooth specification defines an inquiry time of 10.24 seconds. In reality it takes up around 2.5 seconds to establish a Bluetooth link [6]. This makes Bluetooth far from optimal for dynamic wireless systems that need fast response and fast connection establishment. Seconds are very long time for computer systems, and this is a major drawback in applying Bluetooth technology for a number of future dynamic distributed systems.

Connection times can be lengthy, as transmitters and receivers need to synchronize before communication can commence. These limitations would have serious consequences if the wireless link were of a critical nature – for example, a 'panic button', a life-dependant medical monitor, or an engine management system. When it comes to sensor networks, for fixed nodes connection establishment should not be a problem, since once the connection is established it remains active. The problem occurs for mobile sensors, when the long duration of the connection establishment and short range may mean that the mobile sensor will be out of reach before the connection is fully set up.

3.3 Power Consumption

The small range of Bluetooth networks enables Bluetooth devices to spend minimal power to communicate. In addition to this, Bluetooth devices can switch to one of the three low power modes to further decrease the power consumption. The low power modes can also be used to form a dynamic piconet with more than 7 slaves – some slaves can be in low power modes while 7 slaves are continuously active.

Bluetooth specification defines three low power modes: Hold, Sniff and Park. Hold mode allows devices to be inactive for a single short period of time. Sniff mode allows devices to be inactive except for periodic sniff slots. Both Hold and Sniff modes are used in the cases when a node does not have continuous data to send, but does not want to disconnect from the network. Park mode is similar to sniff node, except that parked devices give up their active member address. A device in park mode does wake up periodically to listen to broadcast packets and it can be unparked in this instances. For additional power saving, Bluetooth specification also provides the option to switch to a less accurate low-power oscillator which then drives the Bluetooth clock in the low power mode.

The existence of low power modes is essential for sensor networks, where the power consumption is critical.

3.4 Scheduling and Packet Types

Bluetooth supports two types of communication links – the Asynchronous Connectionless (ACL) links and the Synchronous Connection-oriented (SCO) links.

The ACL link provides a packet-switched communication between a master and a slave when data arrives from the upper layers. Essentially, the data communication in Bluetooth is achieved over ACL links. On these links the master always transmits on even slots and the slave replies on odd-numbered slots, where a slot time is $625\,\mu s$.

Master only transmits data when there is something to send. A slave may only transmit if it has been transmitted to. In the simulation analysis that will follow we assume that all the links are ACL. The wireless sensor network using Bluetooth will be data communication networks and will use ACL Bluetooth links.

SCO links are essentially voice links. They provide reserved link bandwidth and regular periodic exchange of data in the form of reserved slots. If an SCO link is in operation, then that slave must be communicated with regularly according to the SCO repetition rate, T_{sco}.

Two packet types are used on ACL links - DM and DH packets. DM stands for Data Medium, these packets use FEC for error correction. DH stands for Data High rate. These connections do not use FEC. There are options to use 1-slot, 3-slot and 5-slot packets. Table 1 gives data rates for asymmetric communication between a master and a slave. In our analysis, we will be looking at the case of asymmetric data flow where the slaves are sending the data to the master, and all the packets will be single-slot DH1 packets.

In a Bluetooth piconet, multiple master-slave connections are served at the master in a round-robin fashion. The slaves are polled one by one, regardless of their traffic rates and queue sizes. Such a scheme is only efficient in the low load environments or when the incoming traffic is symmetric in all slaves. The master will transmit to and

Table 1. Data rates and packet sizes

Packet type	Max payload (B)	Max Asymmetric data rate FWD (Kbps)	Max asymmetric data rate reverse (Kbps)
DM1	17	108.8	108.8
DH1	27	172.8	172.8
DM3	121	387.2	54.4
DH3	183	585.6	86.4
DM5	224	477.8	36.3
DH5	339	723.2	57.6

receive from each of the slaves that are active at the time. If there is nothing to send, the master may either omit that slave or transmit a NULL packet.

A lot of work is being done in the research community to design and implement new scheduling mechanisms. The schemes presented in research literature all have an objective to avoid empty slots in Bluetooth scheduling – i.e. to avoid the cases when the master is polling a slave that has no data to send. The schemes differ on the basis of whether the polling is done in cycles – whether each slave is guaranteed at least one polling packet in a cycle, or the cycle does not exist.

For example, Capone et al [7] analyse a number of polling schemes in an attempt to find an optimal exhaustive scheme. The schemes they present differ in the process of identifying the next slave to be served. They use the queue lengths at both the master and the slaves to identify the next slave. Das et al in [8] analyse three different scheduling schemes based on a definition of a maximum time limit in which the master has to serve a slave. In their analysis, each slave has a polling interval and is served within that interval. The polling interval for each slave is dynamic – it is longer for slaves with empty queues and shorter for slaves with high traffic load. Lapeyrie and Turletti [9] analyse efficiency and fairness of polling schemes. They calculate the slave priority as a linear combination of the probability the queues are non-empty and the number of slots since the slave has last been polled.

All of these schemes have in common the need to prioritise among the slaves, and to use the queue length and traffic load to calculate the slave priority. Average waiting time (average packet delay) is the main metric used for the evaluation of these schemes. With respect to the analysis given in section 2, we can say that in wireless sensor networks the network Quality of Service should not be assessed on the basis of average packet delay but rather on the transfer delay of a sequence (burst) of packets. This is the main reason why in this paper we experiment with schemes that are optimised for dealing with bursts of data.

In the remainder of this section, we present three schemes that are used in the simulation. In the network model used for our simulation, sensor nodes are modelled as Poisson sources of constant-length bursts of packets.

Scheme 1: *Exhaustive scheduling (ES)*. In this scheme, the master polls a slave for as long as there is data in the slave's buffer. After that, the master moves on to serve the next slave in the cycle. The order of slaves is always the same. This is a standard polling scheme and we are using it to assess the benefit of introducing scheduling schemes optimised for constant-length burst traffic.

Scheme 2: *Limited exhaustive scheduling (LES)*. In this scheme, master polls the slave for as long as the slave buffer contains packets belonging to one data burst.

Once a burst is served, master moves to the next slave in the cycle and, in the case of a non-empty queue, serves the next-in-line burst from that slave. We expected this scheme to outperform the exhaustive scheme in terms of fairness in the presence of asymmetric traffic load.

Scheme 3: *Maximum burst delay first (MBDF)*. In this scheme the master serves the slave that has a data burst that has been waiting the longest. The master serves that slave until the burst of data is transferred, recalculates the delays and starts serving the slave with the maximum burst delay. This is different from Scheme 2 because the slaves are not being served in a cycle, but there is a recalculation of the waiting time after every data burst. We expected this scheme to outperform the other two. The only problem with this scheme is the implementation, since the master can know the burst arrival times only in the case when the queues are never empty. If one queue becomes empty, there is no way master can find out about the arrival of a new data burst. The MBDF scheme as defined here is therefore ideal and will have to be modified to be successfully implemented in a real system. However, the scheme is very important in the QoS provision in sensor node since it is based around the idea that the priority in the service order should be given to the sensor node with the highest probability of the data being flushed (arriving too late, becoming old and useless).

4 Simulation Results

We used discrete event simulator written in C++ to evaluate the performance of the three schemes. The simulated network consists of 7 slaves that generate the traffic in bursts of 50 DH1 packets. We assume that we are dealing with a network of temperature sensors or surveillance cameras that send periodic fresh information about the process or phenomenon monitored. The bursts are generated according to a Poisson process. The master does not have any of its data to send, it is only polling the slaves and receiving data from them. We use simple independent error model in which bit errors happen at random with probability P_e. Das et al in [8] present a more accurate two state Markov error model, but in the preliminary simulations given here a simpler model has been used.

In terms of performance evaluation metrics, in accordance to the analysis given in section 2, we are looking at per-burst performance and use average burst delay as the main performance parameter. Another parameter we use is the *flush rate* – the percentage of bursts that arrived at the master with the overall queuing and transmission delay greater then some Tmax, where we have used 0.5sec and 1 sec as values for Tmax in our simulations. It is critical for wireless sensor network that fresh information arrives at the end-hosts. The information about the flush probability is very important for assessing network availability in the sensor systems. If flushing of data happens often, the Quality of Service in the system decreases.

The simulation experiments observe the cases of symmetric and asymmetric traffic load. The traffic load in the experiments is high, which creates the environment where scheduling becomes critical. In the first experiment (Fig.2 and Fig.3), the average traffic load is 0.8 in all slaves, the bit error probability varies from 10^{-2} to 10^{-5}, and the flush time Tmax is 1sec. In the second experiment (Fig.4 and Fig.5), the traffic load remains symmetric, but varies from 0.6 to 0.9 with constant error rate of 10^{-3}, and Tmax=0.5sec. In the third experiment (Fig.6 and Fig.7) one of the slaves has load of 0.9 while other slaves have load of 0.6, the error rate varies and Tmax=0.5sec.

For the asymmetric load experiment, a metric called *fairness index* is used to measure the performance of the slave with asymmetric load. For example, let the average burst delay for slave i be d_i, $i = 1,...,7$. Then the fairness index for slave 1 is

$$f_i = 1 - \frac{d_i}{\frac{1}{N}\sum_{i=1}^{7} d_i} \tag{1}$$

Since the objective for both the average delay and the flush probability is to be minimal, the smaller the fairness index, the worse the performance. If the fairness index is negative, this means that the slave in question performs worse than the average. This means that the scheduling scheme ignores the high load of that slave and does not prioritise it in any way.

Overall, the results show that, when it comes to average burst delay, exhaustive scheduling outperforms the other two schemes even though they are supposed to be designed for constant-length burst traffic. This happens in both asymmetric and symmetric loading. When it comes to the flush rate, the MBDF scheme is the best. The Limited exhaustive scheme gives better delay performance than the MBDF scheme for symmetric load.

Fig. 6 and Fig. 7 show that Limited Exhaustive scheme performs poorly in the asymmetric load conditions. The other two schemes perform in a similar way, with MBDF better in the flush rate and Exhaustive scheme better in the average delay. It is important to note that both Exhaustive and MBDF scheduling generate positive fairness index, i.e. they both provide priority for the high-loaded slave. While this is just a consequence of the queue sizes for the Exhaustive scheme, the MBDF scheme clearly prioritises the high-loaded slave by serving it much more frequently than other slaves. This is a very interesting conclusion that deserves to be further investigated in the future.

For symmetric load, the difference in the performance between the Exhaustive and Limited Exhaustive scheme is minimal. Additional experiments will show what is the impact of longer bursts on this. For symmetric load, MBDF scheme performs much worse than the other two. The only aspect of performance where the MBDF scheme shows good results is in giving more priority to the heavy-loaded slave. More experiments are needed to find out how relevant is this and how beneficial is it.

Another interesting point that was noticed is that for very high symmetric load (>0.9), Limited Exhaustive scheme starts performing worse than the MBDF scheme, although in terms of the flush rate it remains dominant.

Overall, the results are somehow surprising, since the 'standard' exhaustive scheme has outperformed the other two in our experiments. Additional work is required and will be done in the further evaluation of this conclusion.

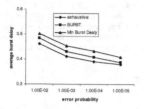

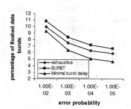

Fig. 2. Average burst delay, symmetric load 0.8

Fig. 3. Percentage of flushed bursts (flush time 1sec), load 0.8

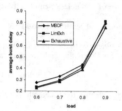

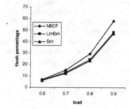

Fig. 4. Average burst delay, symmetric load

Fig. 5. Flush rate, symmetric load, Tmax=0.5sec

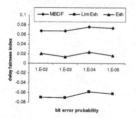

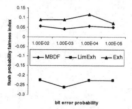

Fig. 6. Average delay fairness index, asymmetric load

Fig. 7. Flush rate fairness index, asymmetric load

5   Conclusion

This paper analysed the use of Bluetooth in sensor networks. Bluetooth is not a clear solution for sensor networks, because of slow connection establishment, high complexity and high power consumption in the process of moving from a low-power mode to a connected mode. However, with some modifications Bluetooth can be used in sensor networks. The paper analysed in detail scheduling schemes that can be used in Bluetooth-based sensor networks and showed some simulation results where the limited benefit of introducing sensor network – specific scheduling schemes was observed. Our preliminary conclusion is that the implementation of a scheduling scheme specially tailored for sensor network does not bring much benefit in terms of QoS. It is only in highly asymmetric loading environments that the new scheduling schemes proved to be beneficial.

References

1. Bluetooth specification, available at http://www.bluetooth.org
2. Bray J, Sturman C. F.: Bluetooth 1.1 Connect Without Cables. Prentice Hall, 2002
3. Akyildiz I. F, Su W. Sankarasubramaniam Y, Cayirci E.: Wireless Sensor Networks: A Survey. Elsevier Computer Networks, 38 (2002), 393-422
4. Leopold M, Dydensborg M. B., Bonnet P.: Bluetooth and Sensor Networks: A Reality Check. in Proc. of SenSys03 (2003)
5. Kasten O, Langheinrich M.: First Experiences with Bluetooth in the Smart-Its Distributed Sensor Networks, in Proc. of Workshop on Ubiquitous Computing and Communications, PACT01, Barcelona, Spain (2001)
6. Kammer D, McNutt G, Senese B, Bray J.: Bluetooth Application Developer's Guide, Syngress Publishing (2002)
7. Capone A., Gerla M., Kapoor R.: Efficient Polling Schemes for Bluetooth Picocells, in Proc. of IEEE ICC 2001, Helsinki, Finland (2001)
8. Das A., et al, "Enhancing Perofmance of Asynchronous Data Traffic over the Bluetooth Wireless Ad-hoc Network", in Proc. IEEE INFOCOM01 (2001)
9. Lapeyrie J-B., Turletti T.: FPQ: A Fair and Efficient Polling Algorithm with QoS Support for Bluetooth Piconet, in Proc. IEEE INFOCOM03, San Francisco, (2003).

A Distributed Algorithm to Provide QoS
by Avoiding Cycles in Routes

Juan Echagüe[1], Manuel Prieto[2], Jesús Villadangos[2], and Vicent Cholvi[1]

[1] Universidad Jaume I,
Campus del Riu Sec s/n,
12071 Castellón, Spain.
telephone: +349728261 fax: +349728486
{echague, vcholvi}@uji.es
[2] Universidad Pública de Navarra,
Departamento de Automática y Computación,
Campus de Arrosadia,
31006-Pamplona-Spain.
{manuel.prieto, jesusv}@unavarra.es

Abstract. We present a novel distributed algorithm which provides QoS by only enabling free-of-cycles routes which are known to ensure network stability. Network stability is synonymous of QoS as allows to deterministically bound maximum delays and backlogs. Cycles are avoided by disabling the use of pair of input-output links around a node (turns). This method improves network utilization compared to previous solutions, which avoids cycles by forbidding the use of whole links.
The algorithm can be applied in joint with any routing algorithm as does not require knowledge of the whole network topology, reducing the communication overhead compared to centralized approaches.
The performance of the proposed algorithm has been compared against a centralized optimization solution. Even though the centralized solution exhibits a lower percentage of prohibited turns, the difference is quite moderate. We have also shown how our protocol can be enhanced to be able to tolerate fail-stop node crashes without the necessity of having to start from the beginning.

1 Introduction

To provide Quality of Service (QoS) in packet switched networks, bounds on the maximum end-to-end delay have to be provided. It is known [1] that when no link is fully loaded, network stability (i.e. bounded backlogs at any node) may deterministically ensure maximum end-to-end delays and maximum backlogs (which avoids packet loss).

In the last few years, much of the analysis of stability has been done by using an *adversarial packet injection approach*, which was initially proposed by Borodin et al. [2] and Andrews et al. [1]. The adversarial model is a worst case packet injection model in which the adversary injects packets and chooses the route for each packet. The adversarial model was an improvement to the use

J. Solé-Pareta et al. (Eds.): QofIS 2004, LNCS 3266, pp. 224–236, 2004.
© Springer-Verlag Berlin Heidelberg 2004

of probabilistic packet injection schemas as reduces the whole set of possible injection scenarios. In a seminal paper, by using and *adversarial model*, Andrews et al. [1] provides a list of universally stable packet scheduling policies (i.e. policies that guarantee network stability if the traffic load at each link is lower or equal to the link's capacity). They showed that policies such as *Farthest-to-Go* (FTG), *Nearest-to-Source* (NTS), *Shortest-in-System* (SIS) and *Longest-in-System* (LIS) are universally stable. In contrast, they also showed that packet scheduling policies like *First-in-First-Out* (FIFO), *Last-in-First-Out* (LIFO), *Nearest-to-Go* (NTG) and *Farthest-from-Source* (FFS) are not universally stable. Furthermore, LIFO, NTG and FFS can be made unstable at arbitrarily low load rates [2]. Recently, it has also been shown that the same applies to FIFO [3,4], the scheduling policy analyzed in this paper which is, by far, the most widely used to schedule packets.

Based on the knowledge of the routes, Echagüe et al. [5] shows that any *work conserving* scheduling policy (i.e. any policy that always schedule packets if there is anyone in the queue) is stable if the load rate is not bigger than $\frac{1}{d}$ (where d is the largest number of links crossed by any packet). They also reduced that bound to $\frac{1}{d-1}$ for *system-wide time priority* scheduling policies (i.e., policies under which, a packet arriving at a buffer at time t has priority over any other packet that is injected after time t).

On the other hand and regarding the case of session oriented networks (i.e. networks where all packets belonging to the same session follow the same route), in [6] Andrews provides an example which shows that FIFO can be unstable under the (σ, ρ)-regulated session–model of Cruz [7]. However, it turns out that, in the case of session oriented networks, if the total load at each link is smaller than 1, then, any network where routes do not create cycles of interdependent packets is stable [7].

A technique used to guarantee that no cycles will appear consists of prohibiting the use of some network resources. A simple approach to transform a graph into a cycle-free graph is to construct a spanning tree and prohibit the use of links not belonging to the tree. Whereas a spanning tree keeps the graph connectivity, it is quite inefficient since it prohibits the use of whole links and links close to the root get congested.

Another approach consists of using the *up/down routing algorithm* [8]. The up/down routing algorithm constructs a graph, based on a spanning tree, and instead of prohibiting the use of determined links, it prohibits the use of pair of links around a node (turn). This algorithm performs better than the spanning tree but also suffers of unbalancing the load; furthermore, the performance of the resulting topology depends on the initial spanning tree.

Starobinski et al. [9] propose an optimization solution (called Turn–Prohibition *TP*) to break cycles by prohibiting the use of network *turns* . Their performance analysis shows that *TP* outperforms the spanning tree algorithm. Also, they show that the maximum number of prohibited turns in the network is a third of the total number of turns. However, implementing *TP* algorithm is centralized and requires knowledge of the whole network topology.

In this paper, we propose a new algorithm that guarantees that routes do not create cycles. Similar to Starobinski et al. [9], it breaks cycles by prohibiting the use of network turns. However and contrary to the *TP* algorithm, our algorithm (which we call Distributed Turn Prohibition *DTP*) does not require that nodes has got whole knowledge of the network topology. Although *TP* performs better than *DTP* in terms of number of prohibited turns (which is foreseeable since one has a global knowledge of the network topology and the other not), we show that the difference is quite moderate (less than 5%). Furthermore, we also present an extension of *DTP* that tolerates network failures in a dynamic fashion.

Our work is arranged as follows. In Section 2 we present the system model and the distributed turn prohibition (DTP) algorithm, as well as an application example. Section 3 provides a formal proof of correctness. In Section 4 we make a performance analysis and in Section 5 we present a self–configuring version of the *DTP* algorithm that tolerates network failures. Finally in Sections 6 we present our conclusions and point some future work.

2   Distributed Turn Prohibition Algorithm (DTP)

We model the network topology as a graph $G = (V, E)$, composed by a set of vertices V and a set of edges E. They represent network nodes and bidirectional links. Thus, we use interchangeably the terms node and vertex, and the terms link and edge.

We denote an edge between node n_i and n_j as $e_{i,j}$. A path is defined as a list of nodes $\{n_i, n_j, n_k \ldots, n_p\}$, such that $e_{i,j}, e_{j,k}, \ldots \in E$. It is said that a packet follows a path with a *cycle* in G if the path contains at least twice the same edge. For instance, a packet following the path $\{0, 1, 2, 4, 0, 1\}$ at the graph of Figure 1, traverses twice edge $e_{0,1}$. Note that a packet following a *path* can traverse several times the same node without creating a cycle.

We say that packets or flows are interdependents if they share at least a link in its respective paths. A *path of interdependent packets* can be obtained following a packet's path and enabling to switch to another packet's path at the shared links of the interdependent packets. A set of interdependent packets can create a *cycle of interdependent packets* in G if, starting from a node belonging to the set of nodes traversed by one of those packets, it is possible to obtain a *path of interdependent packets* which contains at least twice the same link. For example in the graph of the Figure 1 a packet from flow F0 follows path $\{0, 1, 2\}$, a packet from flow F1 follows path $\{1, 2, 3\}$, a packet from flow F2 follows path $\{2, 3, 4\}$ and a packet from flow F4 follows path $\{4, 0, 1\}$. The four packets create a *cycle of interdependent packets* with path $\{0, 1, 2, 3, 4, 0, 1\}$.

From now onwards, when we refer to cycles, we mean cycles composed by a single packet or cycles of interdependent packet flows.

Note that the graph shown in Figure 1 is not unstable, although it contains a cycle. The existence of cycles does not imply network instability, as it has been shown that a ring topology is universally stable. The non-existence of cycles implies network stability [7].

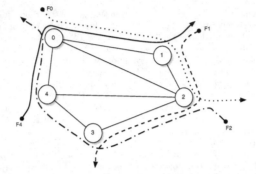

Fig. 1. *Cycle of interdependent packet flows* the path which contains the cycle is $\{0, 1, 2, 3, 4, 0, 1\}$. There are four flows. Flow $F0$'s path is $\{0, 1, 2\}$, flow $F1$'s path is $\{1, 2, 3\}$, flow $F2$'s path is $\{2, 3, 4\}$ and Flow $F4$'s path is $\{4, 0, 1\}$.

A pair of input-output links around a node is called a *turn*. We represent a turn around n_j by the three-tuple (n_i, n_j, n_k), where $n_i, n_j, n_k \in V$ and $e_{i,j}, e_{j,k} \in E$. The prohibition of turn (n_i, n_j, n_k) means that no packet can be forwarded from node n_i to node n_k by means of node n_j, or from node n_k to node n_i by means of node n_j. That is, no packet can traverse link $e_{j,k}$ after traversing $e_{i,j}$ or, in the other direction, no packet can traverse link $e_{i,j}$ after traversing link $e_{j,k}$.

An efficient approach for breaking cycles in a network is based on the *prohibition* of turns in opposite to the prohibition of full links. Next we provide a novel distributed algorithm which provides network stability by the prohibition of turns in a network topology.

2.1 Distributed Turn Prohibition Algorithm

In this section, we propose a distributed turn–prohibition algorithm, called DTP, that guarantees that (i) all cycles within the network will be broken and (ii) that the network will remain connected after algorithm completion.

Our algorithm performs two independent tasks: (i) it builds a spanning tree ensuring that all nodes execute the algorithm, and, (ii) it forbids some turns so that the resulting network topology is free of cycles. In order to guarantee that nodes will remain connected, those forbidden turns cannot be present in the spanning tree.

The DTP algorithm (see Figure 2) uses the data structures shown in Table 1 and acts as follows: the "initiator" or $ROOT$ node (which can be any node) issues a GO message to initiate a depth search. When a node receives a GO message for the first time, it sets the sender as its father and marks it as "explored".

Table 1. DTP algorithm data structures

Variable	Description
V_i	Set of nodes directly connected to node i.
$father_i$	Variable which stores a node identity or the value $ROOT$ for the algorithm's initiator node.
$explored_i$	Boolean variable. Takes the value *true* if i has already received a GO message, and *false* otherwise.
in_i	Node that sent the last GO message to i.
out_i	Last selected node to forward the GO message from i.
$revisedNodes_i$	Set of nodes that have sent or received a GO message from i. Initially $\emptyset$.
$selectOut(S)$	Selects an element from the set of nodes S.
$backList_i$	List of nodes. Initially $\emptyset$.
Q	Ordered set of type $\langle path, turn \rangle$. Initially $\emptyset$.
$update(Q, list\_of\_nodes, turn)$	Appends "$list\_of\_nodes$" to the first element of the last element in Q, and "turn" to the second element of the last element in Q.
$newTopology(Q)$	Creates a topology with the information of the paths contained in Q. That is, the topology contains all the nodes and links present in Q's paths.
$checkCycle(Q, L, i, n)$	Looks for cycles in $newTopology(Q)$, not already broken by the turns belonging to Q, that connects the nodes in the list $L \cdot i \cdot M \cdot n$, where M is a list of path that can be empty. It returns a set of type $\langle path, turn \rangle$, where each path exhibits a cycle broken by its associated turn.

Furthermore, it updates the elements of the GO message and forwards it to one of its adjacent "non-explored" nodes. If it is not the first time it receives a GO message (a cycle has been detected), it replies with a BACK message.

When a node receives a BACK message it forbids the turn formed by the node that sent him the BACK message and the node that sent him the last GO message. Then, it continues forwarding a GO message toward a link that has not been explored yet. When all links have been explored, the node sends back to its father a SEARCH message.

A SEARCH message can be interpreted as the fact that the son has finished to explore its successors and returns the topology it has gathered. Such information is useful in order to check for possible cycles present in the topology recollected by its son. Once the node has checked the topology, and forbidden the cycles found, it continues with the transmission of GO messages, as explained above (in the case where there are still nodes to be explored), or sends a SEARCH message to its father (if all nodes have been explored).

The algorithm ends when the initiator has explored all its successors. That is, when all links present in the topology have been explored.

Regarding the number of messages required to fully complete the DTP algorithm, it linearly depends on the number of links of the network (i.e. the cost of DTP in terms of number of messages is $O(|E|)$). This can be readily seen

```
initiate_i
    father_i ← ROOT;
    explored_i ← true;
    in_i ← ∅;
    out_i ← SelectOut(V_i);
    revisedNodes_i ← out_i;
    send message [GO, i, {{i}{∅}}] to out_i.


rcv_GO_j(γ, Q):: activated when node i receives
the message [GO, γ, Q] from node j
    in_i ← j;
    revisedNodes_i ← revisedNodes_i ∪ in_i;
    IF (explored_i = false) THEN
        explored_i ← true;
        father_i ← j;
        backList_i ← γ;
        out_i ← selectOut(V_i − revisedNodes_i);
        IF (out_i ≠ ∅) THEN
            revisedNodes_i ← revisedNodes_i ∪ out_i;
            update(Q, i, ∅);
            send message [GO, backList_i · i, Q] to out_i;
        ELSE
            send message [SEARCH, Q] to father_i;
    ELSE
        update(Q, i, ∅);
        send [BACK, Q] to j.
```

```
rcv_BACK_j(Q):: activated when node i receives
the message [BACK, Q] from node j
    forbid_turn (in_i, i, out_i);
    update(Q, ∅, (in_i, i, out_i));
    IF (V_i − revisedNodes_i = ∅) THEN
        IF (father_i ≠ ROOT) THEN
            send message [SEARCH, Q] to father_i;
    ELSE
        out_i ← selectOut(V_i − revisedNodes_i);
        revisedNodes_i ← revisedNodes_i ∪ out_i;
        Q = Q ∪ (∅, ∅);
        update(Q, backList_i · i, ∅);
        send message [GO, backList_i · i, Q] to out_i;


rcv_SEARCH_j(Q):: activated when node i receives
the message [SEARCH, Q] from node j
    IF (V_i − revisedNodes_i = ∅) THEN
        for all n ∈ (V_i − {father_i, out_i}) do
            T = checkCycle(Q, backList_i, i, n)
            for each < PATH, TURN >∈ T do
                forbid_turn (TURN);
                Q = Q ∪ (∅, ∅);
                update(Q, PATH, TURN);
        IF (father_i ≠ ROOT) THEN
            send message [SEARCH, Q] to father_i;
    ELSE
        out_i ← selectOut(V_i − revisedNodes_i);
        revisedNodes_i ← revisedNodes_i ∪ out_i;
        Q = Q ∪ (∅, ∅);
        update(Q, backList_i · i, ∅);
        send message [GO, backList_i · i, Q] to out_i;
```

Fig. 2. DTP algorithm

by observing Theorem 1 in the next section which shows that the maximum number of messages exchanged when applying the DTP algorithm is at most of $2 \cdot |E|$. Note that the TP algorithm really solves an optimization problem whose implementation requires a global knowledge of the system. Consequently, its implementation will require a higher number of messages than DTP.

2.2   Example of Application of the DTP Algorithm

To illustrate how the DTP algorithm works, we provide an example of application to the graph of Figure 3-a. We focus our attention on the messages exchanged by nodes into the network. Note that the given algorithm behavior is one of all the possible ones.

We can find the following cycles in the graph of Figure 3-a: $\{0, 3, 2, 0, 3\}$, $\{0, 3, 1, 0, 3\}$, $\{0, 3, 2, 1, 0, 3\}$, $\{0, 3, 1, 4, 0, 3\}$, $\{0, 1, 4, 0, 1\}$, $\{0, 1, 2, 0, 1\}$, $\{0, 4, 1, 2, 0, 1\}$ and $\{1, 3, 2, 1, 3\}$.

Let us assume that the algorithm is started by $node\ 1$. After updating its local data structures, it sends the message $GO(\{1\}, \{\langle 1, \emptyset \rangle\})$ to $node\ 0$, which will send the message $GO(\{1, 0\}, \{\langle\langle 1, 0\rangle, \emptyset\rangle\})$ to $node\ 4$. In turn, it will send message $GO(\{1, 0, 4\}, \{\langle\langle 1, 0, 4\rangle, \emptyset\rangle\})$ to $node\ 1$. As $node\ 1$ has already been explored, it sends a BACK message to $node\ 4$. On the reception of such a message, $node\ 4$ will prohibit the turn $(0, 4, 1)$; $node\ 4$ will not route any message received from link $e_{0,4}$ through link $e_{4,1}$, nor from link $e_{4,1}$ to link $e_{0,4}$. The prohibition of this

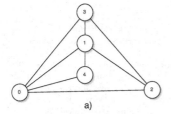
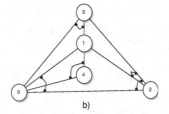

a)   b)

Fig. 3. Example of utilization of the DTP algorithm. The graph in **a)** can produce paths with cycles. After applying DTP algorithm the topology shown in **b)** is free of cycles keeping network connectivity. Curved lines with dots at both sides represents the prohibited turns around a node (each dot touches a link of the prohibited turn). For example it is not possible that a packet follows path $\{0, 4, 1\}$.

turn does not isolate node $node\ 4$ as it is still possible to receive packets targeted to it and send packets from it.

For simplicity, from now onwards, we omit the data contained in the exchanged messages. As $node\ 4$ has already explored all its links, it sends a SEARCH message to $node\ 0$ (who is its father). $Node\ 0$ will send a GO message to $node\ 3$. $Node\ 3$ will send a GO message to $node\ 1$. As $node\ 1$ has already been explored, it will reply to $node\ 3$ with a BACK message. On the reception of such a message, $node\ 3$ will prohibit the turn $(0, 3, 1)$.

$Node\ 3$ continues the depth search by sending a GO message to $node\ 2$, which sends a GO message to $node\ 0$, that in turn responds (to $node\ 2$) with a BACK message. On the reception of such a message, $node\ 2$ prohibits turn $(3, 2, 0)$.

There are yet unexplored links in $node\ 2$. So, it sends a GO message to $node\ 1$, which responds with a BACK message, prohibiting turn $(3, 2, 1)$.

As $node\ 2$ has explored all its links, it sends a SEARCH message to $node\ 3$ (its father). The value of Q received by $node\ 3$ is $Q = \{< \{1, 0, 4, 1\}, \{0, 4, 1\} >, < \{1, 0, 3, 1\}, \{0, 3, 1\} >, < \{1, 0, 3, 2, 0\}, \{3, 2, 0\} >, < \{1, 0, 3, 2, 1\}, \{3, 2, 1\} >\}$. At the reception of the SEARCH message, $node\ 3$ executes function $checkCycle(Q, < 1, 0 >, 3, n)$,for $n \in \{1, 2\}$. It has to find a cycle containing the pattern: $P =< \delta_x \cdot 1 \cdot 0 \cdot 3 \cdot M \cdot n \cdot \delta_y >$, where δ_x, δ_y and M are any valid subpath obtained from Q's paths and P is a valid path. In this case no cycle is found. $Node\ 3$ sends a sends a SEARCH message to $node\ 0$.

$Node\ 0$ performs function $checkCycle(Q, < 1 >, 0, n)$ for $n \in \{2, 3, 4\}$, detecting cycle $\{1, 0, 2, 1\}$ which is avoided by the prohibition of turn $(1, 0, 2)$. Following it sends a SEARCH message to $node\ 1$ which performs checks to test if there still are undetected cycles and concludes the algorithm.

The algorithm has forbidden a total of 5 turns as shown in Figure 3-b. In this new topology cycles can not be formed and connectivity is kept.

3   Correctness Proof

The *DTP* algorithm satisfies two important properties: (1) by forbidding turns it avoids cycles in the network and (2) the network remains connected after algorithm completion.

To verify the above properties, we check some general considerations that the algorithm must verify. First, we prove that the algorithm is of the traversal kind described by G. Tel [10]. Then, it will be easy to prove that the algorithm execution defines a spanning tree in the system. In order to verify that the system remains connected, we prove that the spanning tree is always a way to transfer messages between the whole nodes of the system. That is, turns can not be formed by links of the spanning tree.

Theorem 1. *The proposed algorithm is a traversal algorithm*

Proof: Because the GO message is sent at most once in each direction through each link, it is obviously sent at most $2 \cdot |E|$ times before the algorithm terminates. Because each process sends the GO message through each link at most once, each process receives a BACK or a SEARCH message through each link at most once. Each time the GO message is held by a non-initiator p, node p has received the GO message once more than p has sent the message. This implies that the number of links incident to p exceeds the number of used links of p by at least one, so p does not decide, but forwards the GO message. It follows that the decision takes place in the initiator. It will be proved in three steps than when the algorithm terminates each node it has forwarded the GO message.

1. *All links of the initiator have been used once in each direction.* Each link has been used by the initiator to send the GO message or to respond to such a message forwarded by one of its neighbors, otherwise the algorithm does not terminate. The initiator has received a SEARCH or a BACK message exactly as often as it has sent an GO message and it has sent a BACK message exactly as often as it has received a GO message; because it has been received through a different link each time, it follows that links have been used once in both directions.

2. *For each visited node p, all links of p have been used once in each direction.* Assuming this is not the case, let us choose p to be the earliest visited node for which this property is not true. We observe that by point (1) p is not the initiator. By the choice of p, all links incident to $father_p$ have been used once in each direction, which implies that p has sent the SEARCH message to its father. This implies that p has used all its links to send or receive the GO message; but as the algorithm terminates when the initiator receives the message and all its links have been explored, p has transmitted a message in both directions through all its links. This is a contradiction.

3. *All processes have been visited and each link has been used once in both directions.* If there are unvisited nodes, there are neighbors p and q such that p has been visited and q has not been visited, contradicting that each link of p has been used in both directions. So, by point (2), all nodes were visited, and all links are used once in both directions. ∎

Traversal algorithms are one kind of *wave algorithms* [10] and verify some important properties. One of them shows that each computation of the algorithm defines a spanning-tree in the network topology, as shown in the following lemma. The root of the tree is the initiator, and each non-initiator p has stored its father in the tree in the variable $father_p$ at the end of the computation.

Lemma 1. *Let C be a computation with one initiator p and, for each non-initiator q, let $father_q$ be the neighbor of q from which q received a depth search message (a GO message in our proposal) in its first event. Then, the graph $T = (N, E_T)$, with N the set of nodes and $E_T = \{qr : q \neq p \wedge r = father_q\}$ is a spanning tree directed towards p.*

Proof: The proof is equivalent to that presented in [10] by considering that the depth search message received in its first event is a GO message.   ∎

The proposed algorithm belongs to the class of traversal algorithms but we must prove that the algorithm breaks all cycles of nodes and that the graph remains connected.

Theorem 2. *The algorithm breaks all cycles of the system.*

Proof: As the proposed algorithm belongs to the traversal kind, theorem 1, the GO message will gather the whole network topology, which is backward transmitted by SEARCH messages to the root node following the spanning-tree defined by the nodes of the system, lemma 1. Each node of the network receives at least once the GO message and sends backward a SEARCH message. Before a SEARCH message is sent, the node has checked if it belongs to cycles —*checkCycle()*— in the network topology that connects the node with all its reachable nodes along the spanning tree. Each node, in case it detects the presence of cycles, will create turns ensuring that the node does not belong to any cycle of nodes. In addition, such new information is included in the SEARCH message backward transmitted.

Each node sends a SEARCH message until it is received at the initiator of the computation and the computation terminates (theorem 1). By handling the SEARCH message at the initiator, it checks for cycles (*checkCycle()*) and prohibits the cycles of nodes present in the network. The initiator will create turns for each cycle of nodes that were not previously detected by other nodes in the network, because the message contains the complete network topology and turns defined in the system.   ∎

Lemma 2. *A node does not create turns between its father and son in the spanning-tree defined by the algorithm computation.*

Proof: For a given node, a turn is created after handling a BACK or SEARCH message. According to the algorithm specification it is impossible that a son sends a BACK to its father. SEARCH messages can only be sent by the sons of the node, but in any case the father of the node is not taken into account to search for cycles (see actions associated upon SEARCH reception) and it is verified for all the nodes of the system.   ∎

Table 2. Percentage of prohibited turns between *TP* and *DTP* algorithms when varying the number of nodes. The network degree is 4.

Number of Nodes	Percentage of prohibited turns of *TP*	Percentage of prohibited turns of *DTP*	Difference of percentages (%*DTP* − %*TP*)
16	18.7	23.0	4.3
32	17.0	21.0	4.0
64	17.4	19.6	2.2
128	16.8	18.7	1.9
255	16.7	18.5	1.8

Theorem 3. *The system remains connected*

Proof: By theorem 2, all cycles will be broken and by lemma 2, after algorithm execution for each node of a cycle there is at least one path to reach each node of the cycle. So, the only possibility to be unconnected is to have a node that has turns defined for all its links, which is impossible by lemma 2 and the theorem is verified. ∎

4 Performance Evaluation

In this section we compare the performance of the *DTP* and the *TP* algorithms by means of performing massive simulations.

Regarding the topologies we consider in our analysis, we have used the GT-ITM graph generator [11,12] since it permits us to generate different models of graphs that match current Internet topologies. For any concrete simulation, we generate 99 different topologies (33 flat random, 33 hierarchical and 33 transit-stub) and average the obtained results.

The simulation experiments compare the percentage of prohibited turns using the *TP* algorithm against the *DTP* algorithm. By forehand, we have in mind that *TP* algorithm will perform better than the *DTP* algorithm, since it has knowledge of the whole network topology.

In our first experiment, we analyze how the increase of the number of nodes affects the percentage of prohibited turns between both algorithms. We have generated topologies with a fixed network degree (i.e. the average of the nodes degree) of 4. The number of nodes ranges from 16 to 255. Table 2 shows the obtained results. As it can be seen, the *TP* algorithm prohibits less than 5% fewer turns than the *DTP* algorithm, which is quite moderate.

In a second experiment, the number of nodes has been fixed to 120 and the network degree has been varied from 4 to 10. The results in Table 3 show that, when increasing the network degree, the number of prohibited turns increases in both algorithms. This is clear since, having a high network degree, increases the number of turns in the system and consequently the number of cycles. The

Table 3. Percentage of prohibited turns between *TP* and *DTP* algorithms varying network degree. Number of nodes fixed to 120.

Network degree	Percentage of prohibited turns of *TP*	Percentage of prohibited turns of *DTP*	Difference of percentages $(\%DTP - \%TP)$
4	16.7	18.7	2.0
6	20.0	22.4	2.4
8	24.5	27.5	3.0
10	25.0	28.4	3.4

results show that, even though the *TP* algorithm prohibits less turns than the *DTP*, the difference is also quite moderate (less than 4%).

5   Self–Configuring Distributed Turn Prohibition Algorithm *SDTP*

One important issue in computer networks is the capability to dynamically adapt to the failure of nodes or links. In this section, we describe how the *DTP* algorithm can be enhanced to tolerate fail-stop node crashes without the necessity of having to start the algorithm from the beginning.

1. When a node handles a GO message for the first time labels itself in an ordered fashion.
2. If a node with label l fails (in a fail–stop fashion) do:
   a) Mark as "invalidated" all nodes with a label higher than l.
   b) Look for the node with the highest label, lower than l, linked with any of the "invalidated" nodes. Apply *DTP* starting at this node.
   c) If there are still "invalidated" nodes, repeat the previous step until there are not "invalidated" nodes (connected with "non–invalidated" nodes).

The correctness proof of this new algorithm is almost the same as in the *DTP* (so, we omit it). Due to the node failures it may happen that nodes get isolated. Clearly, in that case it is not possible to guarantee that the network will remain connected.

In Figure 4, we present an application that shows how *SDTP* behaves. For the sake of clarity, we have labeled each vertex with two labels, the original vertex number and the label given by the *SDTP* algorithm as a letter (letters are in alphabetical order).

In a similar way, one can provide an algorithm to add new nodes (or subgraphs) to a *SDTP* conformant graph. An intuitive first approach will consist in: (1) determine which is the node with the lowest label sharing a link with any of the new nodes, (2) invalidate all the nodes with label equal or higher that this node and (3) apply *SDTP* from the above mentioned node.

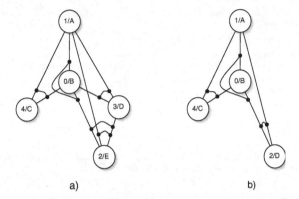

Fig. 4. Example of the behavior of the *SDTP* algorithm. Figure 4.a represents the topology of Figure 3.b. In case of failure of the node $3/D$, invalidate node $2/E$ (its label is higher than node $3/D$). Since the node with the highest label connected to node $2/E$ is node $0/B$, apply *SDTP* starting from node $0/B$. The resulting topology after recovery from node $3/D$ failure is showed in Figure 4-b. Note that the label given to node 2 has changed to $2/D$.

6   Conclusions and Future Work

We have presented a novel distributed algorithm which guarantees network stability by means of forbidding turns without loosing network connectivity. The performance of the algorithm has been compared against the *TP* protocol and we found that the difference between them (in terms of prohibited turns) is quite moderate. However, on the contrary than *TP* whose implementation requires a global knowledge of the system, *DTP* is a fully distributed algorithm that does not require a global knowledge of the network. Finally, we have enhanced the *DTP* protocol to tolerate fail-stop node crashes without the necessity of having to start the algorithm from the beginning.

At this moment we are working on a version of the algorithm which permits multiple initiator nodes.

References

1. Andrews, M., Awerbuch, B., Fernandez, A., Leighton, T., Liu, Z., Kleinberg, J.: Universal-stability results and performance bounds for greedy contention-resolution protocols. In: Journal of the ACM. (2001) 39–69
2. Borodin, A., Kleinberg, J., Raghavan, P., Sudan, M., Williamson, D.: Adversarial queueing theory. Journal of the ACM (2001) 13–38

3. Bhattacharjee, R., Goel, A.: Instability of fifo arbitrarily low rates in the adversarial queueing model. Technical Report Technical Report No. 02–776, Department of Computer Science, University of Southern California (2002)
4. Koukopoulos, D., Mavronicolas, M., Spirakis, P.: Fifo is unstable at arbitrarily low rates (even in planar networks). Technical Report Revision 1 of ECCC Report TR03–016, Electronic Colloquium on Computational Complexity (2003)
5. Echagüe, J., Cholvi, V., Fernández., A.: Universal stability results for low rate adversaries in packet switched networks. IEEE Communication Letters **7** (2003)
6. Andrews, M.: Instability of FIFO in session-oriented networks. In: Proceedings of the Eleventh Annual ACM-SIAM Symposium on Discrete Algorithms, N.Y., ACM Press (2000) 440–447
7. Cruz, R.: A calculus for network delay. Parts i and ii. IEEE Trans. Inform. Theory, vol 37-1 (1991) 114–131 and 131–141
8. Shoreder, M.: Autonet: A high-spedd, self-configuring local area network using point-to-point links". IEEE JSAC **9** (1991) 1318–1335
9. Starobinski, D., Karpovsky, M., Zakrevski, L.A.: Application of network calculus to general topologies using turn-prohibition. IEEE/ACM Transactions on Networking (TON) **11** (2003) 411–421
10. Tel, G.: Introduction to Distributed Algorithms. Ed. Cambridge Press (1994)
11. Ken Calvert, M.D., Zegura, E.W.: Modeling internet topology. IEEE Communications Magazine (1997)
12. Ellen W. Zegura, K.C., Donahoo, M.J.: A quantitative comparison of graph-based models for internet topology. IEEE/ACM Transactions on Networking (1997)

The Case for Source Address Dependent Routing in Multihoming[*]

Marcelo Bagnulo, Alberto García-Martínez, Juan Rodríguez, and Arturo Azcorra

Universidad Carlos III de Madrid
Av. Universidad, 30. Leganés. Madrid. España
{marcelo,alberto,jrh,azcorra}@it.uc3m.es

Abstract. Multihoming is currently widely adopted to provide fault tolerance and traffic engineering capabilities. It is expected that, as telecommunication costs decrease, its adoption will become more and more prevailing. Current multihoming support is not designed to scale up to the expected number of multihomed sites, so alternative solutions are required, especially for IPv6. In order to preserve interdomain routing scalability, the new multihoming solution has to be compatible with Provider Aggregatable addressing. However, such addressing scheme imposes the configuration of multiple prefixes in multihomed sites, which in turn causes several operational difficulties within those sites that may even result in communication failures when all the ISPs are working properly. In this note we propose the adoption of *Source Address Dependent* routing within the multihomed site to overcome the identified difficulties.

1 Introduction

Since the operations of a wide range of organizations rely on communications over the Internet, access links are a critical resource to them. As a result, sites are improving the fault tolerance and QoS capabilities of their Internet access through *multi-homing*, i.e. the achievement of global connectivity through several connections supplied by different Internet Service Providers (ISPs). However, the extended usage of the currently available IPv4 multi-homing solution is jeopardizing the future of the Internet, since this use has become a major contributor to the post-CIDR growth in the number of global routing table entries [1]. Therefore, in IPv6 the usage of *Provider Aggregatable (PA)* addressing is recommended for all sites, included multihomed ones, in order to preserve inter domain routing system scalability. While such addressing architecture reduces the amount of routing table entries in the *Default Free Zone* of the Internet, its adoption presents a fair amount of challenges for the end-sites, especially for those who multihome. Essentially, when PA addressing is adopted, a multihomed site will have to configure multiple addresses, one per ISP, in every node of the site, in order to be reachable through all its providers. Such configuration pose quite a few number of challenges for its adoption, since current

---

[*] This work has been partly supported by the European Union under the E-Next Project FP6-506869 and by the OPTINET6 project TIC-2003-09042-C03-01

J. Solé-Pareta et al. (Eds.): QofIS 2004, LNCS 3266, pp. 237–246, 2004.
© Springer-Verlag Berlin Heidelberg 2004

hosts are not prepared to deal with multiple addresses per interface as it is required. In this note, we will present how *Source Address Dependent (SAD)* routing can be adopted to deal with some of the difficulties present in this configuration.

The rest of this paper is structured as follows: First we will present the rationale for adopting SAD routing within multihomed sites. Then, we will detail the different configurations of SAD routing that may be required in different sites, including some trials performed, and next we will present the capabilities of the resulting configuration. Finally we will present the conclusions of this work.

2   Rationale

2.1   Current IPv4 Multihoming Technique and Capabilities

As mentioned above, a site is multi-homed when it obtains Internet connectivity through two or more service providers. Through multi-homing an end-site improves the fault tolerance of its connection to the global network and it can also perform *Traffic Engineering* (hereafter *TE*) techniques to select the path used to reach the different networks connected to the Internet.

In IPv4, the most widely deployed multi-homing solution is based on the announcement of the site prefix through all its providers. In this configuration, the site S obtains a *Provider Independent (PI)* prefix allocation directly from the Regional Internet Registry. Then, the site announces this prefix to its providers using BGP [2]. Then the multihomed site providers announce the prefix to its own providers and so on, so that eventually the route is announced in the *Default Free Zone*.

This mechanism provides fault tolerance capabilities, including preserving established connections throughout an outage. In addition, the following TE tools are available to the multihomed site: The multihomed site can define which one of the available exit paths will be used to carry outgoing traffic to a given destination by proper configuration of the LOCAL_PREFERENCE attribute of BGP [3]. For incoming traffic, the multihomed site can influence the ISP through which it prefers to receive traffic by using the *AS prepending* technique, which consists in artificially making the route through one of the providers less attractive to external hosts by adding AS numbers in the AS_PATH attribute of BGP [3] (it should be noted that in this case, the ultimate decision of which ISP will be used to forward packets to the site belongs to the external site that is actually forwarding the traffic).

While the presented IPv4 multihomed solution provides fairly good features regarding to fault tolerance and TE, it presents very limited scalability properties with respect to the interdomain routing system. Because of the usage of PI addressing by the multihomed sites, each multi-homed site using this solution contributes with routes to the Default Free Zone routing table, imposing additional stress to already oversized routing tables. For this reason, more scalable multi-homing solutions are being explored for IPv6 [4], in particular solutions that are compatible with the usage of PA addressing in multihomed sites, as it will be presented next.

2.2 Provider Aggregation and Multi-homing

In order to reduce the routing table size, the usage of PA addressing is required. This means that sites obtain prefixes which are part of their provider's allocation, so that its provider only announce the complete aggregate to their providers, and they do not announce prefixes belonging to other ISP aggregates, as presented in figure 1.

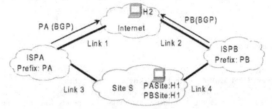

Fig. 1. Provider aggregation of end-site prefixes

When provider aggregation of end-site prefixes is used, each end-site host interface obtains one IP address from each allocation, in order to be reachable through all the providers and benefit from multi-homing capabilities (note that ISPs will only forward traffic addressed to their own aggregates).

This configuration presents several concerns, as it will be presented next.

- Difficulties in the communication in case of failure. When *Link1* or *Link3* becomes unavailable, addresses containing the *PASite* prefix are unreachable from the Internet.
- Ingress filtering [5] is a widely used technique for preventing the usage of spoofed addresses. However, in the described configuration, its usage presents additional difficulties for the source address selection mechanism and intra-site routing systems, since the exit path and source address of the packet must be coherent with the path, in order to bypass ingress filtering mechanisms.
- Established connections will not be preserved in case of outage. If *Link1* or *Link3* fails, already established connections that use addresses containing *PASite* prefix will fail, since packets addressed to the *PASite* aggregate will be dropped because there is no route available for this destination. Note that an alternative path exists, but the routing system is not aware of it.

The presented difficulties show that additional mechanisms are needed in order to allow the usage of PA addresses while still provide incumbent multi-homing solution equivalent benefits. In this note, we will explore the possibility of using Source Address Dependent routing as a tool to help to overcome the identified difficulties.

3 Source Address Dependent (SAD) Routing

Source Address Dependent (SAD) routing essentially means that routers maintain as many routing tables as source address prefixes involved, and packets are routed according to the routing table corresponding to the source address prefix that best matches the source address contained in the packet header.

SAD routing can be used to provide ingress filtering compatibility for routing packets flowing from the multihomed site to the Internet. In this case, the source address of the exiting packets has been determined by the host that initiated the communication (the host in the multihomed site, or the external host through the selection of the destination address of the initial packet) and then the routing system will forward the packet to the appropriate exit router in order to guarantee ingress filtering compatibility. The source address selection determines the ISP to be used for routing packets, since, because of address filtering, the source address determines the forward path from the multihomed site to the rest of the Internet, and it also determines the ISP to be used in the reverse path, since the source address used in the initial packets will become the destination address of the reply packets.

Since source address selection implies ISP selection, the adoption of SAD routing will also affect the mechanisms to be used in multihomed sites to define TE. In particular, it will shift TE capabilities from the routing system to the hosts themselves.

We will next evaluate the adoption of SAD routing in two typical multihomed configurations: sites running BGP but without redistributing the BGP information into an IGP, and sites running an IGP to select the exit path. There is an additional possible configuration using static routes in the multihomed site. However, this last configuration is fairly simple and several commercial routers already support it, so we won't provide a full description of it. Nevertheless, it should be noted that when SAD routing is used, it is possible to obtain fault tolerance and TE capabilities without requiring dynamic routing, since those features are now supported by the hosts themselves and not by the routing system.

In order to enable SAD routing within a site, SAD routing support is not required in all the routers within the site, but it has to be adopted in a connected SAD routing domain that contains all the exit routers [6], as presented in the figure below.

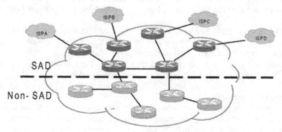

Fig. 2. SAD routing domain

Note that it is not necessary for the generic routing domain to be connected, i.e. it can be formed by a set of disconnected domains, all connected to the SAD routing domain.

3.1 Sites Running BGP but Without Redistribution of BGP Information into IGP

Current IPv4 multihomed sites usually run BGP with their providers. Through BGP, they obtain reachability information from each of their ISPs. However, because of operational issues, some sites do not redistribute the information obtained through BGP into the IGP [3]. So, in order to be able to properly select the intra site path towards an external destination, they include all the routers that are required to properly select the exit path in the IBGP mesh, including not only site exit routers, but also other internal routers that have access to multiple exit routers. This means that the IBGP cloud is wrapping the non-BGP aware routing domain, as presented in figure 3.

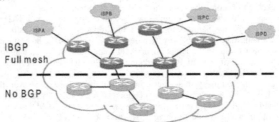

Fig. 3. IBGP full mesh

It should be noted that only the IBGP mesh must be connected, and that the non-BGP aware region may be formed by multiple disconnected domains, only linked by the IBGP domain. It is clear that only the routers included in the IBGP mesh need to implement SAD routing in order to properly select the site exit path. So, since all these routers are running BGP, we can use BGP capabilities to provide SAD routing support.

In order to implement SAD routing, each exit router that is running EBGP has to attach a color tag to the routes received from the ISP, so that it is possible to identify the routes learned through each different ISP. Additionally, once the routing information is colored, it is necessary to map each of the colors to a source address prefix. Once that the information of both a given color and its correspondent prefix is available, it is possible to construct SAD routing tables, containing routing information per source prefix.

SAD routing can be implemented in this scenario using the BGP Communities [7] attribute to color the routing information. So, we assume a multihoming scenario where a multihomed site has n ISPs, each one of them has assigned $Pref\_i$ to the multihomed site, with $i=1,...,n$. In order to adopt SAD routing it is required that:
- First, a private community value is assigned to each different ISP. Therefore, Com_i value is assigned to the routes obtained from $ISPi$, being $i=1,..,n$
- Second, n routing tables are created in each of the routers involved, so that each router has one routing table per prefix in the site (i.e. per ISP). Additionally each router is configured to route packets containing a source address matching $Pref\_i$ using the routing table i.

- Third, BGP processing rules are configured in each router, so that routes containing a community attribute value equal to *Com_i* only affect routing table *i*.
- Finally, each exit router that is peering with an external router in *ISPi* is configured to attach the community value *Com_i* to all the routes received from *ISPi*, when announcing them through IBGP.

The resulting behavior is that each router within the IBGP mesh will have separate routing tables containing the information learned through each ISP. Packets containing a source address with the prefix of the *ISPi* will be routed using the corresponding routing table.

3.2 Sites Using IGP

In this scenario, the multihomed site is using an IGP to inform about both internal and external destinations. The IGP learns about external destinations in one of the following three ways:

- Manually configured routes are imported into the IGP
- BGP redistribution into the IGP
- IGP exchange with the providers

As in the previous case, the whole multihomed site routing system is not required to support SAD routing but only a connected domain that has to contain all the exit routers. However, while BGP provides mechanisms to tag routing information so that the same protocol instance can be used to propagate information with different scopes, as presented in the previous section, current IGPs do not provide such capability.

In order to provide SAD routing support, different instances of the routing protocol run in parallel, each one of them associated with a source address prefix. In this way, different instances of the IGP will update different routing tables within the routers. The main difficulty with this approach is to differentiate messages corresponding to the different instances of the IGP. Normally, different instances of the IGP run in different interfaces, so that each instance only receives its own messages. But in this case we want to run multiple instances of the IGP in the same interfaces, so we need a way to separate messages according to the instance of the IGP they belong to.

A possibility would be to send IGP messages using global addresses as source addresses. Usually, IGP messages are sent using link local addresses. But, since each router can be configured with multiple IP addresses, one per prefix, the router includes different source addresses in the messages corresponding to different instances of the IGP. This ships-in-the-night strategy would allow each IGP instance to believe that they are running alone in the link

In particular OSPF for IPv6 [8] explicitly supports running multiple instances in the same link and packets belonging to different instances are identified using the Instance_ID field in the OSPF header.

3.3 Experimenting with SAD Routing

We will next analyze the deployability of the approach by evaluating the available support for SAD routing in current implementations. In order to asses the deployment effort required to adopt the proposed solution, we have built a testbed with widely available commercial routers and we have performed some trials in the framework of the Optinet6 research project. The testbed evaluated the capabilities to support SAD routing of Cisco 2500 routers, Cisco 7500 routers and Juniper M10 routers.

All the tested routers support static SAD routing, i.e. routing based on the source address of the packets according to statically-defined routes. However, the implementation of the SAD routing support differs considerably between them. Cisco IOS supports static SAD routing through manually defined rules, called route-maps, that affect the processing of packets. In order to enable SAD routing, route-maps corresponding to each source address dependent route have to be defined. On the other hand, Juniper routers support multiple routing tables, so that it is possible to create as many routing tables as source address prefixes are involved, and then define the required rules so that the router will forward packets according to the routing table associated with the prefix contained in the source address. In the case of static SAD routing, the multiple routing tables are configured manually with the desired static routes.

Regarding dynamic SAD routing, the support provided by Cisco routers is very limited. Because SAD routing is supported as a manually defined route-map, and because route-map definition is mainly a manual process performed by the router operator, Cisco routers cannot update the routing information (i.e. route-maps) involved in the SAD routing. This means that neither the BGP nor the IGP case are supported by this router vendor.

Because Juniper routers support multiple parallel routing tables, the support for dynamic SAD routing is provided more naturally. In the case of BGP, it is needed that different routing tables are updated depending on the values of the community attribute contained in the BGP route. While this seems pretty straightforward, it is not currently supported by Juniper routers because of the existent constraint that imposes that a given instance of a routing protocol can only update a single routing table, making not viable that the BGP instance can update different routing tables based on the value of the community attribute. Such limitation does not apply for the IGP case, since the considered approach proposes the usage of multiple instances of the IGP running simultaneously, one per source prefix involved, and that each instance of the IGP updates its corresponding routing table. This configuration is currently supported in Juniper routers for OSPFv2 and also for BGP. It should be noted that this approach, i.e. running multiple instances of BGP in parallel, can be used as a temporary solution for the BGP case while the community based approach is not available.

4   Resulting Capabilities

4.1 Fault Tolerance Capabilities

Since the basic assumptions behind adopting SAD routing for multihoming support are that the source address is determined by the initiating host, and that each source address prefix determines an exit ISP, fault tolerance capabilities will be provided by the hosts themselves. As described in the Host Centric Approach [6], such mechanisms are based on a trial and error procedure. Considering that each source address available in a host is bound to an exit path, the host can try different exit paths by changing the source address. The main difference between the approaches is how fast the host can learn that a destination address is unreachable through the selected path.

When external routes are static, the intra site routing system has no external reachability information, so the packet will be forwarded outside the site and only when it reaches routers that have richer knowledge about the topology it will be possible to determine whether the requested destination is reachable through the selected path. In the worst case, the initiating host will timeout and will retry with a different path.

When the multihomed site runs BGP or an IGP with its providers, reachability information is available closer to the host, i.e. in the site's routers, so in some cases, unreachability will be discovered faster than in the general case, where unreachability information is learned through timeouts. So, the host will attempt to use one of its source addresses to reach a certain destination. The packet will be routed through the generic routing domain to the SAD routing domain. Once there, the routers will determine whether the selected destination is reachable with the selected source address. This means that a route to the selected destination exists in the routing table associated with the selected source address prefix. The possible resulting behaviors are:

- If the selected destination is reachable through the selected source address, then the packet is forwarded towards the site exit router that leads to the ISP corresponding to the source address prefix selected.
- If the selected destination is not reachable through the selected source address, but it is reachable through an alternative source address, then the packet is discarded and an *ICMP Destination Unreachable* with *Code 5* which means *Source Address Failed Ingress Policy* [9] is sent back to the host. The information about the proper source address prefix can be included in this message, for instance in the source address of the ICMP message. The host will then retry using the suggested source address.
- If the selected destination is unreachable, the packet is discarded and an *ICMP Destination Unreachable* is sent back to the host. In this case, the host may retry if an alternative destination address is available.

4.2   Traffic Engineering (TE) Capabilities

As a consequence of using multiple prefixes in multihomed sites in conjunction with SAD routing, the party selecting the address of the multihomed host to be used during the communication is the party that determines the ISP to be used for the packets involved in this communication. So, TE mechanisms will have to influence such selection. It must be noted, that the addresses used in a communication are determined by the party initiating the communication, so in this environment, policy mechanisms will not affect incoming and outgoing traffic separately as in the IPv4 case, but they will affect packets belonging to externally initiated communications and packets belonging to internally initiated communications differently. This is the first difference with the previous case.

4.2.1 TE for Externally Initiated Communications

When a host outside the multihomed hosts attempts to initiate a communication with a host within the multihomed site, it first obtains the set of destination addresses, then it selects one according to the Default Address Selection procedure [10]. It seems then that the only point where the multihomed site can express TE considerations is through the DNS server replies. The DNS server can be configured to modify the order of the addresses returned to express some form of TE constraint.

This mechanism can work fine to provide some form of load balancing and load sharing. The DNS server can be configured so that x% of the queries are replied with an address with prefix of ISPA first and the rest of the times (100-x %) are replied with an address with prefix of ISPB first. In addition SRV [11] records can be used to provide enhanced capabilities by those applications that support them. When the host receives the list of addresses, it will process them according to RFC3484. If none of the rules described works, the list is unchanged and the first address received is tried first. Note that the list may be changed by the address selection algorithm because of the host policies.

4.2.2 TE for Internally Initiated Communications

For internally initiated communications, the exit ISP is determined by the source address included in the initiating packet. This means that the source address selection mechanism [10] will determine the exit ISP. RFC 3484 defines a policy table that can be configured in order to express TE considerations. The policy table allows a fine grained policy definition where a source address can be matched with a destination address/prefix, allowing most of the required policy configurations.

5 Conclusions

In this note we have presented the case for the adoption of SAD routing in multihomed environments. The scalability limitations of the current multihoming solution based on the usage of Provider Independent addressing have been largely acknowledged by the Internet community, and there is a consensus that only a new multihoming solution compatible with PA addressing will preserve IPv6 inter domain routing system scalability. However, the adoption of PA addressing in multihomed environments implies that multihomed sites need to internally configure as many prefixes as providers they multihome to, causing several difficulties, such as incompatibilities with ingress filtering, incapability to preserve established connections through outages and so on and so forth. This is basically due to the fact that when multiple PA prefixes are present in the multihomed site, the source address selection process determines the ISP to be used in the communication. This is so because in order to preserve ingress filtering compatibility, the packet has to be forwarded through the ISP that is compatible with the selected source address. Current destination address based routing does not take into account the source address of the packet, making it unsuitable to provide ingress filtering compatibility, that is source address related. SAD routing is then the natural option to overcome the difficulties caused by ingress filtering. Moreover, once that SAD routing is available on the multihomed site, it is possible to obtain additional benefits such as fault tolerance and traffic engineering capabilities with a reduced complexity. SAD routing is not a new technology and it is available in some form in current router implementation, which facilitates its adoption and deployment. However, SAD routing is currently a special feature whose applicability was limited to very specific scenarios. But, if SAD routing is adopted as a fundamental part of the IPv6 multihoming solution as it proposed in this note, it would imply a massive adoption of SAD routing technology, based on the expected number of multihomed sites.

References

[1]   G. Huston, "Commentary on Inter-Domain Routing in the Internet", RFC 3221, 2001.
[2]   Y. Rekhter, T. Li, "A Border Gateway Protocol 4 (BGP-4)", RFC 1771, 1995.
[3]   I. Van Beijnum, "BGP", Oreilly, 2002.
[4]   J. Abley, B. Black, V. Gill, "Goals for IPv6 Site-Multihoming Architectures", RFC 3582, 2003.
[5]   P. Ferguson, D. Senie, "Network Ingress Filtering: Defeating Denial of Service Attacks which employ IP Source Address Spoofing", RFC 2827, 2000.
[6]   C. Huitema, R. Draves, M. Bagnulo, "Host-Centric IPv6 Multihoming", Internet-Draft (Work in proress), 2004.
[7]   R. Chandra, P. Traina, T. Li, "BGP Communities Attribute ", RFC 1997, 1996.
[8]   R. Coltun, D. Ferguson, J. Moy, "OSPF for IPv6", RFC 2740, 1999.
[9]   Conta, A. and S. Deering, "Internet Control Message Protocol (ICMPv6) for the Internet Protocol Version 6 (IPv6) Specification", RFC 2463, 1998.
[10]   Draves, R., "Default Address Selection for Internet Protocol version 6 (IPv6)", RFC 3484, 2003.
[11]   A. Gulbrandsen, P. Vixie, L. Esibov, "A DNS RR for specifying the location of services (DNS SRV)", RFC 2782, 2000.

QoS Routing in DWDM Optical Packet Networks

Franco Callegati[1], Walter Cerroni[2], Giovanni Muretto[1], Carla Raffaelli[1], and
Paolo Zaffoni[1]

[1] DEIS – University of Bologna
viale Risorgimento, 2 – 40136, Bologna – ITALY
{fcallegati,gmuretto,craffaelli,pzaffoni}@deis.unibo.it
[2] CNIT – Bologna Research Unit
viale Risorgimento, 2 – 40136, Bologna – ITALY
walter.cerroni@cnit.it

Abstract. This paper considers the problem of service differentiation in an optical packet switched backbone. We propose and analyze a QoS routing approach based on different routing and congestion management strategies for different classes of service. Congestion resolution is achieved by using the wavelength and time domain and QoS differentiation in the single node is achieved by resource reservation in the wavelength domain. This is combined with alternate routing at the network level. In the paper we show that the proposed strategy guarantees very good performance to the high priority traffic with very limited impact on low priority traffic.

Keywords: Optical packet switching, adaptive routing, optical buffer, QoS.

1   Introduction

One of the emerging needs in present day networking is the support of multimedia applications, which demands real time information transfer with very limited loss to provide the end-users with acceptable quality of service (QoS). At the same time economics require an efficient use of the network resources.

Assuming that internetworking will be provided by the IP protocol, and accounting for its inability to manage QoS, techniques for QoS differentiation must be implemented in the transport networks. Significant effort has been developed to define QoS models. In backbone networks the most interesting solutions proposed to solve the QoS problem deal with a limited number of service classes collecting aggregates of traffic flows with similar requirements. This approach can greatly improve scalability and reduce the operational complexity [1].

In the near future high-capacity circuit-switched optical backbones will provide a huge bandwidth capacity but with limited flexibility in terms of bandwidth allocation and QoS management. Optical burst switching (OBS) and optical packet switching (OPS) are respectively a medium and a longer term networking solutions that promise more flexibility and efficiency in bandwidth usage combined with the ability to support diverse services [2].

In this paper we will focus on an OPS network even though, as explained in the following, the results presented here may be meaningful also with reference to OBS networks that implement queuing at intermediate nodes.

J. Solé-Pareta et al. (Eds.): QofIS 2004, LNCS 3266, pp. 247–256, 2004.
© Springer-Verlag Berlin Heidelberg 2004

Aggregate QoS solutions such as DiffServ are a viable approach also for OPS networks although, because of the limitations of the optical technology, the number of QoS classes must be kept small to minimize operational efforts. In fact, complex scheduling algorithms are not applicable because of the peculiarity of queues in the optical domain, which usually provide a very limited queuing space being implemented by means of delay lines that do not allow random access [3]. This means that traditional priority-based queuing strategies are not feasible in OPS network, and QoS differentiation can be achieved only by means of resource reservation strategies.

We have shown in previous works that QoS differentiation in an OPS network can be provided with good flexibility and limited queuing requirements by means of resource reservation both in the time and wavelength domains [4][5]. These works deal with algorithms for QoS management implemented at the switching node level. Other opportunities arise when considering the routing decisions.

In this paper we extend the study of QoS differentiation mechanisms at the network level, by investigating how the network topology properties can be exploited together with suitable QoS algorithms in order to differentiate the quality along the network paths.

First, the QoS management issues in standalone optical packet switches are reviewed in section 2. Then the QoS management approach for the whole network is described in section 3. The same approach is then applied, in section 4, to a sample network to analyze the influence of different alternatives. In section 5 a network design procedure given the topology and the traffic matrix is presented and, finally, in section 6 some conclusions are drawn.

2   QoS Management in OPS Networks

We assume a network capable of switching asynchronous, variable-length packets or bursts. Therefore the results presented in the following may refer both to an OBS network implementing queuing in the nodes [6] or to an OPS network [7]. In the following we will generally refer to an OPS network and assume that two classes of traffic exist, namely high priority (HP) and low priority (LP).

We consider optical switches that resolve congestions by means of the wavelength and time domains. We do not deal with switching matrix implementation issues and consider a general switching node with N input and N output fibers, carrying W wavelengths each. The switch control logic reads the burst/packet header and chooses the proper output fiber among the N available.

Packets contending for the same output are multiplexed in the wavelength domain (up to W packets may be transmitted at the same time on one fiber) and, if necessary, in the time domain by queuing, implemented with fiber delay lines (FDLs). The FDL buffer stores packet waiting to be transmitted but does not allow random access to the queue. Therefore the order of packets outcoming from the buffer can not be changed and priority queuing is not applicable. Thus QoS management must rely upon mechanisms based on a-priori access control to the optical buffers.

In general, after the output fiber has been determined, the switch control logic must face a two-dimensional scheduling problem: choose the wavelength and, if necessary, the delay to be assigned to the packet. This problem is called the wavelength and delay selection (WDS) problem. An optimal solution to the WDS problem is hardly feasible, because of computational complexity and heuristics have been proposed in the past [5][8][9]. Here we will use the minimum gap algorithm (MING) [8] that has been shown to realize a good trade-off between complexity and performance. This algorithm performs the wavelength assignment by selecting the queue which introduces the smallest gap between the new packet and the last buffered one.

In this scenario QoS differentiation is achievable in the node by differentiating the amount of resources to which the WDS algorithms is applied. We have already shown in [5] that this can be done adopting either a threshold-based or a wavelength-based technique. In the former case, the reservation is applied to the delay units. The WDS algorithm drops incoming LP packets if the current buffer occupancy is such that the delay required is greater than or equal to the threshold, while HP packets have access to the whole buffer capacity. In the latter case the reservation is applied to the wavelengths. A subset of K≤W wavelengths on any output fiber is shared between HP and LP packets while the remaining W-K wavelengths are reserved to HP packets. Generally speaking, wavelength reservation is more promising because of the larger amount of resources available that provide more flexibility to the algorithms. This is because WDM systems are continuously improving and the number of available wavelengths per fiber is getting larger and larger. On the other hand FDL buffers are bulky and should be kept as small as possible, therefore the number of delays that can be implemented is fairly limited and is probably not going to improve much in the future.

The aforementioned approach provides QoS differentiation at the single network node, but does not tackle the problem at the whole network level. A further extension is to define QoS routing algorithms to obtain even further service differentiation by combining QoS management at the routing level with QoS management in the WDS algorithms.

This paper assumes a meshed network topology and shortest path routing. Traffic is normally forwarded along the shortest path but alternate paths of equal or longer length are also identified and can be used. We define two possible routing strategies:
- Single Link Choice (SL), that implements a conventional shortest path routing based on minimum hop count and do not use alternate paths;
- Multiple Alternative (MA), besides the shortest path calculates alternate paths that are used by the network nodes when the link along the shortest path (also called default link) becomes congested.

QoS management is achievable by differentiating the concept of congestion and/or providing different alternatives to LP and HP traffic.

The proposal analyzed in this work is as follows.
- The WDS algorithm works with wavelength reservation according to a partial sharing approach; H out of the W wavelengths available are reserved for HP traffic while the W-H remaining are shared between HP and LP traffic. Two options are considered:
  - The H reserved wavelength may be fixed, namely the W wavelengths available are ordered and the reserved wavelengths are λ_i with i=1,...,H (FIX strategy).
  - Any H wavelengths are reserved based on the actual occupancy, namely when at least W-H wavelengths are available both LP and HP packets may be transmitted, otherwise when less than W-H wavelengths are available only HP packets can be transmitted (RES strategy).

- In the routing algorithms congestion is defined according to the wavelength occupancy to determine wavelength availability, when at least T out of W wavelengths are busy the fiber (and the path to which the fiber belongs to) is considered congested. The value of T is different for different classes of service; for LP traffic T_{LP} = W-H < W, while for HP traffic T_{HP} = W. This means that for the LP class congestion arises before and alternative path, if any, should be used more frequently.
- Alternate routing is used for LP traffic but not for HP traffic. Therefore HP traffic is always routed with a SL choice, while LP traffic is routed with a MA choice, and alternate paths are used when congestion is present.

The basic idea behind this approach is that the HP traffic stream should be preserved intact as much as possible. Congestion and alternate routing will modify the traffic stream, because of loss, delay, out of sequence delivery etc. Therefore we reserve resources to HP traffic to limit congestion phenomena and do not rely on alternate routing to avoid as much as possible out of order packets.

3   Network Performance Analysis

In this section we provide numerical results to evaluate that the proposed techniques for QoS management may provide service differentiation at the whole network level. Performance is evaluated in terms of packet loss probability. Due to lack of space evaluation of other performance parameters such as delay and out-of-order packet delivery are not shown here. Numerical results were obtained by using an ad-hoc, event-driven simulator. The reference network topology is shown in Fig. 1 and consists of 5 nodes interconnected by 12 fiber links carrying 16 wavelengths each. Traffic enters the network at any node and is addressed to any other node according to a given traffic matrix.

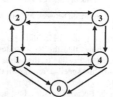

Fig. 1. The reference network topology

The network adopts a connectionless transfer mode, with traffic generated by a Poisson process. The packet size distribution is exponential with average value equal to the buffer delay unit D measured in bytes. This choice minimizes the packet loss at the node level when adopting the MING algorithm [8]. The number of simulated packets is up to 10^8. The traffic matrix has been set up as follows, with two alternatives.

- Balanced traffic matrix (B): the traffic distribution in the network is uniform since each wavelength carries the same average load (80%). With this approach the input load at different ingress points of the network may clearly not be the same.
- Unbalanced traffic matrix (U): in this case the traffic load at the ingresses of the network is assumed to be the same. By making this choice the links have a different average load per wavelength, with the only constraint that the maximum value cannot overtake a fixed value (80% in our simulations).

Since in the balanced case each link is loaded in the same way, we can consider the average loss probability of the whole network as an evaluation parameter. On the other hand, in the unbalanced case this parameter may not be representative for performance evaluation, therefore the worst loss probability among all links will be taken into account.

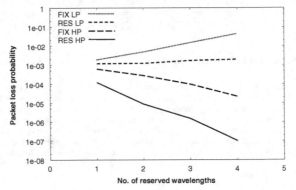

Fig. 2. Comparison between FIX and RES.

At first in Fig. 2 we compare the FIX and RES strategies for wavelength reservation. The graph clearly shows that the RES strategy performs better for both traffic classes. This was expected and is due to the better exploitation of the network resources (the wavelengths in this specific case). In the RES case the reserved wavelength pool is dynamically adjusted to the present state of traffic requests, with a sort of "call packing" approach. Because of the clear advantage of this reservation strategy, in the rest of the paper we will always assume that RES will be used.

In Fig. 3 the packet loss probability is shown for different routing algorithms (SL and MA) and different traffic matrices (B and U) considering undifferentiated traffic.

It is clear that MA performs better than SL even though the gain is not that big. This can be explained by considering that the network topology is only composed by a limited number of hops and not so many routing alternatives may be actually exploited. In [10] it is proved that dynamic algorithms perform much better than the static ones in presence of a bigger and more complex network. However, it is important to take into account that MA keeps packets within the network for longer and then the transmission delay becomes bigger than the SL case. This is why the

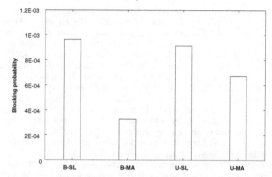

Fig. 3. Packet loss probability for SL and MA algorithms for undifferentiated traffic and for both cases of balanced and unbalanced traffic matrix.

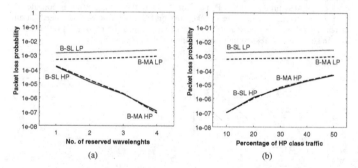

Fig. 4. Packet loss probability for high and low priority classes with balanced traffic matrix, varying the number of reserved wavelengths (a) or the percentage of HP traffic (b).

routing of HP packets always adopts SL. Therefore the choice between SL and MA is relevant only to the routing of LP packets. Since results for the balanced and unbalanced traffic matrix are very similar, only the balanced case is shown in the following due to the limited space available.

In order to understand how different choices affect the network behavior, Fig. 4 shows the results obtained by different approaches. First we evaluate the performance assuming a fixed percentage of the HP class set to 20% while the number of wavelengths reserved to HP class varies from 1 to 4 out of 16 on each link (Fig. 4a). Then, the percentage of HP input traffic varies between 10% and 50% while the number of wavelengths reserved to HP class is fixed to 3 for each link (Fig. 4b).

As expected, the higher the number of dedicated wavelengths, the higher the gain in terms of loss probability that can be obtained for the HP class. When 1 to 4 wavelengths are reserved, the loss probability improves by three orders of magnitude, while the performance of the LP class is barely affected. Packet loss probability remains nearly constant at one order of magnitude worse than the undifferentiated case. HP class reaches very low packet loss probability (10^{-7}) when resource reservation is equal to 25% (i.e. 4 wavelengths) of the whole set. Moreover, it is also not affected by the fact that LP class can be routed in different ways.

On the other hand, for a very low percentage of HP traffic, good level of performance may be achieved. When HP traffic grows over the 10% performance starts getting worse quite rapidly, while the LP class again seems to be slightly affected. It follows that in case a given loss probability is required by HP traffic, the admission to the network has to be kept under control in order to avoid performance degradation due to the limited resources reserved to HP class.

A good degree of differentiation between the two classes may be obtained in both cases reaching up to four orders of magnitude, while the adaptive routing strategy for LP traffic allows a further performance improvement.

The results presented in Fig. 5a let us understand how the amount of reserved resources and the percentage of HP traffic are related when a given value of the HP class packet loss probability (PLP) is required. Only the SL algorithm is considered here. As expected, in case a given packet loss probability has to be guaranteed for an increasing percentage of HP traffic, more resources must be reserved. Furthermore, Fig. 5b shows the corresponding performance of the LP class.

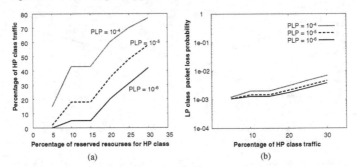

(a)   (b)

Fig. 5. Relation between the HP traffic percentage and the percentage of resources reserved for given packet loss probability (a) and corresponding LP traffic performance (b).

4   Network Design

In this section a network design procedure is presented. The reference network is the same as above but the aim now is to calculate, with relation to the SL routing algorithm and to the traffic matrix adopted, the number of wavelengths required per

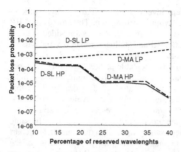

Fig. 6. Packet loss probability for high and low priority classes resulting from the network design procedure as a function of the percentage of resources dedicated to the HP class with 20% of HP class traffic.

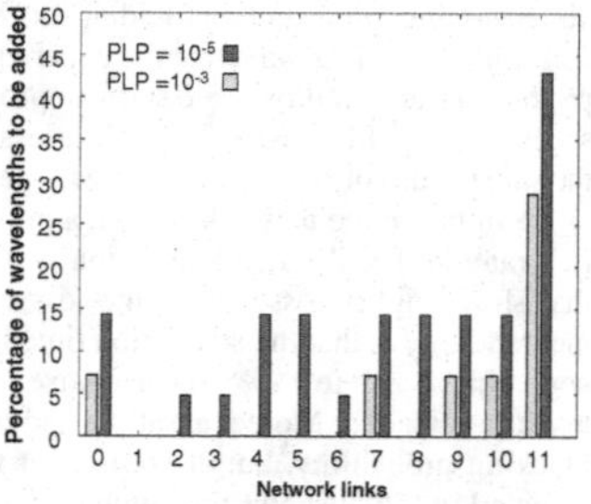

Fig. 7. Percentage of additional wavelengths required by each link for different loss constraints.

Table 1. Number of wavelengths required to achieve different packet loss probabilities.

PLP	No. of iterations	Total no. of wavelengths required by each link											
		0	1	2	3	4	5	6	7	8	9	10	11
10^{-1}	0	14	28	21	21	14	14	21	14	14	14	14	7
10^{-2}	1	14	28	21	21	14	14	21	14	14	14	14	8
10^{-3}	2	15	28	21	21	14	14	21	15	14	14	15	9
10^{-4}	3	15	28	21	21	15	15	21	15	15	15	15	9
10^{-5}	4	16	28	22	22	16	16	22	16	16	16	16	10

fiber so that a given average load per wavelength is obtained. The main assumption is that all nodes generate the same input traffic which is uniformly distributed to the other nodes.

The input traffic value is chosen so that the total number of wavelengths is very close to 16×12 = 192 as in the previous case, each wavelength being loaded at 80%. This allows a better comparison with almost the same cost in terms of wavelengths. The resulting resource distribution varies from 7 to 28 wavelengths per link. In the design procedure it is important to adopt the MA approach, otherwise performance decreases. This is due to the fact that SL, not sharing the wavelength resources, does not achieve load balancing. With the MA approach performance is the same as the balanced case with the advantage that the traffic matrix is now imposed by user needs instead of network configuration as before.

In Fig. 6 the performance of SL and MA is shown for both classes varying the percentage of wavelengths reserved to HP traffic (set to 20%). Obviously, when the percentage of reserved wavelengths is low there is a bad service differentiation between the two classes. Moreover the trend of the curve for HP class is not as smooth as before. This is because losses are not uniformly distributed over the links, varying in a range between 10^{-2} and 10^{-6}. The reason why this happens is because different numbers of wavelength are available on different links due to the dimensioning procedure, providing different levels of wavelength multiplexing. In fact, links with less wavelengths experience worse performance. Thus the overall HP packet loss probability curve starts improving when more resources are added to these specific links. LP class seems to be less affected and its loss probability remains of the same order of magnitude with MA performing better than SL as usual.

To improve the network design, a maximum acceptable packet loss probability per link may be fixed. Then simulation is iterated by adding wavelengths to those links that show higher losses until the loss constraint is satisfied. The drawback of this methodology is that the simulation time increases. The average wavelength load at the beginning is set to 80% but of course, when more resources are added, some links result less loaded. Moreover, at the end of the process links still do not have the same blocking probability, but at least all of them satisfy the loss requirement. The chart depicted in Fig. 7 shows the number of additional wavelengths (as a percentage of the starting number) that must be added to each link in order to meet different packet loss requirements.

In Table 1 the number of iterations and the corresponding number of wavelengths for each link required to achieve different loss constraints are shown.

5 Conclusions

In this paper the problem of quality of service differentiation in DWDM packet-switched optical networks has been addressed. The effects of quality of service routing have been shown by applying dynamic wavelength management on each link jointly with static or dynamic routing strategies. Different quality of service algorithms have been analyzed and then applied to a network dimensioning procedure. The sharing effect produced by the dynamic routing algorithm proved to be particularly effective in this situation. An iterating procedure has then been applied to achieve loss balancing over network links with relation to design constraints. The main conclusion is that the use of assigned wavelength results optimized in relation to the performance target. Both delay and out-of-order problems as well as the application of the algorithms to more complex network topologies are currently under investigation and will be the subject of future works.

Acknowledgments. This work is partially funded by the Italian Ministry of Education and University (MIUR) under the projects "INTREPIDO - End-to-end Traffic Engineering and Protection for IP over DWDM Optical Networks" and "GRID.IT - Enabling platforms for high-performance computational grids oriented to scalable virtual organizations". The authors wish to thank Mr. Michele Savi for his help in programming and running the simulations.

References

[1] S. Blake, D. Black, M. Carlson, E. Davies, Z. Wang, and W. Weiss, "An architecture for differentiated services", IETF RFC 2475, December 1998.

[2] W. Vanderbauwhede, D.A. Harle, "Design and Modeling of an Asynchronous Optical Packet Switch for DiffServ Traffic", Proc. ONDM 2004, Gent, Belgium, pp. 19-35, February 2004.

256   F. Callegati et al.

[3]   D.K. Hunter, M.C. Chia, and I. Andonovic, "Buffering in Optical Packet Switching", IEEE/OSA Journal of Lightwave Technology, Vol. 16, No. 10, pp. 2081-2094, December 1998.
[4]   F Callegati, G. Corazza, and C. Raffaelli, "Exploitation of DWDM for optical packet switching with quality of service guarantees", IEEE Journal on Selected Areas in Communications, Vol. 20, No. 1, pp. 190-201, January 2002.
[5]   F. Callegati, W. Cerroni, C. Raffaelli, and P. Zaffoni, "Wavelength and time domain exploitation for QoS management in optical packet switches", Computer Networks, Vol. 44, No. 4, pp. 569-582, 15 March 2004.
[6]   C.M. Gauger, H. Buchta, E. Patzak, andJ. Saniter, "Performance meets technology - An integrated evaluation of OBS nodes with FDL buffers", Proc. WOBS 2003, Dallas, TX, October 2003.
[7]   M.J. O'Mahony, D. Simeonidou, D.K. Hunter, A. Tzanakaki, "The application of optical packet switching in future communication networks", IEEE Communications Magazine, Vol. 39 , No. 3, pp.128-135, March 2001.
[8]   F. Callegati, W. Cerroni, and G. Corazza, "Optimization of wavelength allocation in WDM optical buffers", Optical Networks Magazine, Vol. 2, No. 6, pp. 66-72, November 2001.
[9]   F. Callegati, W. Cerroni, C. Raffaelli, and P. Zaffoni, "Dynamic Wavelength Assignment in MPLS Optical Packet Switches", Optical Network Magazine, Vol. 4 No. 5, pp. 41-51, September 2003.
[10] F. Callegati, W. Cerroni, G. Muretto, C. Raffaelli, and P. Zaffoni, "Adaptive routing in DWDM optical packet switched network", Proc. ONDM 2004, Gent, Belgium, pp. 71-86, February 2004.

A Proposal for Inter-domain QoS Routing Based on Distributed Overlay Entities and QBGP[*]

Marcelo Yannuzzi[1], Alexandre Fonte[2,3], Xavier Masip-Bruin[1], Edmundo Monteiro[2], Sergi Sánchez-López[1], Marilia Curado[2], and Jordi Domingo-Pascual[1]

[1] Departament d'Arquitectura de Computadors, Universitat Politècnica de Catalunya (UPC)
Avgda. Víctor Balaguer, s/n – 08800 Vilanova i la Geltrú, Barcelona, Catalunya, Spain
{yannuzzi, xmasip, sergio, jordid}@ac.upc.es
[2] Laboratory of Communications and Telematics, CISUC-DEI, University of Coimbra,
Pólo II, Pinhal de Marrocos, Postal Address 3030-290 Coimbra, Portugal
{afonte, edmundo, marilia}@dei.uc.pt
[3] Polytechnic Institute of Castelo Branco,
Av. Pedro Álvares Cabral, n°12, Postal Address 6000-084, Castelo Branco, Portugal

Abstract. This paper proposes a novel and incremental approach to Inter-Domain QoS Routing. Our approach is to provide a completely distributed Overlay Architecture and a routing layer for dynamic QoS provisioning, and to use QoS extensions and Traffic Engineering capabilities of the underlying BGP layer for static QoS provisioning. Our focus is mainly on influencing how traffic is exchanged among non-directly connected multi-homed Autonomous Systems based on specific QoS parameters. We provide evidence supporting the feasibility of our approach by means of simulation.

Keywords: Inter-domain QoS Routing, Overlay, BGP.

1 Introduction

At present, nearly 80% of the more than 15000 Autonomous Systems (ASs) that compose the Internet are stub ASs [1], where the majority of this fraction is multi-homed. For these ASs the issue of Quality of Service Routing (QoSR) at the inter-domain level arises as a strong need [2]. Whereas some research groups rely on QoS and Traffic Engineering (TE) extensions to BGP [3-4], others tend to avoid new enhancements to the protocol and propose Overlay networks to address the subject [5-6]. While the former approach provides significant improvements for internets under low routing dynamics, the latter results more effective when routing changes occur more frequently. The main idea behind the overlay concept is to decouple part of the policy control portion of the routing process from BGP devices. In this sense, the two approaches differ in how policies are controlled and signaled. BGP enhancements tend to provide in-band signaling, while the overlay approach provides out-of-band signaling.

---

[*] This work was partially funded by the MCyT (Spanish Ministry of Science and Technology) under contract FEDER-TIC2002-04531-C04-02, the CIRIT (Catalan Research Council) under contract 2001-SGR00226 and the European Commission through Network of Excellence E-NEXT under contract FP6-506869.

J. Solé-Pareta et al. (Eds.): QofIS 2004, LNCS 3266, pp. 257–267, 2004.
© Springer-Verlag Berlin Heidelberg 2004

The Overlay Architecture is mostly appropriate when communicating domains are multi-homed, and thus may need some kind of mechanism to rapidly change their traffic behavior depending on network conditions. In fact, multi-homing is the trend that most stub ASs exhibit in nowadays Internet, which mainly try to achieve load balance and fault tolerance on the connection to the network [5]. In addition, present inter-domain traffic characteristics reveal that even though an AS will exchange traffic with most of the Internet, only a small number of ASs is responsible for a large fraction of the existing traffic. Moreover, this traffic is mainly exchanged among ASs that are not directly connected; instead they are generally 2, 3 and 4 hops away [4].

Therefore, the combination of all these features made us focus on QoSR among strategically selected non-peering multi-homed ASs. The approach to inter-domain QoSR we propose in this paper is to supply a completely distributed Overlay Architecture and a routing layer for dynamic QoS provisioning, while we use QoS extensions and TE capabilities of the underlying BGP layer for static QoS provisioning. Within the overlay inter-domain routing structure reside special Overlay Entities (OEs), whose main functionalities are the exchange of Service Level Agreements (SLAs), end-to-end monitoring, and examination of compliance with the SLAs. These functionalities allow the OEs to influence the behavior of the underlying BGP routing layer, to take rapid and accurate decisions to bypass network problems such as link failures, or service degradation for a given Class of Service (CoS). The reactive nature of this overlay structure acts as a complementary layer conceived to enhance the performance of the underlying BGP layer containing both QoS aware BGP (QBGP) routers and non-QoS aware routers.

The remaining of this paper is organized as follows. Section II presents an overview of our overlay approach. In Section III the main functionalities required from the underlying BGP and overlay layers are analyzed, while Section IV presents our simulation scenario and results. Finally, Section V concludes the paper.

2   Overview of the Proposed Overlay Approach

As stated in the Introduction, we propose in this paper a combined QBGP and Overlay Architecture for inter-domain QoSR. The main ideas behind the Overlay Architecture are:

- The OEs should respond nearly two orders of magnitude faster than the BGP layer in the case of a network failure.
- The OEs should react and try to reroute traffic when non-compliant conditions concerning QoS parameters previously negotiated for a given CoS are detected.
- The underlying BGP structure does not need modifications, and remains unaware of the QoSR architecture running on top of it.

The next figure depicts a possible scenario were our proposal could be applied. In this model, two peering OEs belonging to different ASs spanning across several AS hops are able to exchange a SLA and agree upon a set of QoS parameters concerning the traffic among them. The intermediate ASs do not need to participate in the Overlay Architecture, and therefore no OEs are needed within these transit ASs. From our perspective, the real challenge is to develop a completely distributed overlay system, where each OE behaves in a reflective manner. In this sense our overlay approach is

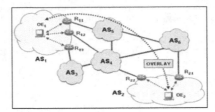

Fig. 1. Inter-Domain QoSR scenario where OEs are used for dynamic QoS provisioning among remote multi-homed ASs

like facing a mirror. Instead of proposing a complex scheme to dynamically and accurately manage how traffic enters a target AS, we focus on how traffic should exit from the source AS. Hence, what we seek is that the OE within the source AS behaves like the image in a mirror of the OE in the target AS. This mirroring scheme allows the OE in the source AS to dynamically manage its outgoing traffic to the target AS, depending on the compliance with the previously established SLA for a given set of CoSs. Then, within each AS, the OE should measure end-to-end QoS parameters along every link connecting the multi-homed AS to the Internet, and check for violations to the SLAs. Henceforth, we assume that the topology has at least two different end-to-end paths between any pair of remote ASs participating in our QoSR model. When a violation is detected, the OE in the source AS is capable of reconfiguring on-the-fly its traffic pattern to the remote AS for the affected CoS. Here, the time scale needed to detect and react to a certain problem is very small when compared with the BGP time scale [7].

The end-to-end measurements are based on active AS path probing among peering OEs. Hence, each OE within an AS spawns probes targeting the remote AS through every available link connecting the source AS to the Internet. We sustain, and we will show by simulation that the AS-AS probing practice is not demanding neither in terms of traffic nor in terms of processing, as long as the number of overlay peering ASs and the number of CoSs remains limited. In fact, the traffic generated between two OEs is negligible. It is worth noting that a non-complying condition may only occur in a single direction of the traffic, which means that the bottleneck is merely on the upstream or the downstream path. For example, in Fig. 1 the OEs in AS_1 and AS_2 measure the same parameters, such as One-Way Delay (OWD) [8] or One-Way Loss (OWL) [9], and react in the same manner due to their mirrored behavior. Therefore, either of them is able to independently decide if it should shift its outbound traffic or not. An advantage of this approach is that BGP updates could be completely avoided if, for example, the LOCAL PREFERENCE (LOCAL_PREF) is used when reallocating this traffic.

Agarwal *et al.* proposed an interesting overlay mechanism to reduce the fail-over time and to achieve load balancing of traffic entering an AS [5]. However, this proposal does not reuse any QoS or TE capabilities from the BGP layer. Moreover, it introduces a centralized and complex server which allows an AS to infer, by means of heuristics, the topology and customer/peer relationships among the multiple ASs that conform all tentative paths known to any given peering AS in the overlay structure.

The complexity introduced is mainly due to the fact that accurately controlling how traffic enters an AS is a very intricate task, particularly when this must be done dynamically. As an alternative, our proposal deals with the allocation of traffic from the source AS, since we strongly believe that simpler approaches such as this one will turn out more attractive to become deployed.

3 Main Functionalities of the Routing Layers

In this Section we describe in detail the overlay routing functionalities as well as the underlay BGP routing functionalities.

3.1 Top Layer: Overlay Routing Functionalities

This layer is composed by a set of OEs:

3.1.1 Basic Set of Components
- At least one OE exists per QoS domain.
- An OE has full access to the border BGP routers within an AS.
- An OE has algorithms for both detecting non-conforming conditions for a given CoS, and deciding when and how to reallocate its traffic.

3.1.2 Main Components

An Overlay Protocol: A protocol between remote OEs is needed. This protocol allows OEs to exchange SLAs with each other, and to exchange substantial information for the Overlay Architecture.

Metric Selection: In order to validate our approach, we choose a simple QoS parameter for the dynamical portion of our QoSR model. The parameter we have selected is a smoothed OWD (SOWD), which defines the following metric:

$$\overline{OWD}(m,n) = \frac{1}{N} \sum_{k=n}^{k=n+N-1} OWD(m,k) \tag{1}$$

in which m and n correspond to the n_{th} probe generated by a source OE and sent towards the m_{th} external link of the AS. This SOWD corresponds to the average OWD through a sliding window of size N. Instead of using instantaneous values of the OWD, we propose to use this low-pass filter, which smoothes the OWD avoiding rapid changes in our metric. A trade-off exists in terms of the size of the window. A large value of N implies a slow reaction when network conditions change and maybe the reallocation of traffic is needed. On the other hand, small values of N could translate into frequent traffic reallocations since it is likely to occur that non-compliant conditions are more frequently met. In this scenario the SLA exchanged by the OEs is simply the maximum SOWD D_j tolerated for each different CoS C_j.

We assume that an OE uses one logical address for each different CoS, and also that specific local policies are applied to Internal BGP (IBGP). Thus, a single OE could be configured to probe a remote OE for any given CoS, through all available egress links of the AS in a round-robin fashion avoiding the hot potato routing problem. Then, the OEs compute a per-CoS cost to reach the remote AS over every external link m based on the previous metric. Furthermore, packets probing a specific CoS belong to that CoS. For instance, in a QBGP framework based on Differentiated Services (DiffServ), when probing a particular CoS which is mapped to an Assured Forwarding (AF) class in each intermediate AS, the probes are tagged under the same AF class [10]. We assume that the OEs are properly synchronized (e.g., by means of GPS) and the details concerning synchronization are out of the scope of this work.

Piggy-Backing mechanism: An important issue is that an active probing technique developed to measure the OWD requires feedback from the remote OE. However, the mirroring scheme implies that the remote OE is already probing the local OE and expects feedback from this latter as well. Thus, the easiest way to avoid unnecessary messages traversing the network is to endow the protocol between the OEs with a piggy-backing technique. Then, feedback for the OWD is carried on the probes itself.

Stability: Another central issue is that the traffic reallocation process should never generate network instability. In order to prevent this from happening, but keeping in mind that we follow a completely distributed architecture design where the OEs should rely on themselves to cope with these problems, we impose the following restriction:

"Traffic targeting a certain CoS C_j should never be reallocated over a link s, if and only if the primary link to reach C_j was s in $[t\text{-}T_b,t]$ or C_j has exceeded its maximum number of possible reallocations $\Rightarrow R_j(t) \bullet R_j^{MAX}$ "

In this way the parameter T_b avoids short-term bounces, while the parameter R_j^{MAX} avoids the long-term ones. Then, each time a traffic reallocation process takes place for a given CoS C_j the variable $R_j(t)$ is incremented. Our approach is to provide a sort of soft penalization similar to BGP damping [11], where the penalty is incremented by a fixed value P with each new allocation, but it decays exponentially with time when no reallocations occur according to:

$$R_j(t) = R_j(T)e^{-\left(\frac{t-T}{\tau}\right)} \qquad (2)$$

where T_b, R_j^{MAX}, P and are configurable parameters, whose values depend on the degrees of freedom in the number of short and long-term reallocations we allow for a given CoS C_j. An additional challenge in terms of stability arises when a path becomes heavily loaded, since several CoSs within the path could experience non-compliant conditions with their respective SLAs. In order to prevent simultaneous reallocations for all the affected CoSs, we endow the OEs with a contention mechanism which prioritizes the relevance of the different CoSs. Then, more relevant CoSs are reallocated faster than less priority classes. The contention algorithm operates as follows:

$\begin{cases} \textit{Let } C_j \textit{ be one of the q affected CoS within link m, where } j = 1,..,q \\ C_j \textit{ will be reallocated in } T_j, \textit{ where } T_j \in [K_{j-1},K_j) \textit{ and } T_j \textit{ is randomly selected, with } K_0 \equiv 0 \end{cases}$

$\Rightarrow$ *Then, the highest priority classes C_1 within link m will be reallocated in a random time $T_1 \in [0,K_1)$, classes C_2 will be reallocated in a random time $T_2 \in [K_1,K_2)$, and so on.*

Clearly, our contention mechanism allows an OE to iteratively reallocate traffic from a loaded path, and to dynamically check if the remaining classes continue under non-compliant conditions. It is likely that as soon as we begin to extract traffic from the path, the remaining classes will start to experience better end-to-end performance. However, a different situation is generated when a link failure occurs. In this case, an OE should react as fast as possible to reallocate all traffic from the affected path. Then, a trade-off exists in terms of both the contention mechanism and the ability to rapidly redistribute all traffic from any given link. Instead of tuning the contention algorithm to efficiently cope with both problems at the same time, we rely on the probing technique since a link failure will cause the complete loss of probes for all the CoSs within the link. Our proposal is based on incrementing the frequency of the probes per-CoS as soon as losses are detected. We maintain that this rising in the frequency does not exacerbate the load on the network, firstly because the fraction of traffic generated by the OE that detects the problem is negligible in terms of the overall traffic exchanged between both ASs. Secondly, this is done for a short period of time and only with the aim of speeding up the re-routing process. Once a CoS is reallocated, the frequency of the probes decreases back to its normal value.

3.2 Bottom Layer: Underlay BGP Routing Functionalities

The set of routes to be tested by the OEs using the probing techniques described in the previous sub-section, are predetermined by the underlying BGP-based layer. In this layer two types of devices can operate; legacy BGP routers and QBGP routers. A QBGP router is able to distribute QoS information and take routing decisions per-CoS constrained to the previously established SLA between different peering domains. In our model, QBGP routers can be seen as the practical tool to establish the overall inter-domain QoSR infrastructure composed by several sub-routing layers, one for each CoS, which in addition could be dynamically influenced by the overlay layer. Interesting approaches and further information on the subject of QBGP could be found in [3, 12].

3.3 Combined QoSR Algorithm

The next scheme (Fig.2) depicts our combined QoSR algorithm. Let m be the external link currently allocating traffic of class C_j. It is important to remark that the approach we follow is that even though an alternative path could have a better cost in terms of SOWD, we avoid reallocating traffic of class C_j from link m until a violation to the SLA is detected. Then, two distinct threads of events occur upon the reception of a probe for class C_j. Initially, the probe (k,l) is separated from the piggy-backed feedback $OWD(m,n)$. In order to accurately reply back to the sender, the first to be processed is the $OWD(k,l)$ which is shown as (I) in Fig. 2. On the other hand, the piggy-backed $OWD(m,n)$ is processed, which is depicted as (II) in the figure.

Once the SOWD is computed, the algorithm checks for violations to the maximum SOWD tolerable, that is D_j. If no violations have occurred the algorithm simply waits for the next incoming probe. However, if a violation is detected in link m the algorithm checks if the maximum number of allowed reallocations R_j^{MAX} is exceeded. In case this is true, the local OE is able to compose a feedback message and warn the remote OE about this situation. The main idea is that the feedback process provides information to the remote OE, and thus it can try to handle the problem by tuning its static QoS provisioning using either QBGP or TE-BGP.

If R_j^{MAX} is not exceeded, then the OE needs to check, within all the external available links p, excepting m, if there exists at least one link i whose cost M_i satisfies the constraint for the class C_j. Moreover, it also needs to check if the link has enough room to handle the class reallocation. Subsequently, and in order to avoid any short term bounce, the OE excludes from the set of capable links those who had allocated traffic of C_j in $[t-T_b, t]$. Once this is done, we rely on QBGP to tiebreak in case two or more links show the same cost in terms of the SOWD. At this step a single link is left as the target for the reallocation of the class. Then, the contention algorithm is executed and T_j seconds later the OE checks if the class still remains in a violating condition. If this is the case, the OE increments $R_j(t)$ by P and reroutes the traffic of C_j.

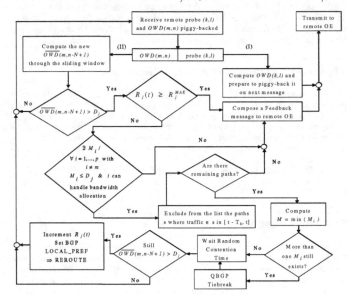

Fig. 2. Combined QoSR Algorithm

4 Simulation Results

The Overlay Architecture proposed in this paper is being evaluated and validated by simulation. In this section some preliminary results are presented to allow a first evaluation of the overall architecture and its capability to support QoS traffic classes in a dynamic way. We are using the J-Sim simulator [13] with the BGP Infonet suite [14] which is compliant with BGP specification RFC 1771 [15]. A set of Java components with the functionalities of the overlay layer was developed. In order to allow the Overlay Entities to have full access to the Adj-RIBs-In and the Loc-RIB of a BGP speaker [15], and to have control over the BGP decision process, it was necessary to add some extensions to the Infonet suite. Furthermore, we have also included the following QoS BGP extensions:

- An optional transitive attribute to distribute the CoS identification (ID), and a set of modifications to BGP tables to allow the storage of this additional information, following a similar approach to the one described in [3].
- A set of mechanisms to: i) allow BGP speakers to load the supported CoSs; ii) allow each local IP prefix to be announced within a given CoS; iii) allow BGP speakers to set the permissibility based on local QoS policies and supported capabilities.

For our simulations, we used the topology presented in Fig. 7. The topology is based in the GÈANT European Academic Backbone with some simplifications to reduce the complexity of the simulation model. In this topology we considered as remote multi-homed AS domains AS_1 and AS_2. All links were assumed to be bi-directional with the same capacity C (C=2Mbps) and propagation delay P_d (P_d=10ms), with the exception of AS_2 links where, in order to have some bottleneck, the capacity chosen was C/2. For complexity concerns, we modeled each AS as a single QBGP router with core DiffServ capabilities configured to support four different IP packet treatments (EF, AF11, AF21 and Best-effort) allowing four different CoSs, namely CoS1, CoS2, CoS3 and CoS4. Thus, on the domain where traffic was injected we used edge DiffServ capabilities to mark packets with a specific DSCP (DiffServ Code Point) depending on its corresponding CoS. These marks were applied to regular IP packets and to the probes generated by the OE. The test conditions are summarized in Table 1. The results obtained are presented in Fig. 3 to Fig. 6. The maximum SOWD tolerated per-CoS (D_j) was heuristically chosen to allow the OEs to take advantage of alternative paths. The SWOD computed when probes were lost was also heuristically chosen. The criterion selected was that 3 consecutive losses imply nearly a rise of 25% in the SWOD. For the tests presented we set $R_j^{MAX} = \infty \ \forall j$. Moreover, no probes were generated for Best-effort traffic (NA=Not Available), and a sliding window of 3 seconds was used in all tests, which is shown as Mov.Average in Table 1.

Table 1. Test conditions

CoS	CBR (Mbps)	Pkt. Size (KB)	PHB	Max. SOWD (ms)	Probing Freq.	Hold (Contention) $\&T_k$ (s)	Mov. Average
CoS1	0,4	1	EF	85	1 s, 1KB	3 & 8	3 s
CoS2	0,8	1	AF11	100	1 s, 1KB	6 & 12	3 s
CoS3	1,0	1	AF21	120	1 s, 1KB	9 & 20	3 s
CoS4	1,6	N.A	BE	N.A	N.A	N.A	N.A

The first objective of the simulation was the validation of the initial assumption that our approach, based on a complementary routing layer, enhances the reaction of the overall routing infrastructure. Then, as a performance indicator, we chose to compare the response time to a link failure. Fig. 3 depicts a set of plots for traffic of CoS1 showing the throughput measured at the destination, the SOWD experienced by

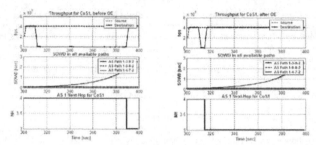

Fig. 3. Link failure reaction with and without OE

probes for all available paths, and the path shifts determined by changes in the next-hop for the source AS, namely AS_1. From these plots, we can observe that a pure QBGP framework (without OEs running on AS_1 and AS_2) needs about 80 seconds to overcome a link failure, but only 5 seconds are needed when OEs are running. This result validates our initial assumption. It is worth mentioning that this last value includes not only the implicit link failure detection condition based on a violation to the maximum SOWD tolerated, but also includes a random contention interval of 3 seconds before re-routing.

Secondly, from figures 4 and 5, we can observe that without OEs there are clear violations to the SLAs established between the end-to-end domains. However with OEs, it becomes clear that the architecture is able to react to SLA violations, and find the best paths to reallocate traffic for the affected traffic classes. Consequently, after a transitory interval of approximately 13 seconds, needed to accommodate the traffic for each CoS, it is visible that a steady state is reached and the SLAs are satisfied for all affected classes. Furthermore, and in order to evaluate overall link utilization, we measured the throughput over all available links at the destination AS (AS_2). Fig. 6 shows that with OEs, in addition to the compliance with the SLAs a better distribution

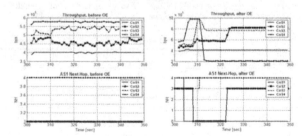

Fig. 4. Throughput for traffic of CoS1-CoS4, with and without OE

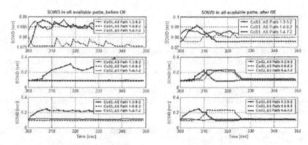

Fig. 5. OWD in all available paths for CoS1-CoS4, with and without OE

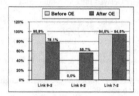   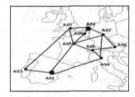

Fig. 6. Remote AS link utilization   **Fig. 7.** Topology based on the GÉANT Network   [16]

of inter-domain traffic is obtained, and thus, resources are more efficiently used. The extra cost in these cases was merely an increment of 8 Kbps, per-CoS, on each link in the remote AS-AS traffic, when oversized probes of 1 KB were spawned.

5   Conclusions

This paper depicts the framework for a combined inter-domain QoSR paradigm based on a completely distributed Overlay Architecture coupled with a QBGP or TE-BGP routing layer. As a first step in our research, and in order to validate our approach we have focused on the coupling of the overlay with a DiffServ QBGP underlying layer. The results obtained show that our distributed Overlay Architecture substantially enhances end-to-end QoS when compared with a pure QBGP model. We believe that whereas significant extensions and enhancements to BGP are certainly going to be seen, the overlay structure arises as a strong candidate to provide flexible and value-added out-of-band inter-domain QoSR. In particular, this becomes perfectly suitable when inter-domain traffic patterns need to dynamically adapt and rapidly react to medium or high network changing conditions, where the former solutions seem impracticable at the present time.

References

1. Olivier Bonaventure, Bruno Quotin, Steve Uhlig, "Beyond Interdomain Reachability", Workshop on Internet Routing Evolution and Design (WIRED), October 2003.
2. E. Crawley, R. Nair, B. Rajagopalan, H. Sandick, "A Framework for QoS-based Routing in the Internet", Internet Engineering Task Force, Request for Comments 2386, August 1998.
3. Cristallo, G., C. Jacquenet, "An Approach to Inter-domain Traffic Engineering", Proceedings of XVIII World Telecommunications Congress (WTC2002), France, September 2002.
4. Olivier Bonaventure, Steve Uhlig, Bruno Quotin, et al, "Interdomain Traffic Engineering with BGP", IEEE Communications Magazine, May 2003.
5. S. Agarwal, C. Chuah, R. Katz "OPCA: Robust Interdomain Policy Routing and Traffic Control", IEEE Openarch, April 2003.
6. Zhi Li, Prasant Mohapatra, "QRON: QoS-aware Routing in Overlay Networks", IEEE Journal on Selected Areas in Communications, June, 2003
7. C. Labovitz, A. Ahuja, A. Bose, and F. Jahanian, "Delayed Internet routing convergence," in Proc. ACM SIGCOMM, 2000.
8. G. Almes, S. Kalidindi, M. Zekauskas, "A One-way Delay Metric for IPPM", Internet Engineering Task Force, Request for Comments 2679, September 1999.
9. G. Almes, S. Kalidindi, M. Zekauskas, "A One-way Packet Loss Metric for IPPM", Internet Engineering Task Force, Request for Comments 2680, September 1999.
10. J. Heinanen, F. Baker, W. Weiss, J. Wroclawski, "Assured Forwarding PHB Group", Internet Engineering Task Force, Request for Comments 2597, June 1999.
11. C. Villamizar, R. Chandra, R. Govindan, "BGP Route Flap Damping", Internet Engineering Task Force, Request for Comments 2439, November1998.
12. IST MESCAL project, "Specification of Business Models and a Functional Architecture for Inter-domain QoS Delivery", Deliverable D1.1, June 2003.
13. J-Sim Homepage, http://www.j-sim.org.
14. Infonet Suite Homepage, http://www.info.ucl.ac.be/~bqu/jsim/
15. Y. Rekhter, T. Li , "A Border Gateway Protocol 4 (BGP-4)", Internet Engineering Task Force, Request for Comments 1771, March 1995.
16. GÉANT Website, http://www.dante.net/server/show/nav.007

Policy-Aware Connectionless Routing[*]

Bradley R. Smith and J.J. Garcia-Luna-Aceves

University of California, Santa Cruz

Abstract. The current Internet implements *hop-by-hop* packet forwarding based entirely on globally-unique identifiers specified in packet headers, and routing tables that identify destinations with globally unique identifiers and specify the next hops to such destinations. This model is very robust; however, it supports only a single forwarding class per destination. As a result, the Internet must rely on mechanisms working "on top" of IP to support quality-of-service (QoS) or traffic engineering (TE). We present the first policy-based connectionless routing architecture and algorithms to support QoS and TE as part of the basic network-level service of the Internet. We show that policy-aware connectionless routing can be accomplished with roughly the same computational efficiency of the traditional single-path shortest-path routing approach.

1 Introduction

The current Internet architecture is built around the notion that the network layer provides a single-class best-effort service. This service is provided with routing protocols that adapt to changes in the Internet topology, and a packet forwarding method based on a single class of service for all destinations. Using one or more routing protocols, each router maintains a routing-table entry for each destination specifying the globally unique identifier for the destination (i.e., an IP address range) and the next hop along the one path chosen for the destination. Based on such routing tables, each router forwards data packets independently of other routers and based solely on the next-hop entries specified in its routing table. This routing model is very robust. However, there are many examples of network performance requirements and resource usage policies in the Internet that are not homogeneous, which requires supporting multiple service classes [2].

Policy-based routing involves the forwarding of traffic over paths that honor policies defining performance and resource-utilization requirements. *Quality-of-service* (QoS) routing is the special case of policy-based routing in the context of performance policies, and *traffic engineering* (TE) is routing in the context of resource-utilization policies. Section 2 reviews previous solutions for supporting QoS and TE. All past and current approaches to supporting QoS and TE in the Internet have been implemented "on top" of the single-class routing tables of the basic Internet routing model. This leads to inefficient allocation of the available bandwidth, given that paths computed based on shortest-path routing within autonomous systems have little to do with QoS and TE

---

[*] This work was supported in part by the Defense Advanced Research Projects Agency (DARPA) under Grant N66001-00-8942 and by the Baskin Chair of Computer Engineering at UCSC.

constraints. Furthermore, current proposals for policy-based routing [1] are connection-oriented, and require source-specified forwarding implemented by source routing or some form of path setup.

There are two key reasons why policy-based path selection (with QoS and TE constraints) has not been addressed as an integral part of the basic routing model of the Internet. Routing with multiple constraints is known to be NP hard [8], and the basic Internet packet-forwarding scheme is based on globally-unique destination identifiers. This paper introduces the first policy-aware connectionless routing model for the Internet addressing these two limitations. It consists of the routing architecture presented in Section 3, and the path-selection algorithms presented in Section 4. The proposed policy-aware connectionless routing (PACR) architecture is the first to extend the notion of label swapping and threaded indices [4] into connectionless packet forwarding with multiple service classes. The path-selection algorithms we introduce generalize Dijkstra's shortest path first (SPF) algorithm to account for *both* TE and QoS constraints. These algorithms have been shown to be correct (i.e., they compute loop-less paths satisfying TE and QoS constraints within a finite time) [12], and are the first of their kind to attain computational efficiencies close to that of SPF for typical Internet topologies.

2   Previous Work

Resource Management: Two Internet QoS architectures have been developed for resource management: The *integrated services* (intserv) architecture [2], and the *differentiated services* (diffserv) architecture. Both work "on top" of an underlying packet forwarding scheme. In Intserv, network resources must be explicitly controlled; applications reserve the network resources required to implement their functionality; and admission control, traffic classification, and traffic scheduling mechanisms implement the reservations. Diffserv provides resource management without the use of explicit reservations. A set of *per-hop forwarding behaviors* (PHBs) is defined within a diffserv domain to provide resource management services appropriate to a class of application resource requirements. Traffic classifiers are deployed at the edge of a diffserv domain, which classify traffic for one of these PHBs. Inside a diffserv domain, routing is performed using the traditional hop-by-hop, single-class mechanisms.

Resource management for TE is quite simple. The desired resource utilization policies are used as constraints to the path-selection function, and traffic classification and policy-based forwarding mechanisms are used to implement the computed paths. Current proposals [1] define resource-utilization policies by assigning network resources to resource classes, and then specifying what resource classes can be used for forwarding each traffic class.

Routing Architectures: Currently proposed policy-based routing architectures are based on a centralized routing model where routes are computed on-demand (e.g. on receipt of the first packet in a flow, or on request by a network administrator), and forwarding is source-specified through the use of source routing or path setup [6] techniques. These solutions are less robust, efficient, and responsive than the original distributed routing method. The forwarding paths in on-demand routing are brittle, because the ingress router controls remote forwarding state in routers along paths it has set up, and must re-establish the paths in the event that established paths are broken.

Path Selection: The seminal work on the problem of computing paths in the context of more than one additive metric was done by Jaffe [8], who defined the multiply-constrained path problem (MCP) as the computation of paths in the context of two additive metrics. He presented an enhanced distributed Bellman-Ford (BF) algorithm that solved this problem with time complexity of $O(n^4 b \log(nb))$, where n is the number of nodes in a graph, and b is the largest possible metric value. Many solutions have been proposed for computing exact paths in the context of multiple metrics for special situations since Jaffe's work. Wang and Crowcroft [14] were the first to present the solution to computing paths in the context of a concave (i.e. "minmax") and an additive metric. Ma and Steenkiste [9] presented a modified BF algorithm that computes paths satisfying delay, delay-jitter, and buffer space constraints in the context of weighted-fair-queuing scheduling algorithms in polynomial time. Cavendish and Gerla [3] presented a modified BF algorithm with complexity of $O(n^3)$ which computes multi-constrained paths if all metrics of paths in an internet are either non-decreasing or non-increasing as a function of the hop count. Recent work by Siachalou and Georgiadis [11] on MCP has resulted in an algorithm with complexity $O(nW \log(n))$. This algorithm is a special case of the policy-based path-selection presented in Section 4 of this paper.

Several other algorithms have been proposed for computing approximate solutions to the QoS path-selection problem. Both Jaffe [8] and Chen and Nahrstedt [5] propose algorithms which map a subset of the metrics comprising a link weight to a reduced range, and show that using such solutions, the cost of a policy-based path computation can be controlled at the expense of the accuracy of the selected paths. Similarly, a number of researchers [8,10] have presented algorithms that compute paths based on a function of the multiple metrics comprising a link weight. These approximation solutions do not work with administrative traffic constraints.

3   Policy-Aware Connectionless Routing (PACR)

Policy-based routing requires the ability to compute and forward traffic over multiple paths for a given destination. For TE, multiple paths may exist that satisfy disjoint network usage policies. For QoS, there may not exist a universally "best" route to a given node in a graph. For example, which of two paths is best when one has delay of $5ms$ and jitter of $4ms$, and the other has delay of $10ms$ and jitter of $1ms$ depends on which metric is more critical for a given application. For FTP traffic, where delay is important and jitter is not, the former would be more desirable. Conversely, for video streaming, where jitter is very important and delay is relatively un-important, the latter would be preferred. Such weights are said to be *incomparable*. In contrast, it is possible for one route to be clearly "better than" another in the context of multi-component link weights. For instance, a route with delay of $5ms$ and jitter of $1ms$ is clearly better than a route with delay of $10ms$ and jitter of $5ms$ for all possible application requirements. Such weights are said to be *comparable*.

The goal of routing in the context of multi-component link weights is to find *the largest set of paths to each destination with weights that are mutually incomparable.*

The weights in such a set are called the *performance classes* of a destination. Supporting policy-based connectionless routing requires three main functions: (a) computing and maintaining routes that satisfy QoS and TE constraints for each destination, (b) classifying traffic intended for a given destination on the basis of TE and QoS constraints, and (c) forwarding the classified traffic solely on the basis of the next hops specified in the routing tables of routers for a given destination and for a given traffic class. The rest of this section outlines our proposed solution, which we call PACR (policy-aware connectionless routing).

3.1 Policy-Aware Route Computation

The routing protocols used in PACR must be designed to carry the link metrics required to implement the desired QoS and TE policies. This requires the use of either a topology-broadcast (also called "link-state") or link-vector routing protocol [7] that exchanges information describing the state of links. The implementation of these routing protocols consists of two main parts: (a) information exchange signaling, and (b) local path-selection.

The signaling component of the protocols is straightforward, because it suffices to re-engineer the signaling of one of many existing routing protocols to accommodate QoS and TE parameters of links. As we discuss subsequently, routing-table information can be exchanged in such signaling, in addition to link-state information. The path-selection component of the protocols is the complex part of PACR, because the path-selection algorithm used must produce paths that satisfy QoS and TE constraints at roughly the same speed with which today's shortest-path algorithms compute paths in a typical Internet topology. Section 4 presents the new path-selection algorithms required in PACR, which arguably constitute the main contribution of the new architecture.

3.2 Packet Forwarding

Solutions for packet classification already exist and can be applied to distributed policy-based routing. However, forwarding packets solely based on IP addresses would require

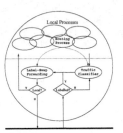

Fig. 1. Traffic flow in PACR

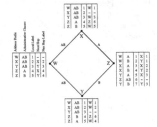

Fig. 2. Forwarding labels in PACR

each relay of a packet to classify the packet before forwarding according to the content of its routing table. We propose using label-swap forwarding technology to require only the first router that handles a packet to classify it before forwarding it. Accordingly, the forwarding state of a router must be enhanced to include local and next hop label information, in addition to the destination and next hop information existing in traditional forwarding tables. Traffic classifiers must then be placed at the edge of an internet, where "edge" is defined to be any point from which traffic can be injected into the internet. Figure 1 illustrates the resulting traffic flow requirements of a router in PACR.

To date, label-swapping has been used in the context of connection-oriented (virtual circuit) packet forwarding architectures. A connection setup phase establishes the labels that routers should use to forward packets carrying such labels, and a label refers to an active source-destination connection [6]. Chandranmenon and Varghese [4] present *threaded indices*, in which neighboring routers share labels corresponding to indexes into their routing tables for routing-table entries for destinations, and such labels are included in packet headers to allow rapid forwarding-table lookups.

The forwarding labels in PACR are similar to threaded indices. A label is assigned to each routing-table entry, and each routing-table entry corresponds to a policy-based route maintained for a given destination. Consequently, for each destination, a router exchanges one or multiple labels with its neighbors. Each label assigned to a destination corresponds to the set of service classes satisfied by the route identified by the label. For example, Figure 2 shows a small network with four nodes with the forwarding tables at each node, two administrative classes A and B, and the given forwarding state for reaching the other nodes.

4   Policy-Based Path-Selection Algorithm

We model a network as a weighted undirected graph $G = (N, E)$, where N and E are the node and edge sets, respectively. By convention, the size of these sets are given by $n = |N|$ and $m = |E|$. Elements of E are unordered pairs of distinct nodes in N. $A(i)$ is the set of edges adjacent to i in the graph. Each link $(i, j) \in E$ is assigned a weight, denoted by ω_{ij}. A *path* is a sequence of nodes $< x_1, x_2, \ldots, x_d >$ such that $(x_i, x_{i+1}) \in E$ for every $i = 1, 2, \ldots, d - 1$, and all nodes in the path are distinct. The weight of a path is given by $\omega_p = \sum_{i=1}^{d-1} \omega_{x_i x_{i+1}}$. The nature of these weights, and the functions used to combine these link weights into path weights are specified for each algorithm.

4.1   Administrative Policies

We use a *declarative* model of administrative policies in which constraints on the traffic allowed in an internet are specified by expressions in a boolean traffic algebra. The *traffic algebra* is composed of the standard boolean operations on the set $\{0, 1\}$, where a set of p primitive propositions (variables) represent statements describing characteristics of network traffic or global state that are either true or false. The syntax for expressions in the algebra is specified by the BNF grammar:

$$\varphi ::= 0 \mid 1 \mid v_1 \ldots v_p \mid (\neg\varphi) \mid (\varphi \wedge \varphi) \mid (\varphi \vee \varphi) \mid (\varphi \rightarrow \varphi)$$

The set of primitive propositions, indicated by v_i in the grammar, can be defined in terms of network traffic characteristics or global state. Administrative policies are specified for an internet by assigning expressions in the algebra to links in the graph, called *link predicates*. These predicates define a set of *forwarding classes*, and constrain the topology that traffic for each forwarding class is authorized to traverse, as required by the administrative policies.

A $SAT(\varphi)$ primitive is required for expressions in the traffic algebra which is the SATISFIABILITY problem of traditional propositional logic. Satisfiability must be tested in two situations: to determine if traffic classes exist that are authorized to use an extension to a known route, and to determine if all traffic authorized for a new route is already satsified by known shorter routes. The first is true *iff* the conjunction of these expressions is satisfiable (i.e., $SAT(\varepsilon_i \wedge \varepsilon_{ij})$). The second is true *iff* the new route's traffic expression implies the disjunction of the traffic expressions for all known better routes (i.e., $(\varepsilon_i \rightarrow \varepsilon_{i_1} \vee \varepsilon_{i_2} \vee ..)$ is *valid*, which is denoted by $(\varepsilon_i \rightarrow \mathcal{E}_i)$ in the algorithms). Determining if an expression is valid is equivalent to determining if the negation of the expression is unsatisfiable. Therefore, expressions of the form $\varepsilon_1 \rightarrow \varepsilon_2$ are equivalent to $\neg SAT(\neg(\varepsilon_1 \rightarrow \varepsilon_2))$ (or $\neg SAT(\varepsilon_1 \wedge \neg \varepsilon_2)$). Satisfiability has many restricted versions that are computable in polynomial time. We have implemented an efficient, restricted solution to the SAT problem by implementing the traffic algebra as a set algebra with the set operations of intersection, union, and complement on the set of all possible forwarding classes.

4.2   Performance Characteristics

Path weights are composed of multi-component metrics that capture all important performance measures of a link such as delay, delay variance ("jitter"), available bandwidth, etc. Our path-selection algorithm is based on an enhanced version of the path algebra defined by Sobrinho [13], which we enhance to support the computation of the best *set* of routes for each destination. Formally, the path algebra $P = <\mathcal{W}, \oplus, \preceq, \sqsubseteq, \overline{0}, \overline{\infty}>$ is defined as a set of weights $\mathcal{W}$, with a binary operator $\oplus$, and two order relations, $\preceq$ and $\sqsubseteq$, defined on $\mathcal{W}$. There are two distinguished weights in $\mathcal{W}$, $\overline{0}$ and $\overline{\infty}$, representing the least and absorptive elements of $\mathcal{W}$, respectively. Operator $\oplus$ is the original path composition operator, and relation $\preceq$ is the original total ordering from [13]. Operator $\oplus$ is used to compute path weights from link weights. The relation $\preceq$ is used by the routing algorithm to build the forwarding set, starting with the minimal element, and by the forwarding process to select the minimal element of the forwarding set whose parameters satisfy a given QoS request.

We add a new relation on routes, $\sqsubseteq$, to the algebra and use it to define classes of comparable routes to select maximal elements of these classes for inclusion in the set of forwarding entries for a given destination. Relation $\sqsubseteq$ is a partial ordering (reflexive, antisymmetric, and transitive) with the additional property that $(\omega_x \sqsubseteq \omega_y) \Rightarrow (\omega_x \succeq \omega_y)$. The relation $\sqsubseteq$ defines an ordering on routes in terms of the containment (subset) of the set of constraints satisfied by one route within the set satisfied by another, i.e., if $\omega_i \sqsubseteq \omega_j$, then the set of constraints that route i can satisfy is a subset of those satisfiable by route j.

algorithm Policy-Based-Dijkstra
   begin

1	$Push(<s, s, \overline{0}, 1>, P_s)$;		
2	**for each** $\{(s, j) \in A(s)\}$	14	$Push(<i, p_i, \omega_i, \varepsilon_i>, P_i)$;
3	$Insert(<j, s, \omega_{sj}, \varepsilon_{sj}>, T)$;	15	**for each** $\{(i, j) \in A(i) \mid SAT(\varepsilon_i \wedge \varepsilon_{ij})\}$
4	**while** $(\mid T \mid = 0)$		**begin**
	begin	16	$\omega_j \leftarrow \omega_i \oplus \omega_{ij}$; $\varepsilon_j \leftarrow \varepsilon_{ij}$;
5	$<i, p_i, \omega_i, \varepsilon_i> \leftarrow Min(T)$;	17	**if** $(T_j = \emptyset)$
6	$DeleteMin(B_i)$;	18	**then** $Insert(<j, i, \omega_j, \varepsilon_j>, T)$
7	**if** $(\mid B_i \mid = 0)$	19	**else if** $(\omega_j \prec T_j.\omega)$
8	**then** $DeleteMin(T)$	20	**then** $DecreaseKey(<j, i, \omega_j, \varepsilon_j>, T)$;
9	**else** $IncreaseKey(Min(B_i), T_i)$;	21	$Insert(<j, i, \omega_j, \varepsilon_j>, B_j)$;
10	$\varepsilon_{tmp} \leftarrow \varepsilon_i$; $ptr \leftarrow Tail(P_i)$;		**end**
11	**while** $((\varepsilon_{tmp} \neq 0) \wedge (ptr \neq \emptyset))$		**end**
12	$\varepsilon_{tmp} \leftarrow \varepsilon_{tmp} \wedge \neg ptr.\varepsilon$; $ptr \leftarrow ptr.next$;	**end**	
13	**if** $(\varepsilon_{tmp} \neq 0)$		

Fig. 3. General-Policy-Based Dijkstra.

Table 1. Notation.

P_n	≡ Queue of permanent routes to node n.
T	≡ Heap of temporary routes.
T_n	≡ Entry in T for node n.
B_n	≡ Balanced tree of routes for node n.

A route r_m is a *maximal element* of a set R of routes in a graph if the only element $r \in R$ where $r_m \sqsubseteq r$ is r_m itself. A set R_m of routes is a *maximal subset* of R if, for all $r \in R$ either $r \notin R_m$, or $r \in R_m$ and for all $s \in R - \{r\}$, $\neg(r \sqsubseteq s)$. The maximum size of a maximal subset of routes is the smallest range of the components of the weights (for the two component weights considered here).

4.3 Path Selection

Path selection in PACR consists of computing the maximal set of routes to each destination in an internet for each traffic class (stated through link predicates and multi-component link weights) for which a path to the destination exists.

The path-selection algorithm in PACR maintains a balanced tree (B_i) for each node i in the graph to hold newly discovered, temporary labeled routes for node i. A heap T contains the lightest weight entry from each non-empty B_i (for a maximum of n entries), and the heap entry for node i is denoted by T_i. Lastly, a queue, P_i, is maintained for each node which contains the set of permanently labeled routes discovered by the algorithm, in the order in which they are discovered (which will be in increasing weight). The general flow of the path-selection algorithm is to take the minimum entry from the heap T, compare it with existing routes in the appropriate P_i, if it is incomparable with existing routes in P_i it is pushed onto P_i, and add "relaxed" routes for its neighbors to the appropriate B_x's.

The correctness of the PACR path-selection algorithm is based on the maintenance of the following three invariants: for all routes $I \in P$ and $J \in B_s, I \preceq J$, all routes to a given destination i in P are incomparable for some set of satisfying truth assignments, and the maximal subset of routes to a given destination j in $P_j \cup B_j$ represents the maximal

Table 2. Operations on data structures.

Notation	Description
	Queue
$Push(r, Q)$	INSERT RECORD r AT TAIL OF QUEUE Q ($O(1)$)
$Head(Q)$	RETURN RECORD AT HEAD OF QUEUE Q ($O(1)$)
$Pop(Q)$	DELETE RECORD AT HEAD OF QUEUE Q ($O(1)$)
$PopTail(Q)$	DELETE RECORD AT TAIL OF QUEUE Q ($O(1)$)
	d-Heap
$Insert(r, H)$	INSERT RECORD r IN HEAP H ($O(\log_d(n))$)
$IncreaseKey(r, r_h)$	REPLACE RECORD r_h IN HEAP WITH RECORD r HAVING GREATER KEY VALUE ($O(d\log_d(n))$)
$DecreaseKey(r, r_h)$	REPLACE RECORD r_h IN HEAP WITH RECORD r HAVING SMALLER KEY VALUE ($O(\log_d(n))$)
$Min(H)$	RETURN RECORD IN HEAP H WITH SMALLEST KEY VALUE ($O(1)$)
$DeleteMin(H)$	DELETE RECORD IN HEAP H WITH SMALLEST KEY VALUE ($O(d\log_d(n))$)
$Delete(r_h)$	DELETE RECORD r_h FROM HEAP ($O(d\log_d(n))$)
	Balanced Tree
$Insert(r, B)$	INSERT RECORD r IN TREE B ($O(\log(n))$)
$Min(B)$	RETURN RECORD IN TREE B WITH SMALLEST KEY VALUE ($O(\log(n))$)
$DeleteMin(B)$	DELETE RECORD IN TREE B WITH SMALLEST KEY VALUE ($O(\log(n))$)

subset of all paths to j using nodes with routes in P. Furthermore, these invariants are maintained by the following two constraints on actions performed in each iteration of these algorithms: (1) only known-non-maximal routes are deleted or discarded, and (2) only the smallest known-maximal route to a destination i is moved to P_i. The details of this proof are presented elsewhere [12].

The PACR path-selection algorithm, presented in Figure 3 computes an optimal set of routes to each destination subject to multiple general (additive or concave) path metrics, in the presence of traffic constraints on the links. The notation used in the algorithms presented in the following is summarized in Table 1. Table 2 defines the primitive operations for queues, heaps, and balanced trees used in the algorithms, and gives their time complexity used in the following analysis. The worst-cast time complexity of Policy-Based-Dijkstra is $O(nW^2A^2)$, where the maximum number of unique truth assignments is denoted by $A = 2^p$ (p is the number of primitive propositions in the traffic algebra), and the maximum number of unique weights by $W = $ min(*range of weight components*). The performance of special-case variants of this algorithm for traffic-engineering and QoS (called the "Basic" algorithms below) are $O(mA\log(A))$ and $O(mW\log(W))$, respectively. Furthermore, for these variants, refinements in the data structures result in algorithms (called the "Enhanced" algorithms below) with $O(mA\log(n))$ and $O(mW\log(n))$ complexity. Details of these variants and the complexity analysis are presented elsewhere [12].

4.4 Performance Results

Figures 4 and 5 present performance results for the path-selection algorithm. The experiments were run on a 1GHz Intel Pentium 3 based system. The algorithms were implemented using the C++ Standard Template Library (STL) and the Boost Graph Library. Each test involved running the algorithm on ten random weight assignments to ten randomly generated graphs (generated using the GT-ITM package [15]).

Fig. 4 show the worst-case measurements for each test of the QoS algorithms. The "Traditional" algorithm is an implementation of Dijkstra's shortest-path-first (SPF) algorithm using the same environment as that used for the other algorithms for use as a reference. The metrics were generated using the "Cost 2" scheme from [11], where the delay component is randomly selected in the range $1..MaxMetric$, and the cost component is computed as $cost = \sigma(MaxMetric - delay)$, where σ is a random integer

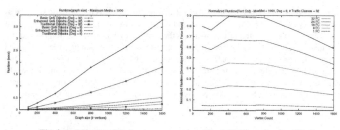

Fig. 4. QoS Runtime(Size)   **Fig. 5.** TE Norm Runtime(Size)

in the range 1..5; this scheme was chosen as it proved to result in the most challenging computations from a number of different schemes considered.

Tests were run for performance (both runtime and space) as a function of graph size, average degree of the graph, and the maximum link metric value. Due to space constraints, only the graphs for runtime as a function of size are shown here with a maximum metric of 1000. These results show that, while costs increase with both graph size and average degree, the magnitude and rate of growth are surprisingly tame for what are fundamentally non-polynomial algorithms.

Fig. 5 shows the performance of the "Basic" traffic engineering algorithm on a similar range of parameters. Each data point represents the worst performance of the algorithm out of 9 runs (3 randomly generated graphs with 3 random link weight assignments each). To control the number of forwarding classes in a graph, each graph was generated as two connected subgraphs. *Bridge* links were then added between 32 randomly selected pairs of vertices from each subgraph to form a single graph with at most 32 paths between any two nodes in different original subgraphs. 32 tests of the algorithm are then run with all traffic classes initially allocated to one bridge link (resulting in one forwarding class for all 32 traffic classes), and successive runs are performed with traffic classes distributed over one additional link for each test, with the final run allowing one traffic class over each bridge link (resulting in a one-to-one mapping of traffic classes to forwarding classes). In each test, the link predicate of all non-bridge links is set to allow all traffic classes (i.e. it is set to *true*). Each plot shows the results for 1, 8, 16, 24, and 32 forwarding classes in terms of the runtime of the algorithm normalized as a fraction of the "Brute Force" runtime required to run the traditional Dijkstra algorithm once for each traffic class. The plots show that the algorithm provides significant savings when the number of forwarding classes is small, and gracefully degrades as the number of forwarding classes grows.

5   Conclusions

We have defined policy-aware routing as the computation of paths, and the establishment of forwarding state to implement paths, in the context of non-homogeneous performance requirements and network usage policies. We showed that a fundamental requirement of policy-aware routing is support for multiple paths to a given destination, and that the address-based, single-forwarding-class Internet routing model cannot support such a requirement. We presented PACR, which is the first policy-based connectionless routing architecture, and includes the first policy-based routing solution that provides integrated support of QoS and TE. The path-selection algorithms introduced for PACR constitute the most efficient algorithms for path selection with QoS and TE constraints known to date. Furthermore, their computational efficiency is comparable to that of shortest-path routing algorithms, which makes policy-aware connectionless routing in the Internet feasible.

References

1. D. O. Awduche, J. Malcom, J. Agogbua, M. O'Dell, and J. McManus, "Requirements for Traffic Engineering Over MPLS," RFC 2702, December 1999.
2. B. Braden, D. Clark, and S. Shenker, "Integrated Services in the Internet Architecture: an Overview," RFC 1633, June 1994.
3. D. Cavendish and M. Gerla. "Internet QoS Routing using the Bellman-Ford Algorithm," *Proc. IFIP Conference on High Performance Networking*, 1998.
4. G. P. Chandranmenon and G. Varghese, "Trading Packet Headers for Packet Processing," *IEEE ACM Trans. Networking*, 4(2):141–152, Oct. 1995.
5. S. Chen and K. Nahrstedt, "An Overview of Quality of Service Routing for Next-Generation High-Speed Networks: Problems and Solutions," *IEEE Network*, pp. 64–79, Nov. 1998.
6. B. Davie and Y. Rekhter, *MPLS: Technology and Applications*, Morgan Kaufmann, 2000.
7. J.J. Garcia-Luna-Aceves and J. Behrens, "Distributed, Scalable Routing Based on Vectors of Link States," *IEEE JSAC*, Oct. 1995.
8. J. M. Jaffe, "Algorithms for Finding Paths with Multiple Constraints," *Networks*, 14(1):95–116, 1984.
9. Q. Ma and P. Steenkiste, "Quality-of-Service Routing for Traffic with Performance Guarantees," *Proc. 4th International IFIP Workshop on QoS*, May 1997.
10. P. Van Mieghem, H. De Neve, and F. Kuipers, "Hop-by-hop quality of service routing," *Computer Networks*, 37:407–423, November 2001.
11. S. Siachalou and L. Georgiadis, "Efficient QoS Routing," *Proc. Infocom'03*, April 2003.
12. Brad Smith, "Efficient Policy-Based Routing in the Internet", PhD Thesis, Computer Science, University of California, Santa Cruz, CA 95064, September 2003.
13. J. L. Sobrinho, "Algebra and Algorithms for QoS Path Computation and Hop-by-Hop Routing in the Internet," *IEEE/ACM Trans. Networking*, 10(4):541–550, Aug. 2002.
14. Z. Wang and J. Crowcroft, "Quality-of-Service Routing for Supporting Multimedia Applications," *IEEE JSAC*, pp. 1228–1234, Sept. 1996.
15. E. W. Zegura, K. Calvert, and S. Bhattacharjee, "How to Model an Internetwork," *Proc. IEEE Infocom '96*, 1996.

Maximum Flow Routing with Weighted Max-Min Fairness

Miriam Allalouf and Yuval Shavitt

School of Electrical Engineering
Tel Aviv University

Abstract. Max-min is an established fairness criterion for allocating bandwidth for flows. In this work we look at the combined problem of multi-path routing and bandwidth allocation such that the flow allocation for each connection will be maximized and fairness will be maintained. We use the weighted extension of the max-min criterion to allocate bandwidth in proportion to the flows' demand. Our contribution is an algorithm which, for the first time, solves the combined routing and bandwidth allocation problem for the case where flows are allowed to be splitted along several paths. We use multi commodity flow (MCF) formulation which is solved using linear programming (LP) techniques. These building blocks are used by our algorithm to derive the required optimal routing and allocation.

1 Introduction

Traffic engineering is a paradigm where network operators control the traffic and allocate resources in order to achieve goals such as, maximum flow or minimum delay. One challenge is to allow different aggregates of flows to share the network, so that the total flow will be maximized while fairness will be preserved. These flows are derived from customers service level agreements (SLAs) and are abstracted here as a list of rate allocation demands to be sent between specific source and destination nodes. Given a network topology and a set of demands, network operators may wish to maximize their profit by routing the traffic so it will maximize the total allocated bandwidth the network can carry, or namely the total assigned net flow. To do this we allow flows to be arbitrarily split inside the network.

One way to maximize the network flow is to formulate it as a maximum multi-commodity flow (MCF) problem which can be solved using linear programming (LP). Each commodity net flow is assigned to a different client. While the solution will maximize the flow, it will not always do it in a fair manner. Flows that traverse several loaded links will be allocated very little bandwidth or non at all, while flows that traverse short hop distances or meet fewer other flows on its way will receive a large allocation of bandwidth.

In an attempt to introduce fairness to the maximum flow problem the concurrent multi-commodity flow problem was suggested. This MCF LP formulation requires that all demands will be equally satisfied and seeks a routing that maximizes network flow in equal portions per demand. However, the end solution under-utilizes the network, sometimes saturating only a small fraction of it.

J. Solé-Pareta et al. (Eds.): QofIS 2004, LNCS 3266, pp. 278–287, 2004.
© Springer-Verlag Berlin Heidelberg 2004

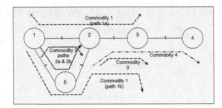

Fig. 1. An example of flow assignment: Maximum Flow, Maximum Concurrent, max-min fair and the WMCM algorithm weighted max-min fair algorithms

In this paper we suggest an algorithm, called WMCM (Weighted Max-min fair Concurrent MCF), that finds routing and resource allocation that maximizes the network link utilization, but, at the same time, adheres to the weighted max min fairness criterion [1]. The WMCM algorithm is based on the maximum concurrent multi-commodity flow problem and bridges the gap between the two other solutions described above.

To clarify the difference between the different algorithms consider the example in Figure 1, which depicts a network with five nodes connected by one unit capacity links. Consider the max-min fairness setting used in [2] where each path represents one flow. We have 6 flows. Two flows from node 1 to node 3: flow $1a$ using path (1-2-3) and flow $1b$ using path (1-5-2-3), two flows from node 1 to node 2: one ($2a$) using link (1,2) and the other, $2b$, via node 5, one flow (flow 3) from node 2 to node 3 and one flow from node 2 to node 4 (flow 4). The max-min fair [2] vector in this case is (1/4,1/4,3/4,3/4,1/4,1/4), where the bottleneck link is shared by four flows, each gets 1/4 unit. The used setting for the MCF flow allocation refers to flows per commodity instead of per path. This example shows 4 commodities: commodity 1 with paths $1a$ and $1b$, commodity 2 with paths $2a$ and $2b$, commodity 3 and 4 with one path each. The maximum MCF problem results in an allocation commodity rate vector (0,2,1/2,1/2) (path vector of (0,0,1,1,1/2,1/2)) starving the two paths ($1a$ and $1b$) of 'commodity 1' flow to achieve the maximum possible flow of 3 units. The WMCM algorithm with equal weights for all commodities will results with the unique commodity rate vector (1/3,5/3,1/3,1/3) for commodities 1,2,3 and 4. Note that in this case there are more than one allocation of flows per path that achieves the max-min vector, e.g., (1/3,0,2/3,1,1/3,1/3) or (1/6,1/6,5/6,5/6,1/3,1/3). In case commodity 1 is given a weight that is double than the rest of the nodes (demand vector (2,1,1,1)), the concurrent MCF problem will allocate it double the bandwidth allocated for the flows in its bottleneck link (link(2,3)) and the weighted max-min vector is (1/2,3/2,1/4,1/4).

2 Related Work

The Max-min fairness bandwidth allocation was mostly examined in the context of one fixed path per session, where a session is defined by a pair of terminals. A simple algorithm that finds the max-min fair allocation where routing is given appears in [2].

Chen and Nahrstedt [3] provide max-min fair allocation routing. They present an unweighted heuristic algorithm that selects the best path so the fairness-throughput is maximized upon an addition of a new flow. Their algorithm assumes the knowledge of the possible paths for each new flow.

Many other distributed algorithms deal with dynamic adjustments of flow rates to maintain max-min fairness when the routes are given [4,5,6,7]. The above algorithms differ by the assumptions on the allowed signaling, and available data. Bartal *et al.* [7] find the total maximum flow allocation in a network for given routes using distributed computations of the global MCF LP problem.

Kelly *et al.* [8] propose the proportional fairness concepts and a convergence algorithm. Mo and Warland [9] generalize the proportional fairness and produce end-to-end flow control of TCP streams by changing the transmission window size, but again they deal with flow allocation without routing.

LP in Traffic Engineering. Many works considered the multi-commodity flow allocation[1] as the mathematical formalism suitable for traffic engineering and path design. Ott *et al.* [16] suggest different off-line LP formulations to optimally spread paths so the excess bandwidth will be maximized in an MPLS networks. In order to avoid link "overflow" caused by flows fluctuation, they calculate how to reduce the assigned throughput per link. Mitra and Ramakrishnan [17] present algorithms that are based on the MCF LP problem allocating bandwidth for the various QoS-type services: QoS and Best-Effort traffic.

Most of the studies, including the ones mentioned above, chose an MCF formulation that considers the demands but they do not discuss the max-min fairness in conjunction with maximum throughput as the WMCM algorithm does.

3   Algorithm

We consider as input a network topology and directional links capacities, a list of ingress-egress pairs, and per-pair traffic average-rate demand. Traffic between ingress-egress pair may be split arbitrarily among different paths. We model the network as a general directed graph where arc label represents link capacity. Each (ingress,egress) peering point pair is represented by a different commodity with some demanded rate.

Our goal is to fulfill clients' demands optimally while keeping a fair sharing of the allocated bandwidth, to lay the set of paths to be used between each pair in the network, and to allocate them bandwidth in a maximal way. The fairness criterion is defined by the weighted max-min fairness.

The WMCM algorithm solves, iteratively, the maximum concurrent MCF LP until network saturation is achieved. Each iteration is computed on the residual capacity from the previous iteration with non-saturated flows left. First, we will provide the formal definition of the weighted max-min criterion (subsection 3.1), then we will describe the maximum concurrent flow (subsection 3.2), and finally we will present our WMCM algorithm (subsection 3.3).

---

[1] Relevant mathematical and algorithmic background on MCF can be found in [10,11,12]. Specific theoretical aspects on the max. concurrent MCF problem and its complexity can be found in [13,14,15].

3.1 The Weighted Max-Min Criterion: Definitions

The weighted max-min fairness criterion is an extension [1] of the max-min fairness criterion [2]. While the max-min definition is stated for the case where each flow takes a single path, it can be also applied to the case where a flow may be split among several paths.

Definition 1. *The Commodity Rate Vector, cr, is a vector whose elements are the rates which were assigned to the commodities.*

Definition 2. *A Flow Rate Vector, f_i, is a vector of the rates assigned to a set of paths of commodity i.*

From the above definitions we can write that for any commodity i, $\sum_{P_{ij} \in P_i} f_{ij} = cr_i$ where P_i is the set of the paths of commodity i. The weighted max-min fair algorithm finds the optimal commodity rate vector cr^* and a flow rate vector f_i per each commodity rate cr_i^* such that cr^* is the lexicographically largest feasible vector. [2]

Note that the algorithm provides the optimal commodity rate vector and a specific flow rate vector f_i that accommodate the optimal cr_i, though the flow rate vector may have several valid realizations (see example 1).

Definition 3. *A commodity rate vector cr is said to be max-min fair if it is feasible and if each of its elements cr_i cannot be increased without decreasing any other element cr_k for which $cr_i \geq cr_k$*

Definition 4. *A commodity rate vector cr is said to be **weighted** max-min fair if it is feasible and if for each commodity i, cr_i cannot be increased (maintaining feasibility) without decreasing any other element cr_k for which $cr_i/dem_i \geq cr_k/dem_k$.*

The two definitions above also hold when traffic may be split to several paths.

3.2 The Maximum Concurrent MCF Problem

The **Maximum Concurrent flow** problem is stated as follows. Let G=(V,A) be a directed graph with nonnegative capacities $c(a), \forall a \in A$. If $a \notin A$ $c(a) = 0$. There are K different commodities: $C_1, \ldots, C_K$ where commodity i is specified by the triplet $C_i = (s_i, t_i, dem_i)$. s_i and t_i are the source and the sink of commodity i, respectively, and dem_i is its rate demand. Each pair is distinct, but vertices may participate in different pairs. The objective is to maximize z so all the $i = 1, \ldots, K$, $z \cdot dem_i$ units of the respective commodities can be routed simultaneously, subject to flow conservation and link capacity constraints. The objective z is the equal maximal fraction of all demands. There are two ways to formulate this problem: path flow and arc flow formulations.

---

[2] For the lexicographical order between two vectors v and u we examine them ordered in increasing order, v and ν, respectively. We say that $v > u$ if there is an index i, such that, $v_i \geq \nu_i$ and $v_j = \nu_j, 1 \leq j \leq i - 1$. Namely we find the longest equal prefixes of the two ordered vectors and define the order according to the first element which is not different.

The maximum concurrent flow: path flow formulation

Let P_i be the set of all the paths of commodity i between s_i and t_i. The Linear program PR assigns the maximum commodity flow to P_i while being restricted by the fairness criterion. The assigned net flow per arc a is the sum of the net flows of the paths passing this arc. PR's solution is composed of the assigned net flow $f(P_{ij})\ \forall P_{ij} \in P_i,\ i = 1, \ldots, K$ and the maximal fairness value z.

LP PR: Path Flow Formulation

maximize z

subject to

$$\forall a \in A, \sum_{i=1}^{K} \sum_{P \in P_i} f(P) \leq c(a)\ /\text{*}a \text{ is an arc on } P\text{*}/ \qquad (1)$$

$$\forall i = 1, \ldots, K, \sum_{P \in P_i} f(P) \geq z \cdot dem_i \qquad (2)$$

$$\forall P \in P_{i=1,\ldots,K} f(P) \geq 0, z \geq 0$$

The size of this linear problem grows with the number of possible paths between any pair of nodes and can be exponentially large when the network is highly connected. For this reason we will reformulate this problem so it could be solved in a polynomial number of steps:

Maximum concurrent flow: arc flow formulation

The variable $acf_k(i, j)\ \forall k = 1, \ldots, K\ \forall a = (i, j) \in A$ holds the assigned net flow per commodity k over arc (i, j). For each vertex $v \in V - \{s_k, t_k\}$, we require that the total flow of commodity k into vertex v is equal to the total flow of commodity k out of vertex v. $\forall a \in A$, the total flow over all commodities is at most $c(a)$. The total flow rate of commodity i out of vertex s_i represents the maximum possible flow per commodity i and should be greater than or equal to the fraction of the respective demand (Eq. 4).

Problem PR' Arc Formulation

maximize z

subject to

$$\forall a = (u, v) \in A, \sum_{k=1..K} acf_k(u, v) \leq c(u, v) \qquad (3)$$

$$\forall k = 1..K, \forall v \in V - s_k, \sum_{v \in V} acf_i(s_k, v) \geq z \cdot dem_i \qquad (4)$$

$$\forall k = 1..K, \forall (u, v, w) \in V - s_k, t_k, acf_k(u, v) - acf_k(v, w) = 0 \qquad (5)$$

$$\forall k = 1..K\ \forall u, v \in V, acf_k(u, v), z \geq 0 \qquad (6)$$

PR' is a reformulation of PR which leads to $O(K \cdot m)$ variables and $O(m + Kn)$ constraints where m is the number of arcs, n the number of vertices, and K the number of commodities, that can be solved in a polynomial number of steps.

The solution of PR' problem gives us the same maximal portion of all the demands restricted by links capacities. The LP variables hold the maximal value of the allocated

---

Flow decomposition algorithm
1. **for** each commodity k **do**
2.   set $P_k = null$
3. P_k holds the paths for commodity k
4.   **while** TRUE
5.   let p be a path between (s_k, t_k).
6.   **if** path p was found **then**
7. $minacf_k = \min_{(u,v) \in p}(acf_k(u, v))$
8.   Add p to set P_k
9. $f(p) = minacf_k$
10. $\forall(w, x) \in p, acf_k(w, x) = acf_k(w, x) - minacf_k$
11.   **else** go to 1

Fig. 2. The flow decomposition algorithm [18].

net flow per commodity. To derive the set of routing paths per each commodity, we need to apply, per commodity, the **flow decomposition algorithm** as described in figure 2 and [18,17]. This algorithm runs in a polynomial number of steps. It decomposes (separately for each commodity) the total net flow per commodity k over all arcs into a set of paths.

3.3 WMCM – Weighted Max-Min Fair Concurrent MCF Algorithm

As noted before, the solution of the maximum concurrent MCF can lead us to an un-saturated network. Assuming two commodities with disjoint paths and the available bandwidth per commodity 1 is larger than that of commodity 2. Restricted by Eq. 4, namely z, PR' will assign both commodities the same net flow size although commodity 2 has more available bandwidth. We have developed a new algorithm that merges the desire for network saturation and fairness. The formal description of algorithm is given in Figure 3.

The main idea behind our WMCM algorithm is to increase the allocated net flow per arc over the residual graph while keeping the weighted fairness criterion at each iteration. We use the maximum concurrent MCF problem solution to increase it optimally. The algorithm recieves as input the list of commodities, $KCOMM$, the vector of demands, dem, and the graph, G. Each iteration starts (see line 4) with a reduced number of commodities (line 19), recalculated demands (line 17) and a residual graph of the unassigned capacities (lines 10-12). $\Delta acf_k(i, j)$ holds the net flow assigned in the iteration, and z holds the fraction of the demands that was fulfilled. Finally, $acf_k(i, j)$ holds the total accumulated net flow per commodity k over arc (i, j). The flow decomposition algorithm is performed (line 21) once and provides the maximum routing with the weighted max-min fair allocation. The strength of the algorithm is due to the optimality and scalability of the maximum concurrent MCF problem. Its running time is $K \cdot T_{concMCF}$ where $T_{concMCF}$ is the running time of solving maximum concurrent MCF LP.

WMCM(KCOMM,dem,G)
1. /* Initialization stage */
2. $\forall (i,j) \in A, k = 1..K, acf_k(i,j) = 0$
3. $G_{RES} = G$ /* G is the original graph */
4. **while** $(KCOMM \neq null)$ **do**
5. /* The calculation of the increased net flow by this iteration*/
6.   **Perform** LP PR' on G_{RES} with $KCOMM$ and dem_k
7.   **Returns**: z and $\Delta acf_k(i,j) \ \forall a = (i,j) \in A$
8.   /* Variable maintenance for the next iteration */
9.   /* G_{RES} calculation: $c(a)$ hold links capacities */
10. $\forall a = (i,j) \in A, k = 1, \ldots, K,$
11. $c(a) = c(a) - acf_k(i,j)$
12.   if $c(a) = 0, A = A \setminus \{a\}$ /* prune saturated links */
13.   /*Commodities and demands calculations */
14. $lastKCOMM = KCOMM$
15.   **for** commodity $k \in lastKCOMM$ **do**
16. $\forall a = (i,j), acf_k(i,j) = acf_k(i,j) + \Delta acf_k(i,j)$
17. $dem_k = dem_k - z \cdot dem_k$
18.   **if** G_{RES} has no connectivity for k **then** /* k has max. possible assignment */
19. $KCOMM = KCOMM \setminus \{k\}$
20. /*end of while*/
21. **Perform** "flow decomposition algorithm" on G and $acf_k(u,v), k = 1..K$.
22. **Returns** per commodity k: set of paths P_k and flows $\forall_{p_i^j \in P_i} f(p_i^j)$

Fig. 3. WMCM weighted max-min fair concurrent allocation algorithm

The WMCM algorithm provides us with a commodity rate vector cr, and a set of flow rate vectors f_k, the rates of the paths $P_k^j \in P_k, k = 1, \ldots, K$ per commodity k, composing each cr_i.

Theorem 1. *The commodity rate vector cr provided by the WMCM is weighted max-min fair.*

Before proving theorem 1, lets state the following lemma and explanations. The rate vector acf^n is composed of the accumulated net flow $acf_k(s_k, j)$ per commodity k where s_k is the source of commodity k, at the end of each iteration n. K^n is a set of the commodities that participate in iteration n.

Lemma 1. $\exists x(n), y(n)$ and $u(n)$ such that $\forall i \in K^n$, we have the following:

- $\Delta acf_i^n = y(n) \cdot dem_i$ *(The increased rate is in proportion to the demand)*
- $acf_i^n = u(n) \cdot dem_i$ *(The accumulated rate is in proportion to the demand)*
- $demRes_i^n = x(n) \cdot dem_i$ *(The residual demand is in proportion to the original demand).*

Proof. We prove by induction on the number of the iterations.

For n=1, the recalculated demand is $(1 - z_1) \cdot dem_i$ (see lines 17), the total increased rate (calculated in lines 7) is $z_1 \cdot dem_i$ and the total rate is $z_1 \cdot dem_i$.

The following three equations are the *induction assumption* for iteration n,

$$\Delta acf_i^n = z_n \cdot \prod_{l=1}^{n-1}(1 - z_l) \cdot dem_i \text{ /*Rate increase */} \tag{7}$$

$$acf_i^n = (\sum_{l=1}^{n} z_l \cdot \prod_{m=1}^{l-1}(1 - z_m)) \cdot dem_i \text{ /*Total rate*/} \tag{8}$$

$$demRes_i^n = \prod_{l=1}^{n}(1 - z_l) \cdot dem_i \text{ /*Residual Demands*/} \tag{9}$$

Proof for iteration $n + 1$: $KCOMM^{n+1}$ holds the commodities that participate in iteration $n+1$. The rate increase (WMCM line 7) for any commodity $i \in KCOMM^{n+1}$ is:

$$\Delta acf_i^{n+1} = z_{n+1} \cdot demRes_i^n = z_{n+1} \cdot \prod_{l=1..n}(1 - z_l) \cdot dem_i \tag{10}$$

Where the first transition is due to Eq. (9).

The accumulated rate for any commodity $i \in KCOMM^{n+1}$ is:

$$acf_i^{n+1} = acf_i^n + \Delta acf_i^{n+1}$$

$$= (\sum_{l=1}^{n} z_l \cdot \prod_{m=1}^{l-1}(1 - z_m)) \cdot dem_i + z_{n+1} \cdot \prod_{h=1}^{n}(1 - z_h) \cdot dem_i$$

$$= (\sum_{l=1..n+1} z_l \cdot \prod_{m=1}^{l-1}(1 - z_m)) \cdot dem_i$$

Where the first transition is due to Eqs. 8 and 10.

The residual demands are calculated at the end of iteration $n + 1$ as follows:

$$demRes_i^{n+1} = (1 - z_{n+1}) \cdot demRes_i^n = \prod_{l=1}^{n+1}(1 - z_l) \cdot dem_i \tag{11}$$

Where the first transition is due to Eq. (9).

The lemma was proved by setting $x(n+1) = \prod_{l=1..n+1}(1 - z_l)$, setting $y(n+1) = z_{n+1} \cdot \prod_{l=1..n}(1 - z_l)$ and setting $u(n + 1) = \sum_{l=1..n+1} z_l \cdot \prod_{m=1..l-1}(1 - z_m)$ as required.

$\square$

We now return to the proof of Theorem 1

Proof. We prove by induction.

Base step: In the first iteration, where $n = 1$, acf^1 is the solution of PR' where for all commodities i and j, $acf_i^1 = z_1 \cdot dem_i$ and $acf_j^1 = z_1 \cdot dem_j$ implying $acf_i^1/acf_j^1 = dem_i/dem_j$. acf_i^1 can not be increased this iteration due to the PR' restrictions (line 7).

Induction Assumption: The weighted max-min fair order holds for iteration n. acf^n is feasible and if for each commodity i, acf_i^n cannot be increased without decreasing any other acf_j^n for some commodity j for which $acf_i^n / acf_j^n \geq dem_i / dem_j$.

Iteration $n+1$: $KCOMM^{n+1}$ is the set of all the commodities that participate in iteration $n+1$. $KCOMM^{SAT}$ is the set of all commodities that were saturated before, in one of the previous iterations. We distinguish among three cases for any commodity i and j:

- Case 1: Both commodities were saturated in the previous iterations, such that $i, j \in KCOMM^{SAT}$. This case holds trivially because of the induction assumption.
- Case 2: Only one of the two commodities was saturated before. W.l.o.g., assume that $i \in KCOMM^{n+1}$ and $j \in KCOMM^{SAT}$. Commodity j cannot increase its flow since it was deleted from the list (see line 19 in WMCM). If it was deleted in the previous iteration, n, then $acf_i^n / acf_j^n = dem_i / dem_j$ holds before starting iteration $n+1$, and thus any increase in commodity i rate (see lines 7 and 16) will imply $acf_i^n / acf_j^n > dem_i / dem_j$. If j was deleted before the previous iteration, n, we know that $acf_i^n / acf_j^n > dem_i / dem_j$ and then any increase in i's rate will keep the relation.
- Case 3: Both commodities participate in iteration $n+1$, thus, $i, j \in KCOMM^{n+1}$. Since both commodities participated in all the previous iterations, they gained rates such that $acf_i^n / acf_j^n = dem_i / dem_j$. As proved in lemma 1, the gain increase in this iteration keeps the same relation between the rates such that $acf_i^{n+1} / acf_j^{n+1} = dem_i / dem_j$.

Finally, $KCOMM^{SAT}$ is reduced in each iteration which ensures termination.

$\square$

4   Concluding Remarks and Consequent Work

We presented a new traffic engineering algorithm for routing demands in a networks in a way that maximizes the flows and maintain fairness. The weighted max-min fair criterion serves better current needs for quality of service regarding clients demands. The algorithm can be used when clients request are above capacity to minimize the loss of revenue and maximize customer satisfaction. It can also be applied when network is not congested to lower the maximum load on the links. The WMCM algorithm provides an optimal max-min fair commodity vector by solving iteratively a linear problem. The run time using linear program techniques can be large, though polynomial. For this reason, we developed, in a consequent work [19], an epsilon-approximation algorithm that is based on some of this work ideas and that solves the problem faster.

References

[1] Marbach, P.: Priority service and max-min fairness. IEEE/ACM Transactions on Networking **11** (2003) 733–746
[2] Bertsekas, D., Gallager, R.: Data Networks. 2nd edn. Prentice Hall, Englewood Cliffs, NJ (1992)

[3] Chen, S., Nahrstedt, K.: Maxmin fair routing in connection-oriented networks. In: Proceedings of Euro-Parallel and Distributed Systems Conference (Euro-PDS '98). (1998) 163–168

[4] Afek, Y., Mansour, Y., Ostfeld, Z.: Phantom: a simple and effective flow control scheme. Computer Networks **32** (2000) 277–305

[5] Awerbuch, B., Shavitt, Y.: Converging to approximated max-min flow fairness in logarithmic time. In: INFOCOM (2). (1998) 1350–1357

[6] Bartal, Y., Farach-Colton, M., Yooseph, S., Zhang, L.: Fast, fair, and frugal bandwidth allocation in atm networks. Algorithmica (special issue on Internet Algorithms) **33** (2002) 272–286

[7] Bartal, Y., Byers, J., Raz, D.: Global optimization using local information with applications to flow control. In: the 38th Ann. IEEE Symp. on Foundations of Computer Science (FOCS). (1997)

[8] Kelly, F.P., Maulloo, A.K., Tan, D.K.H.: Rate control for communication networks: Shadow prices, proportional fairness and stability. Operational Research Society **49** (1998) 237–252

[9] Mo, J., Warland, J.: Fair end-to-end window-based congestion control. IEEE/ACM Transactions on Networking **8** (2000) 556–567

[10] Hochbaum, D.S.: Approximation Algorithms for NP-Hard Problems. PWS Publishing Company (1997)

[11] Vazirani, V.V.: Approximation Algorithms. Springer-Verlag (2001)

[12] Cormen, T.H., Leiserson, C.E., Rivest, R.L.: Introduction to Algorithms. MIT Press, Cambridge, Massachusetts (2001)

[13] Shahroki, F., Matula, D.W.: The maximum concurrent flow problem. Journal of the ACM **37** (1990) 318–334

[14] Plotkin, S.A., Shmoys, D.B., Tardos, E.: Fast approximation algorithms for fractional packing and covering problems. Mathematics of Operations Research **20** (1995) 257–301

[15] Cheriyan, J., Karloff, H., Rabani, Y.: Approximating directed multicuts. In: FOCS. (2001) 291—301

[16] Ott, T., Bogovic, T., Carpenter, T., Krishnan, K.R., Shallcross, D.: Algorithm for flow allocation for multi protocol label switching. Research report, Telcordia (2001) Telcordia Technical Memorandum TM-26027.

[17] Mitra, D., Ramakrishnan, K.: A case study of multiservice, multipriority traffic engineering design for data networks. In: IEEE GLOBECOM. (1999) 1077–1083

[18] Ahuja, R.K., Magnanti, T.L., Orlin, J.B.: Networks Flows. Prentice-Hall, New Jersey (1993)

[19] Allalouf, M., Shavitt, Y.: Fast approximation algorithm for weighted max-min fairness with maximum flow routing. Technical Report EES2004-1, Tel Aviv University (2004)

Survivable Online Routing for MPLS Traffic Engineering

Krzysztof Walkowiak

Chair of Systems and Computer Networks, Faculty of Electronics, Wroclaw University of
Technology, Wybrzeze Wyspianskiego 27, 50-370 Wroclaw, Poland
tel. (+48)713202877, fax. (+48)713202902
Krzysztof.Walkowiak@pwr.wroc.pl

Abstract. Traffic engineering capabilities defined in MPLS enables QoS online
routing of LSPs. In this paper we address issues of network survivability in
online routing. We define a new link weight LFL (Lost Flow in Link) that can
be applied for dynamic routing of LSPs in survivable MPLS network. We propose to use the LFL metric as a "scaling factor" for existing dynamic routing
algorithms that make use of additive metrics. Results of simulations show that
the new metric can substantially improve the network survivability defined as
the lost flow function after KSP local rerouting. However, the application of
LFL metric slightly increases the number of rejected calls.

Keywords: Online routing, survivability, QoS.

1 Introduction

In recent years, there has been an increasing demand for network survivability. In
response to this demand, the research community has been intensively investigating
issues of static and dynamic optimization of survivable networks. In this paper we
propose a new approach to improve the effectiveness of online routing from the perspective of network survivability. We define a link weight that can be applied as an
enhancement of existing constraint-based online routing algorithm. Our interest focuses on MultiProtocol Label Switching (MPLS) technique [11]. The main idea of
MPLS survivability is as follows. Each circuit, LSP (label switched path), has a working route and a backup route. The working route is used for transmitting of data in
normal, failure-free state of the network. After a failure of the working route, the
failed circuit is switched to the backup route. In this work we focus on local restoration (called also local repair) [3], [12]. We assume that the backup route is found only
around the failed link. The origin node of the failed link is responsible for rerouting.
In modern computer networks a single-link failure is the most common and frequently reported failure event [3]. Therefore, in most of optimization models a single-
link failure is considered as the basic occurrence.

We make an assumption that all LSPs are not known a priori. We apply explicit
routing of MPLS proposed in [11]. Since the ingress node of the LSP makes the rout

J. Solé-Pareta et al. (Eds.): QofIS 2004, LNCS 3266, pp. 288–297, 2004.
© Springer-Verlag Berlin Heidelberg 2004

ing decision, links that are congested can be avoided. In addition, the use of dynamic link weight instead of static link weight is enabled. Therefore, selection of paths that meet the required QoS parameters is achievable. We assume that each new LSP's route is determined by the shortest path algorithm applying selected link weight. The following link state information either is flooded by the routing protocol or known administratively: total flow on link, link capacity and network topology. Two first items require traffic engineering extensions provided by extended OSPF [6], [13].

In this paper we apply the LFL (Lost Flow in Link) function defined by the author in [15]. LFL can be effectively applied for assignment of working routes in connection-oriented networks using local restoration. LFL is used to define a new link weight that can be applied for dynamic routing of LSPs in survivable MPLS network. We present and discuss results of extensive simulations run on various networks.

2 Related Work

In this section we describe briefly several constraint-based dynamic routing algorithms developed for connection-oriented networks. The most common routing algorithm applied in computer networks is the shortest path first (SPF) algorithm based on an administrative weight (metric). A popular metric is the number of hops applied in Min Hop Algorithm (MHA). The SPF method applies additive weights - the path's length is calculated as the sum of the link-cost of all links along a path. A valuable enhancement of SPF method is the feasible network approach. The residual capacity of a link is defined as the difference between the capacity of the link and the current flow of the link, which is calculated as a sum of the LSPs' bandwidths that are routed on that link. The feasible network for a new call consists of all routers and links, for which residual capacity exceeds bandwidth requirement of the new path. Thus, routing in the feasible network guarantees that allocation of a new call will not violate the capacity constraint.

Another dynamic routing algorithm is minimum interference routing algorithm (MIRA) proposed in [7]. The major idea of routing in MIRA system is to prevent selecting "critical links" that may "interfere" with potential future paths. The objective of MIRA is to find a feasible path that contains the least number of critical links. The approach similar to MIRA was also applied in papers [4-5], [8].

Authors of [14] proposed a routing algorithm called dynamic online routing algorithm (DORA). The main goal of DORA is to effectively utilize existing network resources and minimize network congestions by carefully mapping paths across the network. Results presented in [14] shows that DORA requires fewer paths to be rerouted and obtains higher successful reroute percentage than either SPF or MIRA. DORA is computationally less expensive than MIRA and has a runtime of SPF.

LIOA (Least Interference Optimization Algorithm) described in [1] reduces the interference among competitive flows by balancing the number of quantity of flows carried by a link. LIOA is based on the SPF method and the feasible network approach. Results of simulations from [1] show that LIOA performs better than MIRA in terms of rejection ration, successful rerouting upon single link failure.

Constraint-based routing research has been typically concentrated on the routing of a single path without taking into account survivability requirements. Recently, there has been much interest in restorable QoS routing, where working and backup paths are setup simultaneously to protect the network against failures. Such an approach was applied in [5], [8-9]. In our work we also address the problem of online routing in restorable MPLS networks. However, we establish only working paths. We assume that when the failure occurs backup paths are established dynamically using the link recovery approach. Backup paths are not established in advance, consequently the capacity has not to be reserved for backup paths. We accept the fact that some of LSPs affected by the failure can be not rerouted due to limited resources of residual capacity. It means that an LSP accepted to the network and allocated to a working path, can be not restored when a failure occurs. Reservation of extra spare capacity for each LSP is not always economically rational. Some LSPs carry traffic of low importance. Such LSPs can be lost in an emergency situation in order to enable better restoration of other high-valued LPSs. The concept of SLA (Service Level Agreement) can be used to facilitate the process of LSP prioritizing. The objective is to minimize the overall lost bandwidth of not rerouted LSPs. Similar approach was proposed in [10].

3   Definition of a New Metric for Survivable Online Routing

In this section we define a new link metric that can be applied for online routing of LSPs in order to improve the network survivability. The metric is based on the LFL function defined and discussed in [15]. We assume that the link rerouting is used a restoration method. The MPLS network is modeled as (G, c), where $G = (V, A)$ is a directed graph with n vertices representing routers or switches and m arcs representing links, $c : A \rightarrow R^+$ is a function that defines capacities of the arcs. We denote by $o : A \rightarrow V$ and $d : A \rightarrow V$ functions defining the origin and destination node of each arc.

To mathematically represent the problem we introduce the following notations

f_a Represents the total flow on arc a, it is a sum of bandwidth requirements of LSPs that uses arc a.

c_a Capacity of arc a.

$$g_v^{out} = \sum_{i:o(i)=v} f_i$$ Aggregate flow of outgoing arcs of v.

$$g_v^{in} = \sum_{i:d(i)=v} f_i$$ Aggregate flow of incoming arcs of v.

$$e_v^{out} = \sum_{i:o(i)=v} c_i$$ Aggregate capacity of outgoing arcs of v.

$$e_v^{in} = \sum_{i:d(i)=v} c_i$$ Aggregate capacity of incoming arcs of v.

The MPLS flow is modeled as a non-bifurcated multicommodity (m.c.) flow denoted by $\underline{f} = [f_1, f_2, ..., f_m]$.

For the sake of simplicity we introduce the following function

$$\varepsilon(x) = \begin{cases} 0 & \text{for} \quad x \le 0 \\ x & \text{for} \quad x > 0 \end{cases} \tag{1}$$

To examine main characteristics of the local restoration we consider an arc $k \in A$. We assume failure of k. In the local rerouting flow on the arc k must be rerouted by the source node of the arc k. Therefore, spare capacity of outgoing arcs of $o(k)$ except k is a potential bottleneck of the restoration process. If spare capacity of arcs leaving node $o(k)$ is relatively small, some flow of the arc k could be lost. We define the function L_k^{out} of the arc k flow lost in the node $o(k)$ in the following way

$$L_k^{\text{out}}(\underline{f}) = \varepsilon\left(g_{o(k)}^{\text{out}} - (e_{o(k)}^{\text{out}} - c_k)\right) \tag{2}$$

Note that L_k^{out} denotes lost flow that cannot be restored using arcs leaving the node $o(k)$ due to limited spare capacity of these arcs. Correspondingly, we define function L_k^{in} of lost flow that cannot be restored using arcs entering the node $d(k)$.

$$L_k^{\text{in}}(\underline{f}) = \varepsilon\left(g_{d(k)}^{\text{in}} - (e_{d(k)}^{\text{in}} - c_k)\right) \tag{3}$$

Function L_k (4) is a linear combination of flow lost in arcs outgoing $o(k)$ and arcs incoming $d(k)$. L_k only estimates the flow of arc k lost after local rerouting.

$$L_k(\underline{f}) = 0.5(L_k^{\text{in}}(\underline{f}) + L_k^{\text{out}}(\underline{f})) \tag{4}$$

In (4) we use functions L_k^{out} and L_k^{in} in the same proportion, however also another combination of these two functions can be applied. Generally, according to our simulations, function $L_k(\underline{f})$ gives similar performance for various values of the proportion ratio. Using $L_k(\underline{f})$ we can define a function $L(\underline{f})$ that shows preparation of the whole network to the local rerouting after a failure of any single arc. We assume that probability of the arc failure is the same for all arcs. Therefore, probability is not included in this function.

$$L(\underline{f}) = \sum_{k \in A} L_k(\underline{f}) = 0.5(\sum_{k \in A} L_k^{\text{in}}(\underline{f}) + \sum_{k \in A} L_k^{\text{out}}(\underline{f})) \tag{5}$$

For more details and comprehensive discussion on the LFL function refer to [15].

In order to make easier the consideration we introduce a new function as follows

$$\varpi(x) = \begin{cases} 0 & \text{for} \quad x \le 0 \\ 1 & \text{for} \quad x > 0 \end{cases} \tag{6}$$

Next, we define the two following functions

$$\tau_v^{\text{out}}(\underline{f}) = \sum_{i:o(i)=v} \varpi\left(g_v^{\text{out}} - (e_v^{\text{out}} - c_i)\right) \tag{7}$$

$$\tau_v^{\text{in}}(\underline{f}) = \sum_{i:d(i)=v} \varpi\left(g_v^{\text{in}} - (e_v^{\text{in}} - c_i)\right) \tag{8}$$

To find a new link weight we use as in [2] the partial derivative $\partial L / \partial f_i$ of the objective function. However, the *LFL* function is not differentiable everywhere [15]. Therefore, combining functions (7) and (8) we define a new link weight as follows

$$l_i^{\text{LFL}} = 0.5(\tau_{o(i)}^{\text{out}} + \tau_{d(i)}^{\text{in}}) \tag{9}$$

Note that l_i^{LFL} is a partial derivative $\partial L / \partial f_i$ for a feasible flow $\underline{f}$ except points $g_v^{\text{out}} = (e_v^{\text{out}} - c_i)$ for all $i:o(i)=v$ and $g_v^{\text{in}} = (e_v^{\text{in}} - c_i)$ for all $i:d(i)=v$. In these points the function l_i^{LFL} is equal to the left-sided derivative of the function L.

The weight l_i^{LFL} of a link i shows the change in the objective function if an incremental amount of a demand is routed on that link or other links adjacent to that link. Note that if the network saturation is relatively low, the value of l_i^{LFL} is 0. When the network load increases and reaches a particular value the weight, l_i^{LFL} starts to grow. The l_i^{LFL} can be used as a link weight in online routing algorithms based on the SPF method. However, we propose to use the metric l_i^{LFL} as a scaling factor for link metric l_i^{METRIC} of any existing online routing algorithm, which applies additive metrics in the following way

$$l_i^{\text{METRIC\_LFL}} = (1 + l_i^{\text{LFL}})l_i^{\text{METRIC}} \tag{10}$$

According to the definition and discussion of l_i^{LFL}, the scaling factor "turns on" when the network is relatively highly saturated. For lightly loaded network the original metric doesn't change. Computational cost of using the new metric in routing algorithm is $O(mn)$ - the same as in algorithms: Constraint Shortest Path First (CSPF), MHA, and second stage of DORA. It means that applying LFL metric doesn't worsen the computation complexity of these algorithms. In contrast, MIRA has complexity of $(O(n^5)+O(m^2))$, the first stage of DORA has complexity of $O(n^3m^2)$ [14].

4 Results

We conducted simulation experiments to evaluate the influence of LFL scaling factor applied to five popular online routing algorithms: MHA, MIRA [7], CSPF, DORA [14] using the bandwidth proportion parameter BWP=0.5 and LIOA [1] using the calibration parameter α=0.5. For each of tested algorithms we use the LFL metric in the same way as in (10). As the major performance indicator we apply the lost flow function using the KSP (k-shortest paths) rerouting method presented in [3], [10]. The KSP lost flow function is calculated as follows. For each link of the network we assume failure of this link. Next, we try to reroute all LPSs that traverse the failed link using the KSP method. Bandwidth requirement of LSPs that cannot be restored due to limited residual capacity is summed up. Obtained KSP lost flow function is an aggregate performance metric that shows the network capability to perform local restoration after a failure of any single link. ·

Results presented in this section are acquired from simulations on the network used as a benchmark in [1], [4] and [7]. The network consists of 15 nodes and 56 directed links. Links capacity is 12 to model the capacity ratio of OC-12. During simulations, the link capacities were scaled by 100 to enable establishing of thousands of LSPs. Each network node is used as an ingress and egress node what results in 210 ingress-egress pairs. Authors [7] considered only LSPs between 4 ingress-egress pairs. We apply the same network as in [1], [4] and [7] in order to enable rational evaluation of our approach against other well-known algorithms proposed in the literature. Moreover, since the MIRA is computationally excessive, we limit our simulations to relatively small network.

Table 1. LFL performance improvement of KSP lost flow aggregated over 10 trails

AVLU	MIRA	DORA	LIOA	MHA	CSPF
0.40	0.00%	25.81%	15.38%	40.64%	16.67%
0.45	0.00%	29.46%	28.69%	38.13%	29.84%
0.50	11.76%	22.45%	25.58%	45.40%	26.95%
0.55	14.97%	21.45%	25.95%	52.43%	26.71%
0.60	16.33%	20.76%	23.40%	50.29%	23.54%
0.65	18.34%	21.32%	19.66%	41.97%	19.56%
0.70	15.00%	18.18%	14.02%	34.88%	14.01%
0.75	6.12%	6.98%	11.70%	26.54%	9.23%
0.80	2.11%	-1.92%	2.21%	18.64%	0.61%

We consider static paths resembling long-lived MPLS tunnels that once established, they stay in the network for a long time [3]. Sets containing LSPs are generated randomly. For comparison of weights in terms of KSP lost flow function, we created such sets of demands that each LSP can be established for each analyzed link weight. In other words, for each tested link weight all LSPs are satisfied, none LSP is rejected. Such simulation scenario enables rational comparison of all link metrics in terms of KSP lost flow function. If for a particular weight some LPSs weren't established, the comparison of metrics' performance would be irrational because the number of demands placed in the network wouldn't be the same for all tested metrics. We assume that the requested bandwidth of LSPs is randomly distributed between 1 to 8 units of bandwidth. The average number of LSPs for each trial is 5600. We repeat the simulations for 10 various sets of demands. We record the KSP function after each LSP setup.

Another objective of our simulation study is to examine the influence of the LFL scaling factor on the path setup rejection ratio. In this case we don't guarantee that all LSPs in a given experiment are established. We examine 10 sets of 10000 static long-lived LSPs and 10 sets of 25000 dynamic short-lived connections.

To present the results we use the AVLU (Average Link Utilization) parameter that shows the average network saturation. For instance, AVLU=0.5 means that in average 50% link capacity is allocated to LSPs in the network. Another parameter is the LFL performance improvement. It is calculated as (RES-RES_LFL)/RES, where RES is the result obtained for standard routing algorithm, RES_LFL is the result obtained for the same algorithm using the scaling factor LFL as in (10).

In Table 1 we report the LFL performance improvement of KSP lost flow function for AVLU between 0.35 and 0.8. We can see that only in one case for strongly loaded network (AVLU=0.8) the LFL scaling factor doesn't improve the KSP lost flow function for metric DORA. In all other cases when LFL is turned on, the network survivability is improved. The best gain is obtained for MHA metric. This can be explained by the fact that MHA uses no information on network congestion and applying the LFL metric introduces some additional traffic engineering information. Fig. 1 shows the performance improvement obtained for AVLU=0.7.

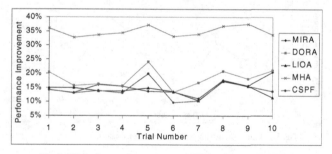

Fig. 1. LFL Performance improvement of KSP lost flow for AVLU=0.7

Generally, good performance of LFL scaling factor is consistent with theoretic considerations presented in Section 3. LFL tries to leave capacity open in regions of the network that are critical for network restoration. The lowest KSP lost flow function is obtained for the metrics MIRA_LFL and LIOA_LFL. The former metric is the best for small and medium saturated network (AVLU<=0.65). The latter link weight

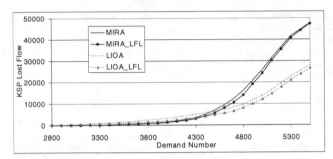

Fig. 2. KSP lost flow function as a function demand number

outperforms other weights for loaded networks (AVLU>0.65). This can be seen on Fig. 2 that shows the KSP lost flow function after every LSP is setup.

Table 2 shows the influence of LFL metric on the performance of tested metrics in terms of rejected calls. We report aggregate results for static long-lived LSPs and dynamic short-lived LSPs. In both cases LFL improves the result only for MHA. For other metrics, LFL scaling factor worsens the result. However, the degradation is not higher than 2.16%. The lowest number of rejected LSPs is obtained for metric LIOA. The worst performance, as reported in many other papers, offers MHA.

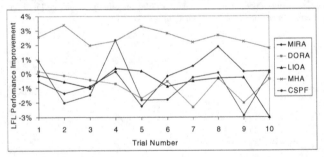

Fig. 3. LFL performance improvement of rejected calls for dynamic LSPs

Table 2. LFL performance improvement of rejected LSPs aggregated over 10 trails

	MIRA	DORA	LIOA	MHA	CSPF
Static LSPs	-1.22%	-2.05%	-2.16%	0.44%	-1.86%
Dynamic LSPs	-0.23%	-0.83%	-0.60%	2.52%	-0.68%

Fig. 3 plots the performance improvement of rejected calls for dynamic LSPs. Fig. 4 shows the number of rejects as a function of demand number for link weights LIOA, LIOA_LFL, MHA and MHA_LFL.

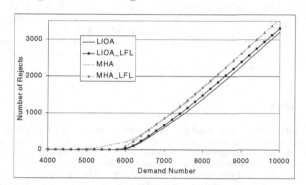

Fig. 4. Number of rejects as a function demand number

In Table 3 we present how the LFL scaling factor changes the capacity usage defined as the overall network capacity allocated by LSPs. We can see that applying LFL leads to selection of longer paths and higher capacity consumption. However, as shown above, this guarantees good performance in terms of network survivability. On the other hand, lower residual capacity obtained for LFL yields blocking of new demands expressed in higher number of rejects. Note that for AVLU<0.5 the performance improvement is 0, since the LFL scaling factor doesn't work for less congested networks.

Table 3. LFL performance improvement of capacity usage aggregated over 10 trails

AVLU	MIRA	DORA	LIOA	MHA	CSPF
0.40	0.00%	0.00%	0.00%	0.00%	0.00%
0.50	-0.01%	-0.05%	-0.02%	-0.12%	-0.04%
0.60	-0.11%	-0.18%	-0.15%	-0.60%	-0.20%
0.70	-0.74%	-0.59%	-0.71%	-0.58%	-0.73%
0.80	-0.60%	-1.76%	-1.85%	0.26%	-2.01%

5   Conclusion

In this paper we have applied the approach proposed in [15] to improve existing online routing algorithms in terms of survivability. We have focused on the local repair – one of methods proposed for recovery of MPLS networks. We have demonstrated that the new metric LFL supports network optimization and protection under single link failure. Simulation results confirm the robustness of our approach. The best improvement of network survivability defined by the KSP lost flow function was obtained for Minimum Hop Algorithm. It is evident, because the classical MHA does not apply any traffic engineering data on network congestion. However, for other algorithms that include information on link saturation the LFL also can improve the results. Applying LFL in considered algorithms does not affect significantly the performance in terms of the path setup rejection ratio. Our approach requires less routing information than in previous works on this field [4], [7].

References

1. Bagula, B., Botha, M., Krzesinski, A.: Online Traffic Engineering: The Least Interference Optimization Algorithm. To appear in the IEEE ICC 2004, 20-24 June 2004, Paris, France
2. Fratta, L., Gerla, M., Kleinrock, L.: The Flow Deviation Method: An Approach to Store-and-Forward Communication Network Design. Networks (1973) 97–133
3. Grover, W.: Mesh-based Survivable Networks: Options and Strategies for Optical, MPLS, SONET and ATM Networking. Prentice Hall PTR, Upper Saddle River, New Jersey (2004)
4. Kar, K., Kodialam, M., Lakshman, T.: Minimum interference routing of bandwidth guaranteed tunnels with MPLS traffic engineering applications. IEEE JSAC, 12 (2000), 2566-2579
5. Kar, K., Kodialam, M., Lakshman, T.: Routing restorable bandwidth guaranteed connections using maximum 2-route flows. IEEE/ACM Trans. on Networking, 5 (2003), 772-781
6. Katz, D., Kompella, K., Yeung, D.: Traffic Engineering (TE) Extensions to OSPF Version 2. RFC 3630 (2003)
7. Kodialam, M., Lakshman, T.: Minimum Interference Routing with Applications to MPLS Traffic Engineering. In Proceedings of INFOCOM 2000, 884-893
8. Kodialam, M., Lakshman, T.: Restorable Dynamic Quality of Service Routing. IEEE Comm. Mag. 6 (2002), 72-80
9. Kodialam, M., Lakshman, T.V.: Dynamic routing of restorable bandwidth-guaranteed tunnels using aggregated network resource usage information. IEEE/ACM Trans. on Networking 3(2003), 399-410
10. Murakami, K., Kim, H.: Virtual Path Routing for Survivable ATM Networks. IEEE/ACM Trans. on Networking, 2 (1996) 22-39
11. Rosen, E., Viswanathan, A., Callon, R.: Multiprotocol Label Switching Architecture. RFC 3031 (2001)
12. Sharma, V., Hellstrand, F. (ed.): Framework for MPLS-based Recovery. RFC 3469 (2003)
13. Smit, H., Li, T.: ISIS Extensions for Traffic Engineering. Internet Draft <draft-ietf-isis-traffic-05.txt>, (2003)
14. Szeto, W., Boutaba, R., Iraqi, Y.: Dynamic Online Routing Algorithm for MPLS Traffic Engineering. Lectures Notes in Computer Science, LNCS 2345 (2002), 936-946
15. Walkowiak, K.: A New Method of Primary Routes Selection for Local Restoration. Lectures Notes in Computer Science, LNCS 3042 (2004), 1024-1035

Smart Routing with Learning-Based QoS-Aware Meta-strategies

Ying Zhang[1], Markus P.J. Fromherz[1], and Lukas D. Kuhn[2]

[1] Palo Alto Research Center, 3333 Coyote Hill Road, Palo Alto, CA 94304, USA
tel: 1-650-812-4654, fax: 1-650-812-4334
{yzhang,fromherz}@parc.com
[2] Ludwig Maximilian University, Munich, Germany

Abstract. Conventional Quality of Service (QoS) routing cannot be applied easily to wireless ad-hoc sensor networks due to the unreliable and dynamic nature of such networks. For these networks, we have proposed a framework of Message-initiated Constraint-Based Routing (MCBR), which consists of a QoS specification and a set of QoS-aware meta-strategies. In contrast to most existing ad-hoc routing with no QoS support, MCBR is able to take QoS specifications into account. In this paper, we focus on learning-based meta-strategies. In contrast to most existing QoS routing approaches, learning-based meta-strategies do not create and maintain explicit routes; instead, packets discover and improve the routes during the search for the destination.

Keywords: Mobile and Wireless Networks, Ad-hoc Sensor Networks, Meta-strategies, Reinforcement Learning.

1 Motivation and Introduction

Large-scale ad-hoc networks of wireless sensors have become an active topic of research. Such networks share the following properties:

- *embedded routers* – each sensor node acts as a router in addition to sensing the environment;
- *dynamic networks* – nodes in the network may turn on or off during operation due to unexpected failure, battery life, or power management; attributes associated with those nodes (locations, sensor readings, load, etc.) may also vary over time;
- *resource constrained nodes* – each sensor node tends to have small memory and limited computational power;
- *dense connectivity* – the sensing range in general is much smaller than the radio range, and thus the density required for sensing coverage results in a dense network;
- *asymmetric links* – the communication links are not reversible in general.

J. Solé-Pareta et al. (Eds.): QofIS 2004, LNCS 3266, pp. 298–307, 2004.
© Springer-Verlag Berlin Heidelberg 2004

Applications of sensor networks include environment monitoring, traffic control, building management, object tracking, etc. Routing in sensor networks, however, has very different characteristics than routing in traditional communication networks. First of all, address-based destination specification is replaced by a more general feature-based specification, such as geographic locations [4] [13] or information gains [3]. Secondly, routing metrics are not just shortest delay, but usually multiple objectives, including energy usage and information density. Thirdly, in addition to peer-to-peer communication, multicast (one-to-many) and converge-cast (many-to-one) are major traffic patterns in sensor networks. Even for peer-to-peer communication, the source/destination pairs often are dynamic (changing from time to time) or mobile (moving during routing).

Various routing mechanisms have been proposed and implemented for sensor networks or wireless ad-hoc networks in general [7]; however, most of them do not have Quality of Service (QoS) support. Distributed QoS routing strategies for mobile ad-hoc networks have also been proposed [2] [1], all of which, however, like most other routing strategies, first establish routes between the source and the sink and then follow up with a route maintenance phase if the route is broken.

We have proposed Message-initiated Constraint-Based Routing (MCBR) [14] for wireless ad-hoc sensor networks. MCBR is a framework of routing mechanisms composed of the explicit specification of constraint-based destinations, route constraints and QoS requirements for messages, and a set of QoS-aware meta-strategies. With the separation of routing specifications from routing strategies, general-purpose *meta* routing strategies can be applied. In contrast to most existing ad-hoc routing strategies with no QoS support, MCBR takes QoS specifications into account. In this paper, we focus on a set of *distributed* routing strategies based on real-time reinforcement learning [11]. In particular, three types of meta-strategies are proposed: real-time search, constrained flooding, and adaptive spanning tree. All of these use the same reinforcement learning core, which estimates and updates the cost from the current node to the destination. The first two strategies have been presented elsewhere [14], while the last one is newly added to this family. The contribution of this paper is twofold: first, the use of MCBR for QoS specification, and second, the introduction of learning-based meta-strategies. The performance evaluations of these strategies and comparisons to AODV [6] are presented as well, using a real application scenario for sensor networks.

The rest of the paper is organized as follows. Section 2 introduces MCBR for QoS routing. Section 3 presents three QoS-aware learning-based meta-strategies. Section 4 discusses performance evaluations for these strategies. Section 5 concludes the paper and points out future directions.

2   MCBR for QoS Routing

MCBR [14] provides a general, flexible, and compositional mechanism for providing QoS message specification and QoS-aware meta-strategies. An MCBR message specification consists of a destination constraint, a route constraint, and a QoS routing objective. An MCBR meta-strategy is QoS-aware using the message specification. This is along the line of Smart Packets for Active Networks [9]; however, in MCBR, packets do not carry code. Only the specification (and possibly an additional selection of a particular meta-strategy) is passed through the network. For networks with small data frames, one can even encode various specifications in nodes and let packets only carry a specification ID with parameters.

A network can be represented as a graph $\langle V, E \rangle$, where V is the set of nodes and E is the set of connections. For an asymmetric network, $(v, w) \in E$ does not imply $(w, v) \in E$. Given a *destination constraint* C_m^d of message m, a node v is a *destination node* for m iff C_m^d is satisfied at v. For example, address-based routing, i.e., sending a message to a node with an address a_d, can be represented using the destination constraint $a = a_d$, where a is the address attribute. Geographical routing, e.g., sending a message to a circular region centered at (x_0, y_0) with radius c, can be represented using the destination constraint $(x - x_0)^2 + (y - y_0)^2 \leq c$, where x and y are location attributes.

Given a *local route constraint* C_m^r of message m, the *active* network of $\langle V, E \rangle$ for m is a subnet $\langle V_m, E_m \rangle$, such that $v \in V_m$ iff C_m^r is satisfied at v and $(v, w) \in E_m$ iff $v, w \in V_m$ and $(v, w) \in E$. For example, a message that should avoid congested nodes while routing to its destination has a local route constraint $l \leq l_m$, where l is the message load attribute (e.g., number of messages in the node's queue) and l_m is the load limit. One can also use geographical constraints (e.g., directional routing) to reduce collision and save energy for a flooding-based strategy. In general, local route constraints redefine the network connectivity on a message-by-message basis.

MCBR explicitly specifies routing objectives. A *local objective function* o is defined on a set of attributes: $o : A_1 \times A_2 \times \ldots \times A_n \to R^+$, where A_i is the domain of attribute i and R^+ is the set of *positive* real numbers. The *value* of o at a node v, denoted $o(v)$, is $o(a_1, a_2, \ldots, a_n)$, where a_i is the current attribute value of attribute i at node v. A local objective function can be a constant such as the unit transmission cost, which induces the shortest path if the objective is minimized. For another example, an energy-aware objective can be defined as $ku + c$, where u is the amount of used energy in the node, and k and c are constants. With this objective, energy-aware routing can be achieved. Similarly, one may use $k/n+c$ as a local objective, where n is the number of neighbors. With this objective, connectivity-aware routing can be achieved. Multi-objectives can be obtained by combining individual objectives, e.g., in a weighted sum.

A local objective can be aggregated over the routing path to form a global routing objective. There are two types of global aggregation, additive and concave. Like general QoS specifications, a *global objective function* O is *additive* if $O(p) = \sum_{i=0}^n o(v_i)$, where o is a local objective function and p consists of a sequence of nodes $v_0, \ldots, v_n$; For the meta-strategies discussed in this paper, only additive objectives are considered.

Problems for MCBR tend to be in one of two classes. One is *anycast*, namely finding an optimal path from the source to *one* of the destination nodes. The other is *multicast*, namely finding an optimal tree from the source to *all* the destination nodes.

An MCBR *specification* for a message m is a tuple $\langle v_m^0, C_m^d, C_m^r, O_m \rangle$. The *goal* of routing is to deliver the message from v_m^0 to one (anycast) or all (multicast) of the destination nodes V_m^d satisfying C_m^d via a sequence or a tree of intermediate nodes $p : v_m^1, \ldots, v_m^{n-1}$ such that C_m^r is satisfied at v_m^i and $\min_p O_m(p)$. Two messages are considered to have the same *type* if they have the same destination and local route constraint as well as the same routing objective.

One should notice that global route constraints are not defined in MCBR. It is well-known that finding an optimal path with an additive objective while satisfying an additive constraint is NP-hard. Unicast MCBR with an additive objective is essentially a *weighted shortest path problem*. Our goal is to make MCBR a simple (in terms of computation) yet still powerful (in terms of representation) mechanism for ad-hoc sensor networks.

3   QoS-Aware Learning-Based Meta-strategies

MCBR separates routing specification from routing meta-strategies. One can modify an existing routing strategy, such as AODV, to be a QoS-aware meta-strategy for MCBR. Most existing strategies, however, establish a route from the source to the destination via flooding the network. In this case, extra control packets are required for repairing broken routes.

Here, we propose QoS-aware learning-based meta-strategies. Real-time reinforcement learning [11] has been studied and applied mostly in agent-based path planning [5]. We apply this powerful technique to develop *distributed* meta routing strategies for sensor networks.

Given a routing specification of a message, including the destination and QoS requirements, one can define a cost function on each node, called *Q-value*, indicating the minimum cost-to-go from this node to the destination. For a distributed sensor network, the cost is initially unknown, and an initial estimation is made according to the type of message. Furthermore, a node also stores its neighbors' Q-values, *NQ-values*, which are estimated initially according to the neighbors' attributes and updated when packets are received from neighbors.

The learning-based meta routing strategies typically consist of an *initialization* phase, a *forwarding* phase, and a *confirmation* phase. Learning happens in all phases. For each packet sent out from a node, the current Q-value of the node for the type of message is attached. All the nodes are set to be in promiscuous listening mode. Whenever a node overhears a packet of type m, whether it is the designated receiver or not, it updates the corresponding NQ-value and re-estimates its own Q-value using the equation

$$Q_m = (1 - \alpha)Q_m + \alpha(o_m + \min_n NQ_m(n)) \tag{1}$$

where α is a learning rate, o_m is the current value of the local objective function, and n is a neighbor of this node.

Using the Q-value, *real-time search* passes the packet to the "best" neighbor according to the estimates, *constrained flooding* decides if and when to re-broadcast the packet according to the cost estimates, and *adaptive spanning tree* forwards the packet to its parent, with parents possibly changing over time pointing to a neighbor with the best Q-value. This approach has a number of attractive properties: (1) explicit use of destination and QoS specifications for finding optimal routes; (2) automatic adaptation with different routes when network conditions change; (3) no need for extra maintenance packets; and (4) no *infinite* looping if a path to the destination exists.

3.1   Meta-strategy 1: Real-Time Search

The pseudo code of real-time search is illustrated in Figure 1, where $Q_m^0(n)$ is an initial estimate for node n, according to the destination and QoS requirement of the message and the attribute values of this node [15]. Please note that although this strategy is *infinite loop* free, it is not loop-free. However, it has been proved that the maximum path length is $O(N^2)$ and it will converge to the optimal path in $O(N^2)$, where N is the number of nodes in the network. For space limitations, readers are referred to [15] for the theoretical bounds and variations of this meta-strategy.

3.2   Meta-strategy 2: Constrained Flooding

In contrast to the search-based methods, where each node decides which of the neighboring nodes to forward the message to, flooding-based strategies decide whether or not to broadcast at each node. A few gradient-flooding type strategies have been developed [12], requiring a cost field to be established beforehand or specialized for geographical routing. We propose a constrained-flooding meta-strategy, where the cost, i.e., the Q-value, can be learned if not known a priori. Figure 2 illustrates the basic idea. Like other gradient-flooding routing protocols [12], the cost is transmitted together with the packet. In addition, the cost at each node is updated every time a packet is received. The update rule is the same as for search-based strategies. Two techniques are used here to control the flood: (1) cost difference – if the receiving node estimates a significantly higher cost than the transmitting node, no action is taken except for updating its cost field; and (2) time difference – the transmit time difference is added to the broadcast, so that nodes with better estimates transmit first, while duplicate packets are suppressed. In this algorithm, a "temperature" variable T is used to control the flood: the higher T, the higher the chance that a packet is broadcast.

If the destination is known a priori, which turns out to be the case for many routing applications in sensor networks, backward constrained flooding from the destination can be used initially to establish the cost field. If there is no initialization and the cost field is flat, one can set T high initially and let it cool down when the cost field is more settled to reduce collisions and save energy. This strategy has been briefly discussed in [14].

Forwarding phase:

> **received** (m, Q) at w from node u **do**
>   **if** new(m) **then**
>   **for all** v with $(v, w) \in E_m$ **do** $NQ_m(v) \leftarrow Q_m^0(v)$; **end**
> $Q_m \leftarrow Q_m^0(w)$;
>   **end**
>   **if** satisfied(C_m^d) **then** $Q_m \leftarrow 0$; **broadcast**$(m, 0)$; **return**; **end**
> $NQ_m(u) \leftarrow Q$;
> $Q_m \leftarrow (1 - \alpha)Q_m + \alpha(o_m + \min_v NQ_m(v))$;
>   **if** designated(m) **then**
> $v \leftarrow \operatorname{argmin}_n NQ_m(n)$; (random tie break)
>   **send**(m, Q_m) to v;
>   **end**
> **end**

Confirmation phase:

> **timeout** $(m$ to $v)$
> $NQ_m(v) \leftarrow \max_n NQ_m(n) + 1$;
>   **if** (resend) **then**
> $v \leftarrow \operatorname{argmin}_n NQ_m(n)$; (random tie break)
>   **send**(m, Q_m) to v;
>   **end**
> **end**

Fig. 1. Real-time search meta-strategy

Forwarding phase:

> **received** (m, Q) at w from node u **do**
>   **if** new(m) **then**
>   **for all** v with $(v, w) \in E_m$ **do** $NQ_m(v) \leftarrow Q_m^0(v)$; **end**
> $Q_m \leftarrow Q_m^0(w)$;
>   **end**
>   **if** satisfied(C_m^d) **then** $Q_m \leftarrow 0$; **broadcast**$(m, 0)$; **return**; **end**
> $NQ_m(u) \leftarrow Q$;
> $Q_m \leftarrow (1 - \alpha)Q_m + \alpha(o_m + \min_v NQ_m(v))$;
>   **if** m is in transmit queue **then** remove m from transmit queue; **return**; **end**
>   **if** $((Q_m - NQ_m(u)) < T)$
>   **broadcast**(m, Q_m) to all neighbors after $k(Q_m - NQ_m(u)) + \delta$ time units;
>   **end**
> **end**

Fig. 2. Constrained-flooding meta-strategy

Initialization phase:

> **for all** v **do** $NQ_m(v) \leftarrow$ inf **end**
> **received** (m, Q) at w from node u **do**
> $NQ_m(u) \leftarrow Q;\ Q_m \leftarrow (1 - \alpha)Q_m + \alpha(o_m + \min_v NQ_m(v));$
> $p'_m \leftarrow \operatorname{argmin}_n NQ_m(n);$ (random tie break)
>   **if** $p_m \neq p'_m$ **then broadcast**$(m, Q_m);\ p_m \leftarrow p'_m;$ **end**
> **end**

Forwarding phase:

> **received** (m, Q) at w from node u **do**
>   **if** satisfied$(\mathcal{C}_m^d)$ **then** $Q_m \leftarrow 0;$ **broadcast**$(m, \mathbf{0});$ **return; end**
> $NQ_m(u) \leftarrow Q;\ Q_m \leftarrow (1 - \alpha)Q_m + \alpha(o_m + \min_v NQ_m(v));$
> $p_m \leftarrow \operatorname{argmin}_n NQ_m(n);$ (random tie break)
>   **if** designated(m) **then send**(m, Q_m) to $p_m;$ **end**
> **end**

Confirmation phase:

> **timeout** $(m$ to $v)$
> $NQ_m(v) \leftarrow \max_n NQ_m(n) + 1;$
> $p_m \leftarrow \operatorname{argmin}_n NQ_m(n);$ (random tie break)
>   **if** (resend) **then send**(m, Q_m) to $p_m;$ **end**
> **end**

Fig. 3. Adaptive spanning-tree meta-strategy

3.3   Meta-strategy 3: Adaptive Spanning Tree

This is a new strategy added to the family, using the same learning core. In cases where destinations (e.g., the base station) are known, it is often more efficient to build an initial spanning tree from the destination. Typical problems with this approach are that the initial tree may be suboptimal due to collisions during tree building, and that an optimal tree may become suboptimal over time due to the dynamic aspects of the network. Rebuilding a complete tree may also result in extra energy consumption and packet loss.

In our framework, an adaptive spanning tree can be built using the same reinforcement learning core as in the previous two meta-strategies. The initialization phase builds an initial spanning tree. The forwarding phase passes the received packet to a node's parent. All the nodes are set to be in promiscuous listening mode. Whenever a node hears a packet, whether it is the designated receiver or not, it updates its corresponding NQ-value and re-estimates its own Q-value, just as in the other two meta-strategies. Similar to real-time search, implicit packet confirmation is used: if the packet is not heard from the forwarded node within a certain time period, the NQ-value is updated to be the largest among the neighbors, and the parent pointer is reset to the neighbor with minimum cost. The pseudo code is illustrated in Figure 3.

4   Evaluations of Meta-strategies

We have simulated the three meta-strategies for several real applications using Prowler [10], a probabilistic wireless network simulator. Prowler provides a radio fading model with packet collisions, static and dynamic asymmetric links, and a CSMA MAC layer.

We use a real application to test the performance of all three meta-strategies and also compare them to AODV. The application, Pursuer Evader Game (PEG) [8], is to use the sensor network to detect an evader and to inform the pursuer about its location. The communication problem in this task is to route packets sent out by one of the sensor nodes to the mobile pursuer. The source is changing from node to node, following the movement of the evader, and the destination is mobile. The network is a 7×7 sensor grid with small random offsets. The maximum radio range is about $3d$, where d is the distance between two neighbor nodes in the grid. Let the source be at the middle and the destination be at the upper right corner initially. Assume that the evader and the pursuer move at about the same speed $0.2d/s$, where d is the grid distance, and the source rate be 1 packet per second. The objective of MCBR in this application is simply the minimum number of hops.

The following performance metrics are used for comparing routing strategies in this paper:

- *latency* – the time delay of a packet from the source to the destination;
- *success rate* – the total number of packets received at the destinations vs. the total number of packets sent from the source;
- *energy consumption* – assuming each transmission consumes an energy unit, the total energy consumption is equivalent to the total number of packets sent in the network;
- *energy efficiency* – the ratio between the number of packets received at the destination vs. the total energy consumption in the network;

Figure 4 shows performance of four meta-strategies: real-time search, constrained flooding, adaptive tree, and AODV. The results are averaged over 10 random runs. We can see that AODV has the shortest latency, but with low success rate, high energy cost, and low efficiency, while constrained flooding has the highest success rate and efficiency. Both constrained flooding and real-time search have been implemented on the Berkeley mote platform, and tested for the real PEG with 49 motes. In practice, constrained flooding also works better. AODV is the worst algorithm in this application, as the source changes very fast, while learning-based strategies adapt to new situations quickly.

5   Conclusion and Future Work

In this paper, we have presented three learning-based meta-strategies for MCBR. MCBR enables X-aware routing strategies, where X can be any attribute of the

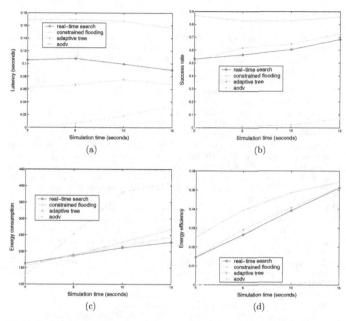

Fig. 4. Performance Evaluation: (a) Latency, (b) Success rates, (c) Energy consumption, (d) Energy efficiency

system, energy, signal strength, connectivity, even sensor readings, or combinations thereof. The three meta-strategies all use the same reinforcement learning core. Even with its minimal overhead (e.g., there are no extra control packets needed other than the initialization process), this routing scheme is highly adaptive to the dynamic changes of a network. We have also implemented ant-routing for sensor network [16]. A comparison between these two types of learning will be presented in the future. We also plan to experiment with different QoS specifications, such as connectivity-aware or reliability-aware routing.

Acknowledgment. This work is funded in part by Defense Advanced Research Project Agency contract # F33615-01-C-1904.

References

1. K. Chen, S. H. Shah, and K. Nahrstedt. Cross-layer design for data accessibility in mobile ad hoc networks. *Wireless Personal Communication*, (21):49–76, 2002.
2. S. Chen. Distributed quality-of-service routing in ad-hoc networks. *IEEE Journal on Selected Areas in Communications*, 17(8), August 1999.
3. M. Chu, H. Haussecker, and F. Zhao. Scalable information-driven sensor querying and routing for ad hoc heterogeneous sensor networks. *Int. Journal on High Performance Computing Applications*, June 2002.
4. B. Karp and H. T. Kung. GPSR: Greedy perimeter stateless routing for wireless networks. In *Proc. 6th Int'l Conf. on Mobile Computing and Networks (ACM Mobicom)*, Boston, MA, 2000.
5. S. Koenig and R. G Simmons. Complexity analysis of real-time reinforcement learning applied to finding shortest path in deterministic domains. In *National Conference on Artificial Intelligence*, pages 99–105, 1993.
6. C. E. Perkins and E. M Royer. Ad hoc on-demand distance vector routing. In *Proc. 2nd IEEE Workshop on Mobile Computing Systems and Applications*, pages 90–100, February 1999.
7. E. Royer and C. Toh. A review of current routing protocols for ad hoc mobile wireless networks. *IEEE Personal Communications*, April 1999.
8. S. Sastry, L. Schenato S. Schaffert, C. Sharp, and B. Sinopoli. Nest challenge problem: Midterm, final demo paln, 2003. DARPA NEST Program Demonstration Plan, http://dtsn.darpa.mil/ixo/nest/day2/UCB1Nest_PL07_10.ppt.
9. B. Schwartz, A. W. Jackson, W. T. Strayer, W. Zhou, R. D. Rockwell, and C. Partridge. Smart packets for active networks. *ACM Transaction on Computer Systems*, 18(1), Feb. 2000.
10. G. Simon. Probabilistic wireless network simulator. http://www.isis.vanderbilt.edu/projects/nest/prowler/.
11. R. S. Sutton and A. G. Barto, editors. *Reinforcement Learning: An Introduction*. The MIT Press, Cambridge, MA, 1998.
12. F. Ye, G. Zhong, S. Lu, and L. Zhang. A robust data delivery protocol for large scale sensor networks. In *Information Processing in Sensor Networks, Lecture Notes in Computer Science*, number 2634, page 658, 2003.
13. Y. Yu, R. Govindan, and D. Estrin. Geographical and energy aware routing: a recursive data dissemination protocol for wireless sensor networks. Technical report ucla/csd-tr-01-0023, UCLA Computer Science Department, May 2001.
14. Y. Zhang and M. Fromherz. Message-initiated constraint-based routing for wireless ad-hoc sensor networks. In *Proc. IEEE Consumer Communication and Networking Conference*, 2004.
15. Y. Zhang and M. Fromherz. Search-based adaptive routing strategies for sensor networks. In *AAAI Sensor Networks Workshop*, July 2004.
16. Y. Zhang, L. Kuhn, and M. Fromherz. Improvements on ant-routing for sensor networks. In *Fourth International Workshop on Ant Colony Optimization and Swarm Intelligence*, 2004.

An Efficient Auction Mechanism for Hierarchically Structured Bandwidth Markets

Marina Bitsaki[1], George D. Stamoulis[2], and Costas Courcoubetis[2]

[1] Department of Computer Science, University of Crete
P.O.Box 2208, Heraklion, Crete, GR-71409, Greece
Tel: +30 2810 393500, Fax: +30 2810 393501
marina@csd.uch.gr
[2] Department of Informatics, Athens University of Economics and Business
76 Patision Str. Athens, GR-10434, Greece
Tel: +30 210 8203693, Fax: +30 210 8203686
{gstamoul, courcou}@aueb.gr

Abstract. In this paper, we formulate a new problem, namely allocation of bandwidth in a two-level hierarchically structured market. In the top level a unique seller allocates bandwidth to intermediate providers [e.g. Internet Service Providers (ISPs)], who in turn allocate their assigned shares of bandwidth to their own customers in the lower level. We present an efficient mechanism comprising auctions in both levels. We prove that, due to the structure of the mechanism and certain rules imposed by the top-level seller, the following dominant strategies apply: a) each of the lower-level customers reveals truthfully his demand in the auction he participates; b) each intermediary reveals truthfully to the top-level seller the aggregate demand in his respective market. Both the mechanism and the results extend to the case of more than two market levels.

Keywords: Auctions, bandwidth markets, efficiency.

1 Introduction

The evolution of network technology has enabled the development of new communication services that demand more and more bandwidth in an unpredictable way. The high competition and the lack of information about the demand for these services motivates the use of auctions to allocate bandwidth to those customers that value it the most. There have been published several studies of allocating network resources by means of auctions. Lazar and Semret propose in [5], [6] the progressive second price (PSP) auction for the allocation of a divisible resource in a link and a network of arbitrary topology respectively. Maillé and Tuffin propose in [7] the one-shot multi-bid auction scheme for the allocation of a divisible resource in a link; this scheme is related to the PSP auction. They extend the multi-bid auction to a special case of a network [8]. Courcoubetis et al present in [3] a descending auction mechanism for bandwidth allocation over paths, where bids are placed simultaneously and independently in each link.

J. Solé-Pareta et al. (Eds.): QofIS 2004, LNCS 3266, pp. 308–317, 2004.
© Springer-Verlag Berlin Heidelberg 2004

In the general case of a business model, there may exist multiple levels in the process of providing bandwidth. Indeed, in the presence of large and distributed sets of potential buyers, their direct transaction with the seller is either impossible, or entails high computational and physical or communication overheads. This motivates a hierarchical business model. In this paper, we formulate a *new* resource allocation problem. In particular, we deal with a two-level hierarchical business model for selling C units of bandwidth in a single link. In the top level the social planner allocates bandwidth to a set of M intermediate providers [e.g. Internet Service Providers], who in turn allocate their assigned shares of bandwidth to a set of $N(N > M)$ customers in the lower level. The social planner imposes certain allocation and payment rules for both levels. Our objective is the efficient overall allocation of the entire supply of bandwidth to the customers, as if the social planner were to assign this bandwidth directly. Hence, the use of the term social planner, rather than profit-seeking seller.

Despite its hierarchical structure, the above allocation problem cannot be solved efficiently in two independent stages, one for each level. The social planner cannot sell the bandwidth to the providers before they acquire knowledge of the demand they will face by their customers, while the providers cannot trade with their customers before they learn the amount of bandwidth they obtain from the social planner. Thus, a dynamic mechanism enforcing coordination and exchange of information between the two levels is required. In light of this necessity, we propose an *innovative mechanism*, comprising an appropriate auction in each level. The auctions are coordinated so that supply is exhausted at the end, as required for attaining efficiency. The service providers and the customers are expected to act according to their own *incentives*, i.e. so that their respective benefits are maximized, without being concerned about social welfare. Nevertheless, our mechanism is specified so that bidders of both auctions have the incentives both to participate uninhibitively and to bid truthfully (incentive compatibility property) in an environment where each player only possesses a certain part of *information* on the entire market. In particular, we assume that customers' preferences are privately known to them, providers have no prior information about their local markets, while the social planner may only interact with providers. Moreover, the rules of the mechanism are announced to all players by the social planner but the bidding process outcomes of both levels are only released after the end of the procedure.

2   Problem Formulation – Analysis of a Simple Case

We assume that the quantity C of bandwidth available in the top level is known to all players, and that customers are partitioned into fixed groups, each of which constitutes the local market of a certain provider. We denote the local market of provider j as the set of customers S_j, for $j = 1, \cdots, M$. Each customer i, receives marginal utility $\theta_{i,k}$ by his k^{th} allocated unit of bandwidth, for $k = 1, \cdots, C$. We assume that marginal utilities of each customer are diminishing and privately known to him. A customer knows neither the utility function of other customers, nor the distribution this utility is drawn from, not even the quantity to be offered in his local market. In contrast to customers, providers do not know their own valuations. Indeed, provider's j marginal utility $u_{j,k}$ is assumed to equal the revenue he would obtain if he were to sell the kth unit of bandwidth after the trade. Thus, it cannot be determined or predicted without knowledge of market demand, which is taken to be completely unknown. The objective of the social planner is to *maximize social welfare*, which measures the overall well-being of the society. In our setting, it is given by the formula: $\text{SW}(\boldsymbol{x}) = (\theta_{1,1}+\cdots+\theta_{1,x_1})+\cdots+(\theta_{N,1}+\cdots+\theta_{N,x_N})$, where $\boldsymbol{x} = (x_1,\cdots,x_N)$ is the vector of the allocated quantities to the N customers. In case of complete information, the optimal allocation is determined by ordering the $\theta_{i,k}$s over all customers and selecting the largest C ones. It is necessary that we only employ efficient mechanisms in both levels. Still, this is not a sufficient condition for efficiency in our problem. To illustrate this we will first analyze the case of selling a single unit of bandwidth, which provides us with considerable insight and understanding of the issues that arise in the general case too. Thus, suppose that the social planner wants to sell a single good (i.e., unit of bandwidth) to one of N customers with valuations (i.e. utilities) θ_i through M providers. Efficiency is attained if the good is ultimately sold to the customer with the highest valuation. Below we examine two combinations of well-known mechanisms, only one of which attains efficiency under the assumption of privately known valuations.

First, we deal with the combination of Vickrey auctions in both levels. The lower-level auction in each market is performed first, so that the respective provider learns his utility. That is, his revenue if he does win the good in the top-level auction; this revenue equals the second highest bid among the bids placed by his customers. Then, the providers participate to the top-level auction according to this utility. All players (in both levels) bid truthfully, due to the Vickrey payment rule, but the outcome may *not* be efficient. That is, the good may not be sold to the provider with whom is associated the customer having the overall highest valuation. For example, assume two providers A and B having two and three customers respectively, with valuations $\theta_1 = 2$, $\theta_2 = 7$, $\theta_3 = 5$, $\theta_4 = 6$, $\theta_5 = 3$. Providers A and B bid 2 and 5 respectively. Provider B wins the good at price 2 and sells it to customer 4 at price 5, which is not the efficient outcome. Even though the auctions in both levels are efficient, overall efficiency is not attained because no local market's valuation (i.e., the maximum valuation among market's customers) is reflected in the top-level auction. Each provider derived his own valuation according to his revenue, as explained above, which differs from his local market's valuation. The same results hold in case of applying the English auction independently in the two levels starting from the lower one: each provider learns his potential revenue, which again equals the second highest valuation of his customers.

Suppose now that we apply the English auction in both levels simultaneously as follows: a common price clock ascends continuously in each level. At every price, each customer either accepts (e.g. by pressing a button) the offer according to his valuation, or withdraws from the auction. Regarding each provider's strategy, it is meaningful to accept the offer at the same price if at least one of his customers accepts it; otherwise, he withdraws from the top-level auction. This auction terminates at the first price where only one of the providers still accepts the offer. This is the winner and pays the current price. His market's auction continues until only one of his customers still accepts the offer. This customer is the final winner and pays the final price of his local market. The winning provider surely ends up with non-negative profits, because his buy price is always less than or equal his sell price. The mechanism is efficient, since as the price increases only the customer with the highest valuation remains active. The provider's strategy described above is a weakly dominant strategy: if he withdraws instead of accepting the offer at a given price where some of his customers are still active, then he ends up with a zero profit; if he accepts instead of withdrawing then he may end up with a negative or zero profit.

It is intuitively clear that a necessary condition for overall efficiency is that each provider submits truthfully his market demand in the top level auction. Thus, the social planner has to design the whole mechanism so that each provider maximizes his profits by transferring his market demand in the top level. A requirement for this, is that all players obtain non-negative profits at the end; negative profits raise participation issues and generate incentives that result in inefficiencies. Since customers know their own valuations and act rationally, there is no possibility of obtaining negative profits in any mechanism. But this is not the case for the providers, since they participate in two different trades. Thus, depending on the mechanism, there may exist a possibility of price inconsistency for a provider; i.e. to buy certain units of bandwidth at prices that are higher than the corresponding selling prices. Such cases do not arise with our mechanism.

3   The Hierarchical Auction Mechanism

We propose a synchronized mechanism in the top and lower levels of our hierarchy for selling C indivisible units of bandwidth to N customers through M providers. The Ascending Clock Auction with Clinching (ACC) introduced by Ausubel in [1] is performed in both levels. In the lower level, however, we introduce a new allocation rule, which makes use of the outcome of the top-level auction; this rule is discussed below. The price clock is common in the two auctions, starts at a reserve price and increases *continuously* until the end of the process at a final price p_{final}. Players of both levels bid for quantities at every price according to their strategies. We assume that no information about his opponents' bids is available to any player during the bidding process. Each customer observes price p and decides whether to submit a new bid at this price or not. A pure strategy for customer i is the function $X_i : \Re \to \aleph$, where $X_i(p)$ denotes the quantity demanded at price p. Each customer's bids have to be non-increasing as price ascends. Each provider observes his own customers' demanded quantities at price p and uses this information to calculate his bid for the top-level auction, as we discuss in Sect. 4. A pure strategy for provider j is a function denoting the quantity this provider demands in the top-level auction at price p. For each price p, this depends on the vector of demanded quantities in the provider's local market. In order to simplify notation, we denote the bid at price p as $Q_j(p)$ and the bidding function as $Q_j : \Re \to \aleph$. Again, provider's j bids have to be non-increasing as price ascends. We will determine the dominant strategies of all players in the next section. The procedure terminates at the first time where demand equals supply in the top-level auction. The evolution of the mechanism is given completely by the set of prices p^l, $l = 1, \cdots, L$, corresponding to the *occasions* on which one or more players (in any of the two levels) strictly decreased his quantity. Next, we will define the allocation and payment rules for the two levels, using the notation of [1].

TOP LEVEL AUCTION: Let q_j^l denote the quantity demanded by provider j at the lth occasion. Due to clinching in the top-level auction, at each price p^l, each provider has already guaranteed a quantity for his respective market. The quantity C_j^l clinched up to (and including) price p^l by provider j is given by:

$$C_j^l = \max\{0, C - \sum_{k \neq j} q_k^l\}, \text{ for } l = 1, \cdots, L \text{ and } j = 1, \cdots, M \ . \quad (1)$$

After the bidding process is completed, the social planner announces to each provider the quantity he wins and his charge. In particular, each provider obtains the quantity he demanded at the final price p_{final} and pays for each unit of bandwidth the standing price at which he clinched this unit, as suggested by the ACC auction. Formally, the outcome of the top-level auction is defined by:

$$\text{Allocation of Provider } j : \ q_j^* = C_j^L = q_j^L, \text{ for } j = 1, \cdots, M \ ; \quad (2)$$

$$\text{Payment of Provider } j \ = \sum_{l=0}^{L} p^l \cdot (C_j^l - C_j^{l-1}), \text{ for } j = 1, \cdots, M \ . \quad (3)$$

LOWER LEVEL AUCTION: Let x_i^l denote the quantity demanded by customer i at the lth occasion. For the allocation and payment by the customers we introduce a new clinching rule: At each price p, and at each local market of provider j, the condition for determining the quantity to be clinched by the various customers employs the already guaranteed supply C_j^l in this market. That is, as long as a provider clinches new units of bandwidth in the top-level auction, he is required to make them available in the lower auction of his own market. This is different than performing the original ACC auction for C_j^L units of bandwidth. We denote as B_i^l the quantity clinched by customer i up to (and including) price p^l; this is given by:

$$B_i^l = \max\{0, C_j^l - \sum_{k \neq j} x_k^l\}, \text{ for } l = 1, \cdots, L \text{ and } i = 1, \cdots, N \ , \quad (4)$$

where C_j^l is the supply offered at the lth occasion at the customer's i local market j. Each customer wins the quantity demanded at the final price p_{final} of

Table 1. Hierarchical auction in the example

	Bidding process						Clinching process					
	Lower A		Lower B		Top		Top		Lower A		Lower B	
Price	C1	C2	C3	C4	PrA	PrB	PrA	PrB	C1	C2	C3	C4
1	3	3	2	3	6	5	3	2	-	-	-	-
2	3	2	2	3	5	5	-	+1	1	-	-	1
3	2	2	2	3	4	5	-	+1	-	1	1	+1
4	1	2	2	3	3	5	-	+1	-	+1	+1	+1
Allocation							3	5	1	2	2	3
Payments							3	11	2	7	7	9
Profits							6	5	8	11	13	15

the top-level auction and is charged according to two restrictions, which are part of the definition of our mechanism: i) each customer pays the standing prices at which he clinched the won units of bandwidth as defined in the modified process above, ii) no unit of bandwidth is sold at a price higher than p_{final}. Formally, the outcome of the lower-level auction is defined by:

$$\text{Allocation of Customer } i: \ x_i^* = B_i^L = x_i^L, \ i = 1, \cdots, N \ , \tag{5}$$

$$\text{Payment of Cust. } i = \sum_{l=0}^{L} \left(\min \left\{ p^l, p_{\text{final}} \right\} \right) \cdot \left(B_i^l - B_i^{l-1} \right), \ i = 1, \cdots, N \ . \tag{6}$$

In the example below, we apply the proposed mechanism assuming that each customer bids truthfully and each provider bids the aggregate demand of his local market observed at each price. In Sect. 4 we will prove that these are indeed the dominant strategies for the customers and the providers respectively.

Example. Suppose there is an amount of $C = 8$ units of bandwidth that is allocated to two service providers A and B that have two customers each, with marginal valuations $\boldsymbol{\theta_1} = (10, 4, 3)$ [$= (\theta_{1,1}, \theta_{1,2}, \theta_{1,3})$], $\boldsymbol{\theta_2} = (12, 6, 2)$, $\boldsymbol{\theta_3} = (11, 9, 1)$, and $\boldsymbol{\theta_4} = (9, 8, 7)$ respectively. The efficient outcome is given by the vector $\boldsymbol{x}^* = (x_1^*, x_2^*, x_3^*, x_4^*) = (1, 2, 2, 3)$ and the optimal social welfare is $SW(\boldsymbol{x}^*) = 10 + 12 + 6 + 11 + 9 + 9 + 8 + 7 = 72$. We apply the proposed procedure letting the price start at price 1. Table 1 presents the bids for both levels at the prices of the various occasions. Note that all auctions terminate simultaneously and the final allocation is the efficient one. After the procedure is terminated, the social planner and the providers calculate their allocations and payments, as shown in Table 1.

4   Derivation of Players' Strategies and Efficiency

In this section, we analyze players' strategies and prove that all the objectives set by the social planner are met when the proposed mechanism is applied.

Regarding players' strategies, first note the following: providers wish to maximize their profits from participation in two trading markets that interact with each other only through their own actions. Their strategy involves a buy process in the top level and a sell process in the lower level. Since it is taken that the social planner has pre-specified the mechanism in both levels including charging in the lower level, the provider's strategy reduces to the bidding strategy of the top level on the basis of the information progressively revealed to him in his own local market auction. The key factor determining his optimal bidding strategy in the top-level auction is his utility function. If the demand in his local market were known and he had the authority to define the payment rule for his own customers, he could calculate his utility function completely and bid truthfully according to this function in the top-level auction. In our setting neither local demand is known nor the payment rule is determined by the provider. Demand is derived step-by-step starting from lower prices (higher quantities) to higher prices (lower quantities). At each price the information available to the provider is the demand up to this level. On the other hand, each customer takes part in the local market auction and wishes to maximize his net benefit according to his known utility function.

We henceforth restrict attention to: i) the following strategy of providers: bid the quantity indicated by local demand if not exceeding the capacity C, otherwise bid C; and ii) the following strategy of customers: bid the quantity indicated by his utility function at each price p. Recalling that $Q_j(p)$ and $X_i(p)$ denote the bidding strategies of providers and customers respectively, we have:

$$Q_j(p) = \min\{C, \sum_{i \in S_j} X_i(p)\} \quad \text{and} \quad X_i(p) = \begin{cases} \max\{k : \theta_{i,k} > p\}, & \text{if } \theta_{i,1} > p \\ 0, & \text{otherwise} \end{cases} \tag{7}$$

We claim that provider j maximizes his profit by adopting Q_j and customer i maximizes his net benefit by adopting X_i, regardless of other players' strategies.

Proposition 1. *Demand revelation by every provider constitutes a weakly dominant strategy.*

Proof. Suppose that all providers but j and customers bid according to arbitrary but fixed strategies. We will prove that provider j maximizes his profit by revealing his local demand. Suppose that at a given price, provider j bids a quantity less than this demand. Since the other providers and provider's j customers do no deviate from their strategies, the procedure will remain the same except for termination, that will be reached earlier, i.e., at a lower price. Indeed, provider j will win less or the same number of bandwidth units than he would if he had revealed his demand. Due to the second restriction of the payment rule in the lower level, all units will be sold at most at this price. This implies that:

1. Provider's j buy price for all bandwidth units won is the same compared to the case where his bid equals his demand at each price $\rightarrow$ no extra profit.
2. His sell price for bandwidth units sold up to the new market-clearing price is the same $\rightarrow$ no extra profit from these units.

3. His sell price for bandwidth units that would be sold at higher prices is now the new market-clearing price, which is lower than the original one → loss.
4. He will probably not win some units of bandwidth that could be sold with non-negative profit → possible loss.

Conversely, suppose that at a certain price provider j bids a quantity higher than his demand. Termination of the process will be delayed which implies that:

1. Provider j will probably obtain more units that he will not be able to sell → possible loss.
2. His buy price for units besides the extra ones does not change since the other providers do not change their bids → no extra profit.
3. His sell price for units besides the extra ones will not be sold in higher prices since his customers do not change their bidding → no extra profit.

Thus, provider j should bid his local demand, regardless of the other players' strategies.   □

Proposition 2. *Truthful bidding by every customer constitutes a weakly dominant strategy.*

Proof. Suppose that all providers bid according to an arbitrary but fixed strategy. We will prove that every customer maximizes his net benefit by bidding truthfully. Indeed, the number of a customer's clinched units of bandwidth at each price is independent of his bids. The units of bandwidth and the prices at which he gains them depend on the bids of his competitors in the same market and on local supply, which is derived by the other providers' bids. In other words, if, at a certain price, customer i reports a higher quantity than the true one, then he might win an extra unit at a price, that is higher than his corresponding marginal value, thus resulting in a loss. Conversely, if customer i reports a lower quantity than the true one, then he faces a loss from not winning an extra unit with positive net benefit, without achieving a lower price for any of his other units won. Thus, customer i bids truthfully.   □

Corollary 1. *If each provider's strategy is demand revelation, then all auctions terminate simultaneously when demand equals supply in the top-level auction.*

Outline of proof. Let p_{final} be the price at which demand equals supply in the top-level auction. All lower-level auctions are terminated no later than price p_{final}, since local market demand equals the respective local market supply at p_{final} Additionally, no lower-level auction is terminated at a lower price than p_{final}, since each provider sells at least one unit at price p_{final} for otherwise the whole procedure would have terminated sooner than p_{final}. □

Remark. It is the second restriction of the payment rule in the lower level (namely, no unit is sold at a price higher than the final price) that renders demand revelation a weakly dominant strategy. Had it been omitted, providers would shade demand to their benefit: Lower level auctions would terminate at higher prices (not simultaneously) yielding more revenues per won unit, while the top-level auction would terminate at a lower price yielding smaller charges.

In the sequel, we explain why efficiency is attained. The selected mechanisms in both levels lead to efficient outcomes if considered independently. This is not enough, as discussed in Sect. 2, for achieving the overall efficiency in our hierarchical allocation problem. Revelation of local demand for every provider is a necessary condition for the mechanism to be efficient. This implies that provider's j marginal utility (revenues for the kth additional unit) should equal the local market's marginal utility $v_{j,k}$. Our mechanism satisfies the condition that the final price (which is the maximum price at which a won unit can be sold) be equal to each provider's marginal utility for the last demanded unit of bandwidth. Indeed, since the providers reveal demand, each of them will sell at least one unit at the final price p_{final} (see proof of Corollary 1). Thus, the final price equals each provider's revenues for the last unit of bandwidth, that is his marginal utility. The provider need not reveal all the demand; a part of it starting from low prices up to the final one is enough to achieve efficiency. Our mechanism has the important property that the social welfare attained in a market with intermediaries is the same as that under direct allocation by the social planner. More interesting is the fact that customers' net benefits are identical in both cases. However, the social planner faces a loss that is conveyed as profit to the providers. This is proved in the next proposition:

Proposition 3. *Allocation and payments for each customer are the same either the trade is performed directly and efficiently by the social planner or hierarchically by means of our mechanism.*

Proof. Assume first that the social planner allocates C units of bandwidth directly to the customers. Let p_1 be the price at which customer 1 clinches his first unit of bandwidth. Then p_1 is the smallest value that satisfies the condition:

$$C - \sum_{i \neq 1} q_i(p_1) = 1 \Leftrightarrow C - \sum_{i=1}^{N} q_i(p_1) + q_1(p_1) = 1 , \qquad (8)$$

where $q_i(p_1)$ is the bid of customer i at price p_1, $i = 1, \cdots, N$. We will prove that at the hierarchical auction we propose, customer 1 clinches his first unit of bandwidth at price p_1 too. Suppose there are two providers A and B, and customer 1 belongs to the set S_A of customers of provider A. Let $Q_A(p), Q_B(p)$ be the bids of providers A and B respectively and $a_A(p)$ the quantity already clinched by provider A at price p. Let p_1' be the price at which customer 1 now clinches his first unit of bandwidth. Then p_1' is the smallest value that satisfies the condition:

$$a_A(p_1') - \sum_{i \in S_A, i \neq 1} q_i(p_1') = 1 \Leftrightarrow a_A(p_1') - Q_A(p_1') + q_1(p_1') = 1 . \qquad (9)$$

Clearly, $a_A(p) = C - Q_B(p)$ for every price p. Combining this with (9), we obtain

$$C - Q_B(p_1') - Q_A(p_1') + q_1(p_1') = 1 \Leftrightarrow C - \sum_{i=1}^{N} q_i(p_1') + q_1(p_1') = 1 . \qquad (10)$$

Combining this with (8) we obtain $p_1 = p_1'$. Similarly, it is easily seen that all the remaining units clinched by customer 1 in a direct ACC are clinched at the same price in the hierarchical auction, thus giving rise to the same allocation and payments for the customer in the two cases.   □

Corollary 2. *The hierarchical auction mechanism yields the efficient outcome.*

Proof. Allocation of bandwidth to customers under our mechanism is the same with that under ACC, which is efficient.Hence, efficiency of our mechanism follows.   □

5   Concluding Remarks

In this article, we focus on strategic interactions among sellers, retailers and potential buyers. We propose and analyze a new auction mechanism to sell bandwidth in a hierarchically structured market efficiently. We take advantage of the distribution of information over the parts involved and coordinate the various trades that have to take place, such that no one has the incentive to deviate from bidding truthfully. The aforementioned mechanism can be applied in other cases too, such the hierarchical trading of units of other services (e.g. call minutes), or the trading of bandwidth of an overprovisioned backbone link (top-level trade) that interconnects with uncongested access networks (lower-level trade), considered by Maillé and Tuffin in [8]. Regarding implementation, recall that according to the presentation of the mechanism in Sect. 3, the auctions of both levels are performed simultaneously. However, the auctions of the lower level can alternatively be performed asynchronously and prior to the top-level auction, until the total demand equals the total available bandwidth C. This implementation is simpler and more appropriate for practical cases.

References

1. L. Ausubel. An efficient ascending-bid auction for multiple objects. Revision, 7 August 2002. To appear in American Economic Review.
2. L. Ausubel and P. Cramton. Demand Reduction and Inefficiency in Multi-Unit Auctions. Revision, 17 July 2002, University of Maryland.
3. C. Courcoubetis, M. Dramitinos, G.D. Stamoulis. An auction mechanism for bandwidth allocation over paths. 17th International Teletraffic Congress (ITC-17), "Teletraffic Engineering in the Internet Era", Salvador de Bahia, Brazil, Dec. 2-7, 2001.
4. P. Cramton. Ascending Auctions. European Economic Review 42:3-5, 1998.
5. A. A. Lazar and N. Semret. Auctions for network resource sharing. Technical Report CU/CTR/TR 468-97-02, Columbia University, 1997.
6. A. A. Lazar and N. Semret. The PSP auction mechanism for network resource sharing. In 8th Int. Symp. on Dynamic Games and Applications, Maastricht, 1998.
7. P. Maillé and B. Tuffin. Multi-Bid Auctions for Bandwidth Allocation in Communication Networks. IEEE Infocom, Hong-Kong, China, 2004.
8. P. Maillé and B. Tuffin. Pricing the Internet with Multi-bid Auctions. Technical Report, Irisa, n. 1630.

Multi-bid Versus Progressive Second Price Auctions in a Stochastic Environment

Patrick Maillé[1] and Bruno Tuffin[2]

[1] GET/ENST Bretagne,
2, rue de la Châtaigneraie CS 17607
35576 Cesson Sévigné Cedex, France
patrick.maille@enst-bretagne.fr
[2] IRISA-INRIA, Campus universitaire de Beaulieu
35042 Rennes Cedex, France
btuffin@irisa.fr

Abstract. Pricing is considered a relevant way to control congestion and differentiate services in communication networks. Among all pricing schemes, auctioning for bandwidth has received a lot of attention. We aim in this paper at comparing a recently designed auction scheme called multi-bid auction with the often referenced progressive second price auction. We especially focus on the case of a stochastic environment, with players/users entering and leaving the game. We illustrate the gain that can be obtained with multi-bids, in terms of complexity, revenue and social welfare in both transient and steady-state regime.

1   Introduction

To cope with congestion in communication networks, it has been proposed to switch from current flat-rate pricing to usage-based or congestion-based pricing schemes (see for instance [3,4] for surveys on pricing in telecommunication networks, describing the range of possibilities; for the sake of conciseness, we do not describe all the schemes here). Among those pricing schemes, auctioning has appeared as a possibility to share bandwidth. The first time auctioning was proposed was in the seminal smart-market scheme of MacKie-Mason and Varian [6], where each packet contains a bid and, if served, pays the highest bid of the packets which are denied service. This scheme requires a high engineering cost, but has pioneered the auction-based pricing activity in the networking community.

Progressive Second Price (PSP) Auction [5,11,12] has recently been proposed as a trade-off between engineering feasability and economic efficiency. In PSP, players submit bids at different epochs, each bid consisting of the required amount of bandwidth and the associated unit-price, until a (Nash) equilibrium is reached. The scheme has been proved to be incentive compatible and efficient. Variants of PSP have been designed in [8,14] in order to fix some of its drawbacks.

In [9], multi-bid auction (a one-shot version of PSP) has been proposed. It consists for each player in submitting multiple bids once only, providing therefore an approximation of her own valuation function. Market clearing price and

J. Solé-Pareta et al. (Eds.): QofIS 2004, LNCS 3266, pp. 318–327, 2004.
© Springer-Verlag Berlin Heidelberg 2004

allocation can be subsequently computed. Here again, incentive compatibility and efficiency are proved (up to a given constant). The scheme presents the advantage, with respect to PSP, that no bid profile diffusion is necessary along the network, and that there is no convergence phase up to equilibrium, then yielding a gain in engineering and economic efficiency, especially when players enter and leave the game randomly.

The goal of this paper is to numerically highlight and illustrate the gain that can be obtained by multi-bids over PSP. We place ourselves in a stochastic environment, with users of different types entering and leaving the game at random times, and investigate the transient (for a given trajectory) behavior and steady-state performance of both multi-bids and PSP. We especially focus on three criteria: network revenue, social welfare and computational complexity.

Note finally that there exist other auction schemes in the literature [2,10, 13], but due to space limitation, and since our main purpose was to emphasize the degree of improvement when using the multi-bids instead of PSP, we do not include them here.

The layout of the paper is as follows. In Section 2, we present the stochastic model that will be used to describe the system behavior. In Section 3 we present the PSP mechanism and its properties; the same is done for the multi-bid scheme in Section 4. Section 5 illustrates the gain that can be obtained by using the multi-bid scheme in a stochastic environment; both transient and steady-state results are provided. Finally, we conclude in Section 6.

2   General Model

In order to look at the auction schemes' behavior in a stochastic environment, we model for convenience the system by a Markov process.

Consider a single communication link of capacity Q. We assume that there exists a finite number T of different valuation (or willingness-to-pay) functions, corresponding for instance to different sets of applications. A player/user i is then characterized by her *type* $t_i \in \{1, ..., T\}$.

Players compete for bandwidth. To model their behavior, we represent their perception/valuation of the service they can get by a quasi-linear utility function of the form

$$U_i(s) = \theta_{t_i}(a_i(s)) - c_i(s), \tag{1}$$

where θ_{t_i} is the *valuation function* of a type-t_i player, which depends on the quantity of resource received a_i. Quantity c_i is for the total cost charged to player i. Both c_i and a_i will depend on the auction scheme used and on the whole set of bids s (where the term "bid" will depend on the auction scheme).

We assume that new players enter the game according to a Poisson process with rate λ, and that the type of a new player is chosen according to a discrete probability distribution $\mathbb{P}_t$, so that the arrival rate of type u is $\lambda_u = \lambda \mathbb{P}_t(u)$. We also assume that each type-u player sojourn time is exponentially distributed with rate μ_u (independent here of the obtained accumulated bandwidth, like for real-time applications). Let $\mathcal{I}(\tau)$ be the set of active players at time τ (i.e. the set of players present in the game at this time) and $I(\tau)$ be the total number of players at time τ.

To ensure that the bandwidth is not sold at a too low level, the seller can thus be seen as a (permanent) player, noted by 0, with valuation function $\theta_0(q) = p_0 q$. p_0, the reserve price, guarantees that no bandwidth will be sold at a unit price under p_0.

Our goal is to compare the behavior of PSP and multi-bids. Let us now recall the basic concepts of both schemes.

3   Progressive Second Price Auction [5,11]

In PSP, a player i submits a 2-dimensional bid $s_i = (q_i, p_i) \in \mathcal{S}_i = [0, Q] \times [0, +\infty)$, where q_i is the desired quantity of resource and p_i the *unit* price player i is willing to pay for that resource. $s = (s_1, ..., s_I)$ will denote the bid profile, and $s_{-i} = (s_1, ..., s_{i-1}, s_{i+1}, ..., s_I)$ will be the bid profile that player i faces, so that $s = (s_i; s_{-i})$ (the dependence on time τ is omitted to simplify the notations).

PSP allocation and charge to player i are

$$a_i(s) = q_i \wedge \left[Q - \sum_{p_k \geq p_i, k \neq i} q_k \right]^+ \tag{2}$$

$$c_i(s) = \sum_{j \in \mathcal{I}(\tau), \; j \neq i} p_j \left[a_j(s_{-i}) - a_j(s_i; s_{-i}) \right], \tag{3}$$

so that players bidding the highest get the bandwidth they request and total charge corresponds to declared willingness to pay of players who are excluded by player i's bid.

Each time a player submits a bid, she tries to maximize her utility, and a bid fee ε is charged to her. Under some concavity and regularity assumptions over functions θ_u, when the number of players is fixed and players bid sequentially, the game is proved to converge to a so-called ε-Nash equilibrium, so that no player can improve unilaterally her utility by more than ε. The scheme is also proved to be incentive compatible (meaning that users' best interest is to truly reveal their willingness to pay), and efficient in the sense that the social welfare $\sum_{i \in \mathcal{I}(\tau) \cup \{0\}} \theta_i(a_i)$ is asymptotically maximized (when the algorithm has converged).

Based on the assumption that users enter or leave the game, efficiency might become an issue. We suppose here that each type-u player in the game has the opportunity to submit a new bid at different times. Inter-bid times are assumed to follow an exponential distribution with parameter ν_u, independent of all other random variables. When a new player arrives, she is assumed to submit an optimal bid (meaning that she knows the bid profile).

4   Multi-bid Auction [9]

In the multi-bid scheme, users, when they enter the game, submit a set of M 2-dimensional bids $s_i = \{s_i^1, ..., s_i^M\}$, where for all m, $1 \leq m \leq M$, $s_i^m = (q_i^m, p_i^m)$ is as in PSP (the seller just submits one 2-dimensional bid $s_0 = (q_0, p_0)$ with $q_0 > Q$ and p_0 the reserve price). We assume without loss of generality that bids are sorted such that $p_i^1 \leq p_i^2 \leq ... \leq p_i^M$. With respect to PSP, the bids are submitted just once, so that users do not submit new bids at given epochs. This reduces the signaling overhead.

From the multi-bids of all competing players at time τ, the so-called pseudo-demand function of user i can be computed as the function $\bar{d}_i : \mathbb{R}^+ \rightarrow \mathbb{R}^+$, defined by

$$
\bar{d}_i(p) = \begin{cases} 0 & \text{if } p_i^M < p \\ \max_{1 \leq m \leq M} \{q_i^m : p_i^m \geq p\} & \text{otherwise.} \end{cases}
\tag{4}
$$

The pseudo-aggregated demand function is the function $\bar{d} : \mathbb{R}^+ \rightarrow \mathbb{R}^+$ defined by $\bar{d}(p) = \sum_{i \in \mathcal{I}(\tau) \cup \{0\}} \bar{d}_i(p)$, where $\bar{d}_0(p) = q_0 \mathbb{1}_{p \leq p_0}$ (apply (4) for $M = 1$) .

From the pseudo-aggregated demand function, we define the pseudo-market clearing price $\bar{u}$ by

$$
\bar{u} = \sup \left\{ p : \bar{d}(p) > Q \right\}.
\tag{5}
$$

Such a $\bar{u}$ always exists since $\bar{d}(0) \geq \bar{d}_0(0) = q_0 > Q$. Moreover for $p > \max_{i \in \mathcal{I}(\tau) \cup \{0\}}(p_i^M)$ we have $\bar{d}(p) = 0$, and therefore $\bar{u} < +\infty$.

Describe now the allocation and pricing rules. First define, for every function $f : \mathbb{R} \rightarrow \mathbb{R}$ and all $x \in \mathbb{R}$, $f(x^+) = \lim_{z \rightarrow x, z > x} f(z)$.

The allocation, recomputed each time a player enters or leaves the game, is

$$
a_i(s_i, s_{-i}) = \bar{d}_i(\bar{u}^+) + \frac{\bar{d}_i(\bar{u}) - \bar{d}_i(\bar{u}^+)}{\bar{d}(\bar{u}) - \bar{d}(\bar{u}^+)}(Q - \bar{d}(\bar{u}^+)),
\tag{6}
$$

meaning that each player receives the quantity she asks at the lowest price $\bar{u}^+$ for which supply excesses pseudo-demand, $\bar{d}_i(\bar{u}^+)$, and the excess of resource is shared among players who submitted a bid with price $\bar{u}$.

The total charge is computed according to the second-price principle [1,15] (but using the pseudo-demand functions instead of the real ones):

$$
c_i(s_i, s_{-i}) = \sum_{j \in \mathcal{I}(\tau) \cup \{0\}, j \neq i} \int_{a_j(s)}^{a_j(s_{-i})} \bar{\theta}'_{t_j}(q) dq,
\tag{7}
$$

with $\bar{\theta}'_{t_j}$ pseudo-marginal valuation function of j, defined by

$$
\bar{\theta}'_{t_j}(q) = \begin{cases} 0 & \text{if } q_j^1 < q \\ \max_{1 \leq m \leq M} \{p_j^m : q_j^m \geq q\} & \text{otherwise.} \end{cases}
\tag{8}
$$

As for PSP, incentive compatibility (each user i should better reveal its bandwidth valuation, i.e., $p_i^m = \theta'_{t_i}(q_i^m)$ $\forall m$), and efficiency are proved, but up to a controlled constant here (see [9] for details).

It is shown in [9] that it is in the players' interest to submit a uniform quantile repartition of their bids, i.e., $(q_i^m, p_i^m = \theta'_{t_i}(q_i^m))$ $\forall 1 \leq m \leq M$ such that

$$\int_{d_i(p_i^{m+1})}^{d_i(p_i^m)} (\theta'_{t_i}(q) - p_i^m)dq = C_i \qquad \forall m, \text{ where } \begin{cases} p_i^{M+1} = \theta'_{t_i}(0) \\ p_i^0 = p_0. \end{cases} \qquad (9)$$

5   Comparison of Performance

Multi-bids present the following advantages with respect to PSP:

- since the bids are submitted exactly once, no convergence phase is required by resubmitting new bids until an equilibrium is reached. It might be argued that the mean number of re-submission up to equilibrium is less that the number M of multi-bids in some cases; this situation is less likely to occur in the situation of customers arriving or leaving the game, meaning that a new re-submission phase is required for each player in PSP, whereas nothing has to be done for already present players when using multi-bids.
- Following the same idea, when submitting a new bid in PSP, each player is assumed to know the bid profile, meaning that it is advertised to all players. This is not required for the multi-bid scheme, saving then a lot of signaling overhead.

We propose to illustrate the above advantages of multi-bids in the following sub-sections. We especially wish to show that this gain in terms of signaling/complexity is not at the expense of efficiency, in terms of seller's revenue or social welfare, both on a trajectory and during the convergence phase of PSP, as well as in steady state, and that it even actually is the converse.

5.1   Transient Analysis

Figure 1 displays the behavior of PSP and multi-bids when the number of players is fixed and until equilibrium is reached for PSP, with two types of players, three type-1 and two type-2 players. The upper left-hand side figure displays the valuation and marginal valuation functions for both types of players, that we used in all our simulations. The lower left-hand side represents social welfare $\sum_{i \in \mathcal{I}(\tau) \cup \{0\}} \theta_i(a_i)$. Since the number of players is fixed during the simulation, multi-bid allocations and charges are fixed, and it can be observed that the social welfare is 60.37, very close to the optimal one 60.71. On the other hand, for PSP auctions the social welfare changes at each re-submission from a user (resulting in the discontinuous curve), reaching equilibrium (with value 60.68) around time $\tau = 26$, but showing a lower social welfare than multi-bids before reaching equilibrium. The lower right-hand side of Figure 1 represents the network revenue for both schemes. Again, multi-bid revenue is constant due to the fact

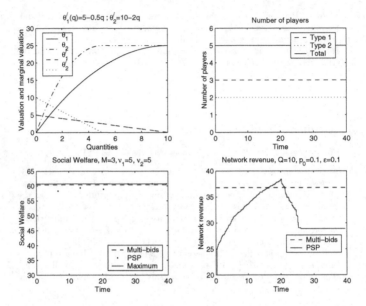

Fig. 1. Comparison of PSP and multi-bids for a fixed number of players, until convergence is reached for PSP

that the number of players is fixed. Also, the revenue for PSP is first increasing, overtaking the one with multi-bids after a while, but then dropping under it just before reaching equilibrium. Actually, we proved in [7] that when the total demand at the unit price p_0 exceeds the available capacity Q, the revenue with PSP in equilibrium tends to $p_0 \times Q$ (i.e. all the resource is sold at the reserve price) when the bid fee ε tends to 0.

Figure 2 illustrates the behavior of both schemes on a trajectory, with players entering and leaving the game[1]. Here the number of players of each type varies, as described on the upper right-hand figure. The curves of social welfare show that, when using multi-bids, the resulting social welfare is always very close to the optimal one, whereas, due to the convergence phase, there is a loss of efficiency when using PSP. Similarly, on this trajectory, the network revenue generated by multi-bids is significantly larger than the one generated by PSP.

---

[1] The parameters we chose are precised in the figure, and were also used for the study of steady-state performance.

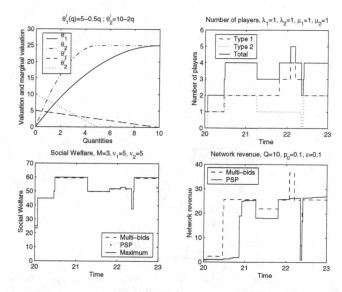

Fig. 2. Behavior of PSP and multi-bids on a trajectory, with players entering and leaving the game

5.2 Steady-State Analysis

Figure 3 illustrates the evolution of the mean efficiency ratio (obtained steady-state social welfare divided by the optimal one), the mean network revenue and the complexity of the algorithm when the number M of multiple bids increases, and compares those performance measures with the ones obtained for PSP. The complexity of computing PSP allocations and prices is of the order $O(I^2)$ [11], and the complexity of multi-bid auction is of the order $O(M \times I^2)$ [9]. We therefore display the mean number of applications of each auction rule by unit of time, multiplying this number by M for multi-bid auction. This curve does not precisely give the number of elementary operations that are conduced, but just gives an idea of how the computational complexity evolves when the parameters vary. Note that this computation of complexity does not include the signaling overhead necessary for PSP. It can be observed then that for small values of M, computational complexity is even smaller also with multi-bids. More important, thanks to the one-shot property of multi-bids (i.e., the fact that no convergence phase is required unlike PSP), steady-state social welfare (for $M \geq 2$) and revenue are larger with multi-bids.

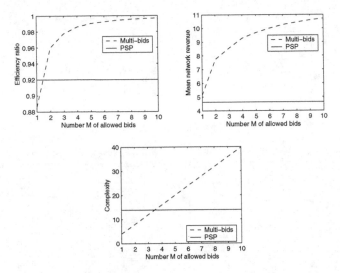

Fig. 3. Steady-state performance of multi-bids for an increasing number M of allowed two-dimensional bids in multi-bid auctions, compared with PSP

Figure 4 displays the evolution of efficiency ratio, network revenue and complexity when the arrival rate increases (with all other parameters fixed).

Again, multi-bids are shown to provide better performance. The difference increases with λ. This is due to the fact that the number of players varies more frequently, so that convergence to optimal values is less likely to occur for PSP, whereas it does not affect multi-bids.

Figure 5 illustrates the three criteria considered in this paper, when the bid-resubmission rate varies for both types of players. Even when this rate increases, leading to a larger computational complexity, we see that the multi-bid auction still outperforms PSP as concerns efficiency and network revenue.

6   Conclusions

The goal of this paper was to compare PSP and multi-bid schemes, two auction mechanisms for bandwidth allocation in telecommunication networks. Based on this purpose, we have considered a model representing a communication link, with players applying for connections at random epochs, and for a random time. Our conclusion is that multi-bid auction scheme significantly reduces the signal-

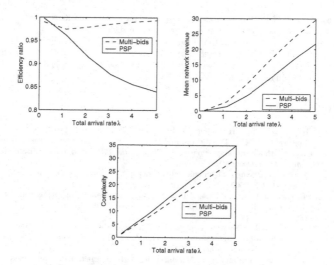

Fig. 4. Steady-state performance comparison when the charge increases

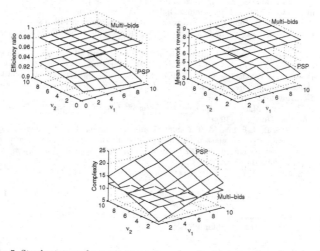

Fig. 5. Steady-state performance comparison when the frequency of re-submission increases in PSP

ing overhead of PSP, but also yields larger social welfare and network revenue (at least for this stochastic context regarding the social welfare).

As future work, we plan to extend the multi-bid auction scheme to a whole network. We already have some results in the case of a tree network, which properly represents the case where the backbone network is overprovisionned and the access networks have a tree structure.

References

1. E. H. Clarke. Multipart pricing of public goods. *Public Choice*, 11:17–33, 1971.
2. C. Courcoubetis, M. P. Dramitinos, and G. D. Stamoulis. An auction mechanism for bandwidth allocation over paths. In *Proc. of the 17th International Teletraffic Congress (ITC)*, Dec 2001.
3. L. A. DaSilva. Pricing for QoS-enabled networks: A survey. *IEEE Communications Surveys*, 3(2):2–8, 2000.
4. M. Falkner, M. Devetsikiotis, and I. Lambadaris. An overview of pricing concepts for broadband IP networks. *IEEE Communications Surveys & Tutorials*, 3(2), 2000.
5. A. A. Lazar and N. Semret. Design and analysis of the progressive second price auction for network bandwidth sharing. *Telecommunication Systems - Special issue on Network Economics*, 1999.
6. J. K. MacKie-Mason and H. R. Varian. Pricing the internet. In *Public Access to the Internet*, JFK School of Government, May 1993.
7. P. Maillé. Market clearing price and equilibria of the progressive second price mechanism. Technical Report 1522, IRISA, Mar 2003. submitted.
8. P. Maillé and B. Tuffin. A progressive second price mechanism with a sanction bid from excluded players. In *Proc. of ICQT'03, LNCS 2816*, pages 332–341, Munich, Sept 2003. Springer.
9. P. Maillé and B. Tuffin. Multi-bid auctions for bandwidth allocation in communication networks. In *Proc. of IEEE INFOCOM*, Mar 2004.
10. P. Reichl, G. Fankhauser, and B. Stiller. Auction models for multiprovider internet connections. In *Proc. of Messung, Modelierung und Bewertung (MMB'99)*, Trier, Germany, Sept 1999.
11. N. Semret. *Market Mechanisms for Network Resource Sharing*. PhD thesis, Columbia University, 1999.
12. N. Semret, R. R.-F. Liao, A. T. Campbell, and A. A. Lazar. Pricing, provisionning and peering: Dynamic markets for differentiated internet services and implications for network interconnections. *IEEE Journal on Selected Areas in Communications*, 18(12):2499–2513, Dec 2000.
13. J. Shu and P. Varaiya. Pricing network services. In *Proc. of IEEE INFOCOM*, 2003.
14. B. Tuffin. Revisited progressive second price auctions for charging telecommunication networks. *Telecommunication Systems*, 20(3):255–263, Jul 2002.
15. W. Vickrey. Counterspeculation, auctions, and competitive sealed tenders. *Journal of Finance*, 16(1):8–37, Mar 1961.

Distributed Multi-link Auctions for Network Resource Reservation and Pricing

Dominique Barth and Loubna Echabbi

Laboratoire PRiSM , Université de Versailles, 45 Av. des Etats-Unis, 78035
Versailles-Cedex, France barth, lechabbi@prism.uvsq.fr

Abstract. In this paper, we propose a Distributed Multi-link Auction mechanism(refereed by DiMA) that deals with request connection establishment in order to provide some End-to-end guarantee of services. The DiMA mechanism determines hop-by-hop the path to be taken by a request while reserving the required resource over it. It consists of consecutive local auctions that each request has to win in order to be satisfied. We consider the problem of determining requests global budgets and bidding strategies. We give some simulative analysis mapping the relation between prices, network utilization and distribution of accepted requests.

Keywords: Packet network, Quality of Services, distributed auction strategies, pricing.

1 Introduction

Today's Internet resources are equally distributed making the only one possible service be a best effort service. With the growth of Internet, several traffic flows belonging to new sophisticated applications have some specific requirements and need some guarantees of Quality of Service (QoS). These services can pay to get a level of QoS when best effort still guarantees connectivity to classical traffic. The question that arises is how to price the different proposed QoS levels in an efficient way.

Some pricing models have been proposed in the context of QoS-enabled architectures including ATM, MPLS and QoS over IP (Integrated services, RSVP). In such models, resources are negotiated and priced at a connection setup timescale in order to meet QoS requirements. Those pricing models can be classified according to the overlay architecture: ATM [12,1,14,13], MPLS[3], Intserv [16, 8]. However, some mechanisms are not specific to a particular architecture and can be applied to any environment where resources are allocated in advance to guarantee QoS requirements. They can be identified by three main concepts: adjustment schemes, effective bandwidth and auctions.

Pricing Models that use adjustment schemes [5,12][19,9] come with different flavors but mainly consist of the same concept; updating prices until an equilibrium point is reached. The objective function to be optimized is the social welfare

J. Solé-Pareta et al. (Eds.): QofIS 2004, LNCS 3266, pp. 328–337, 2004.
© Springer-Verlag Berlin Heidelberg 2004

(a combination of network benefit and user surplus) under some capacity constraints (bandwidth and/or buffer) that guarantee the QoS level. The problem is decomposed into a network problem of updating prices and a user problem of choosing adequate resource demand given its price. In [9], an arbitrager layer is introduced to remove direct consideration of network resources while focusing in the QoS parameters. Indeed, the arbitrager negotiates with users optimal amount of loss probability for instance, and then purchases the needed bandwidth from the network. In [6] an intelligent agent is introduced in order to replace the user in choosing the willingness to pay based on user learned preferences.

An important concept to support QoS over IP is effective Bandwidth. It corresponds to the notion of equivalent bandwidth in ATM. It maps the QoS required to satisfy a request into an amount of bandwidth that can be used by the admission control rule. In [10], a simple linear tariff which is tangent to the bandwidth evolution curve is proposed. This is an incentive to the users to declare (the more precisely) their real requirements.

Incentive compatibility is an important issue about pricing. Indeed, to be efficient, a pricing model should include some incentives to users to declare their real valuations of services. Auctions have been studied [4,11] as an interesting alternative to pricing network resources since they deal with this issue.

In [7], we have introduced the Distributed Multi-link Auction mechanism(referred by DiMA) that deals with request connection establishment by reserving hop by hop the required resource on a possible path. We have studied the complexity of considering more than one possible path for each request. We have proved that this problem is NP-hard and proposed a heuristic approach to solve the problem. When in [7] we were interested in complexity and theoretical properties of the model, here we address some implementation and technical issues about connection establishment and bidding strategies.

Layout: We first describe the communication model we have adopted and remind the concept of DiMA. We address the problem of bidding strategies then we provide some simulation results highlighting how the DiMA mechanism can be deployed to allocate network resources efficiently.

2   Communication Model

2.1   Managing Quality of Services

Many architectures have been proposed to provide QoS support while saving scalability.

Integrated services have been considered as an alternative to provide some QoS guarantees. Resources are reserved to real time applications before the data transfer by using a signaling protocol(RSVP). Statistical multiplexing can be deployed for aggregated flows. However each router needs to reserve resources for each flow. Hence a flow-state must be maintained at each crossed router. This addresses a problem of scalability in addition to important requirements in routers [20].

With Diffserv [15,20] some differentiation can be given to services in scalable way without any resource reservation but then it is impossible to give any deterministic guarantees on the service. Intserv architecture can be implemented at lower levels (access network) to manage individual flows. Diffserv can be then implemented on the backbone by adding flow admission control capabilities to the edge nodes of a statistically provisioned Diffserv network region [2].

A second possibility [21] is to use a Bandwidth Broker(BB) that has sufficient knowledge of resource availability and network topology. BB can then make Admission Control and resource reservation to aggregated flows. Core routers are then freed from performing admission control decisions. However, in this model, BB needs to manage the overall network domain and to store informations about flows and paths. ,This addresses a problem of scalability.

This can be removed as in [17] by distributing the functions assigned to the BB between all nodes in the network domain in a scalable way. Although nodes need to maintain state information for every reservation, author claim that available routers have sufficient memory. Indeed aggregation and thus classification and scheduling is made on a class basis. In addition, a label switching mechanism can be deployed to simplify forwarding packets. The proposed architecture includes per-flow signaling with enhanced scalability and resource reservation for aggregated flows at both core and access nodes.

We adopt this approach for our model. Indeed, we suppose that the traffic is aggregated to some defined class of traffic with specific requirements. Some effective bandwidth techniques can be used to determinate amount of bandwidth which is necessary to meet the QoS requirements. Hence each request corresponds to an inelastic demand and requires then a fixed amount of bandwidth.

2.2   Connection Establishment

The DiMA consists in establishing connections for traffic classes on demand by reserving required resources on appropriated paths before data transmission. Such demand-driven connections requires an on-line treatment. Moreover, due to delay and control constraints, a centralized process seems to bee unrealistic. Hence, in our model request acceptance and routing decisions are decentralized and made by the crossed routers.

The model can be mainly described as follows: Each node of the network corresponds to a router and can also be the origin and/or the end of a connection. For instance, consider that a node u wants to establish a connection with a node v in the network. The node u emits a Connection Demand (CoD), $i.e.$, a packet containing the destination address and the required capacity. Each router that the CoD reaches, by using its local routing function, determines the outgoing link(s) that the CoD could take to the next router with respect to its destination.

At each time step, each router performs the following steps:

○ Collecting CoD that arrive.
○ Selecting CoD to be accepted by satisfying their demand while respecting links capacity constraints and local routing tables.

o Storing accepted requests and sending them to the corresponding outgoing links.
o Updating the amount of available capacity on each outgoing link.
o Destroying the non accepted CoD.
o Sending on each corresponding incoming link a Destruction of Connection (DoC) packet.

Thus, at each time step, the first operation realized by each router is to deal with the DoC it receives by:

o Liberating the allocated capacity on the outgoing link used by the corresponding connection.
o Sending back the DoC to the incoming link of the connection.

When a communication is end-to-end accepted, it remains reserved for the two communicating nodes until one of them cuts the communication by sending a DoC instruction to the other node.

2.3   The DiMA Auctions

In our model, auctions are used to select requests to be accepted while holding capacity constraints. As explained before, the request acceptance and routing decisions are distributed over the involved routers. Hence only local informations can be considered to make such decisions. Moreover, the path assigned to a request (if accepted) is discovered hop by hop based on local routing tables and subject to capacity constraints and auction results.

Each request has a global budget that it is willing to pay to get the desired resource on the entire route to its destination. A partial budget (by link) is derived (see 3.2) to participate at local auctions. In a distributed management setting, only such format is feasible to deal with consecutive auctions. Each request has to win all local auctions on its route to be accepted. Auctions take place at a fast time-scale considering only new requests since bandwidth assigned to already accepted requests is not renegotiated.

We have proposed in [7] a heuristic that solves the auction problem at a router since the problem is NP-Hard in the general case. This heuristic can solve the problem exactly for linear and tree topologies where only one path is possible to each destination. It consists of three points:

o Sort requests that are present at this step in a decreasing order of their unitary budgets i.e (partial budget/ demand).
o Accept requests in this order on one corresponding output link while keeping capacity constraints held.[1]
o Accepted requests on an output link pay for a unit of bandwidth the first non accepted unitary bid on that link.

---

[1] Among the possible output links, we can choose either the first one with available capacity (first fit) or the one with the highest available capacity (best fit).

For a request, a corresponding output link is the following link on a shortest path to the destination[2]. In the auction algorithm, we choose the output link with the highest capacity among the corresponding ones (best fit heuristic). Accepted requests are then routed to the following corresponding router.

The auction rule adopted here consists on a second price rule adapted to inelastic demands. The incentive compatibility is still saved. Indeed, the obtained unitary price of the resource is the minimum such that accepted requests are insured that the refused ones could not bid higher. If a request is refused then the bandwidth that was previously reserved to it is freed. It can be allocated by requests that arrive later. If a request win all auctions on the determined path then it is accepted and the required resource is reserved during the connection duration. It pays for each time unit of its duration, the sum of all prices that results of local auctions on its path. Due to the temporal dimension of the problem, this tariff is efficient if all requests durations are in the same range (see 4.1). Otherwise, a multi-session auction problem can be considered where at each time interval auctions are rerun. This problem has been studied in [18].

3   Bidding Strategies

3.1   Global Budgets

Global budgets can be function of demand, duration and distance between the origin and the destination of the corresponding request. In a mono network manager context, unitary budget can represent a kind of priority that can be given to a traffic class. Let us consider unitary budgets given by the following functions:

Function1: An increasing function of demand. For instance $bid = demand$. A such function gives priority to big requests.
Function2: $bid = 1$. This function does not give any priority since unitary bids are the same. Hence, the auction mechanism is not involved.
Function3: A decreasing function of demand. For instance $bid = \lceil log(10 * demand)/demand \rceil$. It gives priority to small requests.

Example 1. Consider a request r_1 with $d^1 = 1$ and a request r_2 with $d^2 = 7$. Let b^i be the unitary budget calculated, then with:

- Function1: $b^1 = 1$ and $b^2 = 7$ priority is then given to request r_2.
- Function3: $b^1 = 3$ and $b^2 = 1$ priority is then given to request r_1.

We have tested those budget functions to investigate whether the priority level is respected. We provide some numerical results in section 4.

---

[2] Non shortest path can be considered but then some rules should be required to ensure routing stability.

3.2 Bid Distribution

The multi-link auction problem is closely related to the multi-session problem. In [18], authors discuss this relation and outline some issues that concern both dimensions particularly bid distribution. They present the Markovian property of such auctions and thus derive some Markovian bidding strategies. Here, a similar Markovian property can be outlined. That is, at a router, a request do not need to know the complete history about previous auctions. Then the decision of the amount of budget to bid depends only in the remaining steps (remaining distance to the destination). Indeed, in the proof given in [18], we only need an inelastic valuation metric on successful connections. Since a request is only interested by winning all link auctions, we can substitute links to periods and then we obtain the same result.

Hence, We adopt a proportional Markovian strategy that consists in bidding at each step the remaining budget on the remaining distance. One point of interest is that we use a second price rule. That is request does not have to pay its offered bid but rather the first non accepted bid. The difference is added to the remaining budget giving more chances to the request to be accepted in the following auctions.

4 Experimental Studies

4.1 Scenario1

Due to the temporal dimension of the problem and the on-line approach adopted here, some high bidders could be denied from getting the resource because it has already been allocated to some previous low bidders. The question that arises is how often such configurations could happen. One could expect that these precedence constraints induced by the on-line approach attempt to the efficiency of the mechanism. we propose a simple scenario to investigate this issue with the assumption that connection durations are in the same range.

We consider a linear topology of 9 links with fixed capacity equal to 500 unit of bandwidth. A set of new requests is generated each 10 time slots between the first and the last node. Each requesting 1 unit of bandwidth for along a mean duration of 60 time slots. We consider two traffic classes with a unitary budget of 1 and 10. We compare the proportion of accepted requests belonging to each type of class traffic for different request arrival rate. For instance, a rate equal to 5 means that 5 new requests arrive at each 10 time slots. With curve in figure 1, it is clear that requests with highest bid have more chances to be accepted than those belonging to the second class. This differentiation is more visible when the network is overloaded and thus auctions are more involved. We can conclude that the precedence problem does not really influence the efficiency of our mechanism.

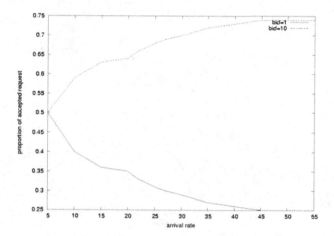

Fig. 1. Accepted request distribution for two type of traffic (two different bids)

4.2   Scenario2

In this scenario, we consider a mono network manager context where some priority level is given to different class of traffic by giving them an appropriate global budget. For instance, the differentiation can be based on request demands (traffic volume).

We consider a network where the topology is a grid with 10*10 nodes. For simplicity, links capacities are fixed to 100 unit of resource. A set of new requests is generated each 10 time slots. Requests origins and destinations are generated with a uniform distribution on the grid nodes. Connection durations are fixed to 50 time slots for all requests. Demands in terms of unit of resource are generated uniformly in [2..7]. Global bids are determined by functions in section3.1.

The curve of figure 2 illustrates the impact of requests budget function on the distribution of requests acceptance. With the function1, the network accepts more big requests than with the 3 and function2. Indeed, requests with big demands should have highest bids to be sure to be accepted. Small requests have more chances to be accepted because they fit easily when the network is high-utilized.

4.3   Scenario3

In this scenario, we simulate a hot spot point in the grid. All requests are destined to this point. We observe prices while approaching the destination.

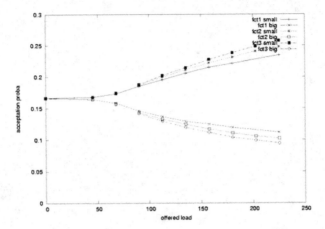

Fig. 2. Accepted request distribution for different bid functions

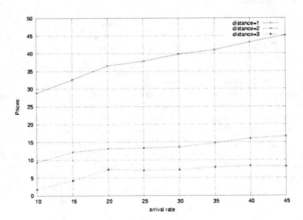

Fig. 3. Resource prices at different distance from the Hot-Spot for different arrival rates

Figure 3 illustrates the evolution of prices on links at different distances from the grid center for different request arrival rates. Prices are more important at one hop from the destination. This is due to the fact that those links are highly utilized since all requests have to use them. Prices are then good indicators of the network utilization.

5   Conclusions and Future Work

The proposed DiMA mechanism is a realistic resource allocation scheme where routing and resource reservation are distributed upon involved network nodes. The calculated prices are good indicators of the network utilization and can be used to propose an efficient pricing scheme.

An open problem is how to propagate information about resource prices on different links to take it into account to optimize the decision making. Some distributed mechanism (an inundation algorithm) can be used by routers to communicate local prices to their neighbors. These informations can be useful to routers while choosing an output link among many possible ones. Indeed, requests can be routed towards less congested directions. Moreover, requests can deduce in each part of the network they need to bid more.

For this purpose, some procedure is required in routers to summarize the different informations that are received and then use it for the decision making.

Acknowledgments. Authors would like to thank their colleague Frank Quessette for his valuable suggestions about the simulation work. Special thanks to Pr Peter Reichl for valuable discussions about the DiMA model. Authors would like to thank also their partners in the Prixnet project. Finally, thanks are addressed to the anonymous referees for their constructive comments which helped to improve the quality of the paper.

References

1. N. Anerousis and A.A. Lazar. A framework for pricing virtual circuit and virtual path services in ATM networks. *ITC-15*, pages 791–802, 1997.
2. Y. Bernet, P. Ford, R. Yavatkar, F. Baker, L. Zhang, M. Speer, R. Braden, B. Davie, J. Wroclawski, and E. Felstaine. A framework for Integrated Services Operation over Diffserv networks. *RFC 2998, Internet Engineering Task Force*, 2001.
3. S. Bessler and P. Reichl. A Network Provisioning Scheme Based on Decentralized Bandwidth Auctions. *Proc. KiVS(Kommunikation in Verteilten Systemen)*, February 2003.
4. C. Courcoubetis, M.P. Dramitinos, and G.D. Stamoulis. An auction mechanism for bandwidth allocation over paths. *International Teletraffic Congress ITC-17, Salvador da Bahia,Brazil*, pages 1163–1174, December 2001.
5. C. Courcoubetis, V.A. Siris, and G.D. Stamoulis. Integration of pricing and flow control for available bit rate services in ATM networks. *ICS-FORTH*, 1996.

6. C. Courcoubetis, G.D. Stamoulis, C. Manolakis, and F.P. Kelly. An intelligent agent for optimizing QoS-for-money in priced ABR connections. *Telecommunications Systems*, 2000.
7. L. Echabbi and D. Barth. Multi-link Auctions for Network Resource Reservation and Pricing. *MCO2004 Proceedings*, 2004.
8. G. Fankhauser, B. Stiller, C. Vogtli, and B. Plattner. Reservation -based charging in an integrated services network, 1998.
9. S. Jordan. Pricing of buffer and bandwidth in a reservation-based qos architecture. 2003.
10. F.P. Kelly. Charging and Accounting for Bursty Connections. *In Lee W. McKnight and Joseph P. Bailey, editors, Internet Economics, MIT Press*, pages 253–278, 1997.
11. A.A. Lazar and N. Semret. The progressive second price auction mechanism for network resource sharing. In *8th International Symposium on Dynamic Games*, pages 5–8, July 1998.
12. S. Low and P.P Varaiya. A new approach to service provisionning in atm networks. *IEEE transaction communications*, 1:547–553, 1993.
13. J. Murphy and L. Murphy. Bandwidth allocation by pricing in atm networks. *Broadband Communications*, pages 333–351, 1994.
14. J. Murphy and L. Murphy. Distributed pricing for embedded atm networks. *In International teletraffic congress ITC-14*, 1994.
15. K. Nichols, V. Jacobson, and L. Zhang. A Two-bit Differentiated Services Architecture for the Internet. 1997.
16. C. Parris, S. Keshav, and D. Ferrari. A framework for the study of pricing in integrated networks. Technical report, International Computer Science Institute, Berkeley, 1992.
17. R. Prior, S. Sargento, S. Cris stomo, and P. Brand o. End-to-End QoS with Scalable Reservations. Technical report, 2003.
18. P. Reichl, S. Bessler, and B. Stiller. Second-chance auctions for multimedia session pricing. In *MIPS2003*, pages 306–318.
19. J. Sairamesh, D. F. Ferguson, and Y. Yemini. An approach to pricing, optimal allocation and quality of service provisioning in high-speed packet networks. *Proceedings of the IEEE INFOCOM*, pages 1111–1119, 1995.
20. X. Xiao and L. M. Ni. Internet QoS : A Big Picture. *IEEE Network*, 13(2):8–18, 1999.
21. Zhi-Li Zhang. Decoupling QoS control from Core routers: A novel Bandwidth Broker Architecture for scalable support of guaranteed services. *SIGCOMM*, pages 71–83, 2000.

A Game-Theoretic Analysis of TCP Vegas*

Tuan Anh Trinh and Sándor Molnár

High Speed Networks Laboratory
Department of Telecommunications and Media Informatics
Budapest University of Technology and Economics
{trinh,molnar}@tmit.bme.hu

Abstract. We use the tools from game theory to understand the impacts of the *inherent congestion pricing schemes* in TCP Vegas as well as the problems of parameter setting of TCP Vegas on its performance. It is shown how these inherent pricing schemes result in a rate control equilibrium state that is a Nash equilibrium which is also a global optimum of the all-Vegas networks. On the other hand, if the TCP Vegas' users are assumed to be selfish in terms of setting their desired number of backlogged packets in the buffers along their paths, then the network as a whole, in certain circumstances, would operate *very inefficiently*. This poses a serious threat to the possible deployment of Vegas-based TCP (such as FAST TCP) in the future Internet.

1   Introduction

The Internet has been a huge success since its creation in the early 70's. It has a big impact on the way we interact and communicate. As the Internet evolves, it is shared, used by millions of end-points and many kinds of applications. They compete with each other for the shared resources and their demand for resources (such as bandwidth) is growing rapidly. As a result, congestion at certain points of the network is inevitable. The TCP protocol suite was originally designed to control congestion in the Internet and to protect it from congestion collapse. Basically, TCP is a closed loop control scheme. Congestion in the network is fed back to the source in the form of losses (Reno-like versions) or delay (such as TCP Vegas) The source then reacts to the congestion signal from the network by reducing its transmitting rate. In other words, we can consider packet loss and high queueing delay as the *cost* of (aggressively) sending packets into the network. The higher the rate, the higher the cost (certainly, the relationship is not necessarily linear in nature), given a fix network. Furthermore, as the Internet has been gradually transforming from a government sponsored project to a private enterprise (or even a commodity), the economics of the Internet becomes more and more important issue. Consequently, Internet connectivity and services will have to confront issues of pricing and cost recovery. In this perspective, the cost of congestion can be in monetary form. Introducing cost

---

* This work is supported by the Inter-University Center for Telecommunications and Informatics (ETIK)

J. Solé-Pareta et al. (Eds.): QofIS 2004, LNCS 3266, pp. 338–347, 2004.
© Springer-Verlag Berlin Heidelberg 2004

of congestion into the network creates balance, stability and high utilization of resource usage.

From what has been discussed so far, the cost of congestion, in our case, can be either price or delay (the application is delay-sensitive). It is suggested in [13] that congestion pricing could be implemented by using "smart market" where price for sending a packet varies on a very short time scale. Specifically, for the TCP implementation, the congestion price is updated every round-trip time.

A natural question arises then: Why TCP Vegas? There is a number of reasons that motivate us to re-examine TCP Vegas. Firstly, it is because TCP Vegas has *inherent pricing schemes* in its design that resemble the congestion pricing schemes proposed in the literature. We believe that by better understanding TCP Vegas' inherent pricing schemes, we will have a better insight into understanding and designing pricing schemes for TCP traffic in general. Secondly, the emergence of very large bandwidth-delay product networks such as the transatlantic link with a capacity in the range of 1 Gbps - 10 Gbps, new transport protocols have been proposed to better utilize the network in these circumstances. One promising proposal is the FAST TCP [18]. Since the design of FAST TCP is heavily based on the design of TCP Vegas, there is a need to reconsider the benefits as well as the drawbacks of TCP Vegas in order to have an insight into the performance and possible deployment of FAST TCP in the future Internet.

Given the pricing schemes (parameter setting), another natural question arises then: Are these schemes efficient? Is there any equilibrium state from where no one has the incentive to deviate? Game theory (see [4] for a comprehensive introduction) provides us the tools to answer these questions as we will illustrate later in the paper. We use these game-theoretic tools to investigate the impact of the pricing schemes and parameter setting in TCP on the performance of each user as well as the network as a whole.

Regarding the related work, we would like to divide it into two classes. The first class mainly deals with game-theoretic analysis of flow control mechanisms in the Internet. Scott Schenker in his pioneering paper [2] used game-theoretic approach to analyze the flow control mechanisms (for Poisson arrivals) with different queueing disciplines at the routers. Korilis *et al* in [3] also used game-theoretic approach to study the existence of equilibria in noncooperative optimal flow control, especially those with QoS constraints. Recently, Akella *et al* in [1] also used the tools from game-theory to examine the behavior of TCP Reno-like (loss-based) flow controls under selfish parameter setting. Our work is different from their work in the sense that we study delay-based versions of TCP. We provide an extensive analysis of the parameter setting problem of the traditional TCP Vegas, the modified version of TCP Vegas (TCP Vegas under REM) as well as FAST TCP. The second class deals with the mechanisms and issues of congestion pricing in the Internet. We would mention here the work of MacKie-Mason *et al* [13], [14]. These papers apply economic theory ("club theory") to study basic issues of congestion pricing in the Internet. Incentive-compatible pricing strategies in noncooperative networks are introduced and analyzed in

[8]. A survey on Internet pricing and charging in general can be found in [6]. These papers deal with the general pricing problems. Our paper, on the other hand, deals specifically, from game-theoretic point of view, with TCP Vegas and its variants.

The main contributions of the paper are the followings. First, we prove that, under the inherent congestion pricing schemes in TCP Vegas, there exists a unique Nash equilibrium of the rates of the TCP Vegas flows sharing a common network and this equilibrium rate vector is also system-wide optimal. Secondly, we provide an extensive game-theoretic analysis of the parameter setting problem in TCP Vegas. We conclude that, in this case, the Nash equilibria (if any), can be very inefficient. This implies that all-Vegas networks are vulnerable to selfish action of end-users posing a serious threat to the possible deployment of all-Vegas based (such as FAST TCP) in the future Internet.

The rest of the paper is organized as follows. The background on TCP is provided in Section 2. The TCP Vegas games are described and analyzed in detail in Section 3. Finally, Section 4 concludes the paper.

2   Background

2.1   TCP Vegas

TCP Vegas was first introduced by Brakmo *et al* in [16]. Basically, it is a delay-based congestion control scheme that uses both queueing delay and packet loss as congestion signal. TCP Vegas tries to control the number of packets buffered along the path with the targeted number to be between α and β ($\alpha \leq \beta$). Let $w(t)$ denote the congestion window at time t, RTT denote the round-trip time and $baseRTT$ is the smallest value of the round-trip time so far (actually, this is an *estimate* of the propagation delay). Denote $\mathsf{diff} = \frac{RTT - baseRTT}{RTT} w$, then the dynamics of the congestion window of TCP Vegas can be expressed as follows:

$$w(t+1) = \begin{cases} w(t)+1 & \text{if } \mathsf{diff} < \alpha, \\ w(t)-1 & \text{if } \mathsf{diff} > \beta, \\ w(t) & \text{otherwise.} \end{cases} \tag{1}$$

In a TCP Vegas/REM network [15], a slight modification is introduced into the updating mechanism of the congestion window. Each link l (with capacity c_l) update the *link price* $p_l(t)$ in period t based on the aggregate input rate $x^l(t)$ and the buffer occupancy $b_l(t)$ as follows:

$$p_l(t+1) = [p_l(t) + \gamma(\mu_l b_l(t) + x^l(t) - c_l)]^+ \tag{2}$$

where $0 < \gamma$ and $0 < \mu_l < 1$ are scaling factors of REM. Each source will estimate the *total* price along its path and update its sending rate accordingly. To feed back the prices to sources, link l marks each arriving packet in period t, that is not already marked at an upcoming stream, with probability $m_l(t)$ defined as:

$$m_l(t) = 1 - \varphi^{-p_l(t)}$$

where $\varphi > 0$ is a constant. Once a packet is marked, its mark is carried to the destination and then conveyed back to the source via acknowledgement, like ECN scheme. The source i estimates the end-to-end marking probability by the fraction $\hat{m}_i(t)$ of its packets marked in period t, and estimates the path price $p_i(t)$ by:

$$\hat{p}_i(t) = -\log_\varphi(1 - \hat{m}_i(t))$$

The dynamics of the congestion window of TCP Vegas/REM can be expressed as follows:

$$w_i(t+1) = \begin{cases} w_i(t) + 1 & \text{if } -\frac{w_i(t)}{RTT_i(t)} < \frac{\alpha}{\hat{p}_i(t)}, \\ w_i(t) - 1 & \text{if } -\frac{w_i(t)}{RTT_i(t)} > \frac{\alpha}{\hat{p}_i(t)}, \\ w_i(t) & \text{otherwise.} \end{cases} \tag{3}$$

2.2   Throughput Models of TCP Vegas

Throughout the paper our game-theoretic analysis uses the models that are previously derived. These are:

Model 1: (Thomas Bonald's model)
In [17], the throughput of multiple flows sharing a bottleneck link is analyzed (by using fluid approximation) both for TCP Reno and TCP Vegas case. Assume N TCP flows sharing a bottleneck link with capacity μ, propagation delay τ and buffer size B. The parameters of TCP are: α and β. The main results in their paper that we use in our analysis are the following:

 – If $N\alpha < B$, there exists a finite time from which no loss occurs. In addition, the window size stabilizes in finite time. If $\alpha \neq \beta$, the congestion windows converge not to a single point (but a region). This implies unfairness among flows even in equilibrium. If $\alpha = \beta$, then $w_1 = w_2 = ... = w_N = \frac{\mu\tau}{N} + \alpha$ and the average rate $\lambda_1 = \lambda_2 = ... = \lambda_N = \frac{\mu}{N}$. Note that in this case the link is fully utilized.
 – If $N\alpha \geq B$, then TCP Vegas behaves exactly like TCP Reno. Let $\omega = \frac{\mu\tau}{B}$ and γ is the multiplicative decrease of TCP Reno (typically $\frac{1}{2}$). If $\omega \geq \frac{\gamma}{1-\gamma}$ then $\lambda_{total} = \frac{(1-\gamma^2)(\omega+1)^2}{2(1-\gamma)(\omega^2+\omega)+1}\mu < \mu$. This implies that in this case the link is not fully utilized.

Model 2: (Steven Low's model)
Steven Low *et al* in [10], [11], [12], [15] described an optimization framework to study the performance of the TCP Vegas in a general network topology and under different queue management schemes at the routers. We would mention the result regarding the throughput of TCP Vegas under REM queue management scheme (TCP Vegas/REM) that we will use in our analysis later in this paper. It is proved in [11] that the equilibrium rate of TCP Vegas can be calculated as: $\lambda_i = \frac{\alpha_i}{p^*}$, where p^* denotes the equilibrium price. Note that this result is true for a general network topology (not restricted to a single bottleneck link).

3   The TCP Vegas Games

In this Section, the games regarding the inherent pricing schemes (for rate allocation) and parameter setting of TCP Vegas are described and analyzed in detail. We also investigate the impact of the results on the performance of TCP Vegas and on the network as a whole.

3.1   Game 1: Rate Allocation of TCP Vegas

We consider a network that consists of a set $\mathcal{L} = \{1, 2, ..., L\}$ of links with capacity c_l, $l \in \mathcal{L}$. Assume that the network is shared by a set of flows (sources). The set of flows is denoted by $\mathcal{N} = \{1, 2, ..., N\}$. The rate of flow i is denoted by x_i, $i \in \mathcal{N}$. Flow i uses a subset ($\mathcal{L}_i$) of $\mathcal{L}$ in its path ($\mathcal{L}_i \subseteq \mathcal{L}$). Let us define the routing matrix as follows:

$$\mathbf{R}_{li} = \begin{cases} 1 & \text{if } l \in \mathcal{L}_i, \\ 0 & \text{otherwise.} \end{cases}$$

The physical capacity constraints of the flows therefore can be defined as :

$$\mathbf{Rx} \leq \mathbf{c} \tag{4}$$

where $\mathbf{x} = (x_1, x_2, ..., x_N)$ is the flow rate vector and $\mathbf{c} = (c_1, c_2, ..., c_L)$ is the link capacity vector. In addition, flow rates cannot be negative:

$$x_i \geq 0, \qquad i = 1, 2, ..., N \tag{5}$$

The set of flow rate vectors Λ that satisfy both conditions 4 and 5 is called a feasible set.

It should be mentioned that our network, as TCP network in general, assumes feedback-based flow control. The feedback can be implicit (e.g. queueing delay) or explicit (by pricing and/or using Explicit Congestion Notification (ECN)). The sources (end-points) use Vegas-style flow control, as defined in [16], [11]. We consider the flows as the players of the game. The strategy space for a player is the range of it sending rate.

Let us define the following generic payoff function for each player

$$B_i(x_i) = \alpha_i \log(x_i) - \sum_{l \in \mathcal{L}_i} \int_0^{x_i} \pi_l(y) dy \tag{6}$$

and $\pi_l = p_l(\sum_{l \in \mathcal{L}_k} x_k)$ is defined as the function of the total flow rates on link l. This function is actually the *price* that is fed back to the player i sending at rate x_i, which is an increasing function. The higher the rate, the higher the price. Hence, the second term in Equation 6 can be interpreted as the bandwidth *cost* fed back to player i when it attempts to transmit at rate x_i. The first term in Equation 6 reflects the *gain* of player i when transmitting at rate x_i, [12] (note that this is a concave function of x_i). As a result, the payoff $B_i(x_i)$

represents the *net* benefit of player i when transmitting at rate x_i. The price (cost) can be communicated to the end-user (the player) by the mean of the total queueing delay of its packets in the path, as in TCP Vegas/Drop-Tail network. The price can also be communicated explicitly to the user by using REM active queue management scheme (with ECN) and here we have a Vegas/REM network. In the first case, we would like to mention that we *implicitly* use the PASTA property for Poisson arrivals with FIFO scheduling principle to derive the proportional relationship between the total rate arrive at the link and the queuing delay. We can also suggest here the Little's formula for this relationship. This is assumed frequently in the literature with or without any mention. In any case, if the aggregate arrival flow is not Poisson (e.g. self-similar traffic), then queue length (queueing delay) is generally larger than the Poisson one. Furthermore, the expression of queueing delay in our model is assumed to be *additive* among links. This is true for a Norton network with Poisson arrivals. So, strictly speaking, our analysis can be considered as a *worst case* analysis for TCP Vegas/Drop-Tail network. For TCP Vegas/REM network, the additive assumption is justified when the mark rates are small. Indeed, let $\pi_l(t)$ be the marking probability at link l at time t and the end-to-end marking probability $q_i(t)$ that the end-point i observes (and to which source algorithm reacts). For small $\pi_l(t)$, $q_i(t) = 1 - \prod_{l \in \mathcal{L}_i}(1 - \pi_l(t)) \approx \sum_{l \in \mathcal{L}_i} \pi_l(t)$.

Under the assumptions mentioned above, our problem can be modelled as a non-cooperative game. The strategy space for a player is its sending rate and is determined by the capacity of the links. The strategy for player i can be defined as $S^{(i)} = \{x_i | 0 < x_i \leq c_{max}^i\}$, where $c_{max}^i = \mathbf{max}\{c_l | l \in \mathcal{L}_i\}$. The strategy space for the game is defined as the Cartesian product $S = \bigotimes_{i=1}^{N} S^{(i)}$, which is equivalent to the feasible set Λ. Strategy $\mathbf{x} = (x_1, x_2, ..., x_N) \in \Lambda$ is called a strategy profile. Each player (e.g. player i) chooses the sending rate (x_i) in the feasible set in order to maximize its *own* payoff function $B_i(x_i)$ in a selfish way. By "in a selfish way" we mean that the player does not care about other players' payoff, as far as the rate vector is in the feasible set.

One of the key questions in a non-cooperative flow control game in general, and our game in particular, is whether the network converges to (or settles at) an equilibrium point, such that no player can increase its payoff by adjusting its strategy unilaterally. In the game-theory terminology such a point is called a *Nash equilibrium*. The Nash equilibrium in our game also reflects the *balance* of the gain and the cost for each player as well as for the network as a whole. A non-cooperative game may have no Nash equilibrium (in its pure strategy space), multiple equilibria, or a unique equilibrium. As for the TCP Vegas game, we can prove the following theorem:

Theorem 1. *There exists a unique Nash equilibrium (in its pure strategy space) for the TCP Vegas game described above.*

We follow the proof methodologies provided in [5] as well as in a recent paper [9].

Proof. First, let's consider the existence of the Nash equilibrium for the TCP Vegas game. Notice that the feasible set $\Lambda = \{\mathbf{x} | \mathbf{Rx} \leq \mathbf{c}, \mathbf{x} \geq \mathbf{0}\}$ is a nonempty, convex and compact set. It is nonempty because $\mathbf{x} = (\epsilon, \epsilon, ..., \epsilon) \in \Lambda$, where $0 < \epsilon < \frac{c_{min}}{N}$, $c_{min} = \mathbf{min}\{c_l | l \in \mathcal{L}\}$. It is bounded because $x_i \leq c_{max}, i \in \mathcal{N}$, where $c_{max} = \mathbf{max}\{c_l | l \in \mathcal{L}\}$. Assume that $\mathbf{x}_1, \mathbf{x}_2 \in \Lambda$ and $0 < \rho < 1$, we have:

$$\rho \mathbf{x}_1 + (1 - \rho)\mathbf{x}_2 \leq \mathbf{R}(\rho \mathbf{x}_1 + (1 - \rho)\mathbf{x}_2) \leq \mathbf{c}$$

This result implies the convexity of Λ.

Now let's consider the payoff functions of the players. Notice that $B_i(x_i)$ is a concave function of x_i. Indeed:

$$B_i''(x_i) = -\frac{\alpha_i}{x_i^2} - \sum_{l \in \mathcal{L}_i} \pi_l' < 0 \tag{7}$$

From what have been discussed so far, our game has the following properties:

1. The joint strategy space is nonempty, convex and compact.
2. The payoff function of each player is concave in its own strategy space.

According to Theorem 1 in [5], there exists a Nash equilibrium in its pure strategy space.

For the uniqueness of the Nash equilibrium, let's consider the (nonnegative) weighted sum of the payoff functions:

$$\sigma(\mathbf{x}, \mathbf{w}) = \sum_{i=1}^{N} w_i B_i(\mathbf{x}), \qquad w_i \geq 0 \tag{8}$$

Denote $g(\mathbf{x}, \mathbf{w})$ the pseudo-gradient of $\sigma(\mathbf{x}, \mathbf{w})$, then the Jacobian of $g(\mathbf{x}, \mathbf{w})$ with respect to $\mathbf{x}$ can be computed as follows:

$$\mathbf{G} = \begin{pmatrix} B_{11} & B_{12} & \dots & B_{1N} \\ B_{21} & B_{22} & \dots & B_{2N} \\ \vdots & \ddots & \ddots & \vdots \\ B_{N1} & B_{N2} & \dots & B_{NN} \end{pmatrix}$$

where

$$B_{ij} = \begin{cases} w_i(-\frac{\alpha_i}{x_i^2} - \sum_{l \in \mathcal{L}_i} \pi_l') < 0 & j = i \\ -w_i \sum_{l \in \mathcal{L}_{(i,j)}} \frac{\partial \pi_l}{\partial x_j} < 0 & j \neq i, \mathcal{L}_{(i,j)} \neq \emptyset \\ 0 & j \neq i, \mathcal{L}_{(i,j)} \equiv \emptyset. \end{cases}$$

where $\mathcal{L}_{(i,j)} = \mathcal{L}_i \bigcap \mathcal{L}_j$. The matrix $\mathbf{G}$ defined above is thus negative definite. As a result, according to Theorem 6 in [5], $\sigma(\mathbf{x}, \mathbf{w})$ is diagonally strictly concave. According to Theorem 2 in [5], the equilibrium point of the TCP Vegas game is unique.

Remark 1. To reach this equilibrium, [5] shows that each player can change its own strategy at a rate proportional to the gradient of its payoff function with respect to its strategy and subject to constraints. This method is in fact equivalent to the gradient projection algorithms described in [10].

Remark 2. The authors of [10], using optimization framework, also showed that, under certain assumptions on the step size, these algorithms converge to a system wide optimal point (which is also proved to be unique). Furthermore, it is proved in [11], [12] that the rate control of TCP Vegas/Drop Tail and TCP Vegas/REM is indeed based on these algorithms. This implies that *the TCP Vegas game described above converges to a unique Nash equilibrium that is system wide optimal.*

3.2   Game 2: Parameter Setting of TCP Vegas

In this game, we consider the parameter setting of TCP Vegas. As described in [16], TCP Vegas tries to maintain the number of backlogged packets in the network between α and β. We examine here the situation when a selfish (and greedy) user tries to increase the number of its backlogged packets in the network in order to grab more bandwidth in the network. If all other players do the same thing (i.e. they are also selfish and greedy), the total number of packets in the network would increase without bound. However, the size of the buffers at routers are bounded and packet loss would occur, reducing the throughput of the connection. We are interested in a situation (i.e. a parameter setting, if at all exists) from where no player would deviate.

We consider a simple topology of N TCP Vegas sources sharing a *single* bottleneck link with a buffer size of B packets. Source i is associated with a set (α_i, β_i). In this paper, we deal with the case when $\alpha_i = \beta_i$. The case when $\alpha_i \neq \beta_i$ is left for future work.

Players: N TCP Vegas flows
Actions: Each player can set its parameter (α_i) in order to control the number of its backlogged packets in the queue of the bottleneck link (with capacity μ and delay τ). The router is assumed to use Drop-Tail mechanism (FIFO principle)
Payoff: $f(\alpha_i) = \lambda_i$ (the average throughput)

If the total number of backlogged packets is smaller than the buffer size at the bottleneck router (i.e. $\sum_{j=1}^{N} \alpha_j < B$) then the payoff function of player i can be expressed as follows:

$$f(\alpha_i) = \lambda_i = \frac{\alpha_i}{\frac{\sum_{j=1}^{N} \alpha_j}{\mu}} = \frac{\mu \alpha_i}{\sum_{j=1}^{N} \alpha_j} = \frac{\mu \alpha_i}{\alpha_i + \sum_{j \neq i} \alpha_j} \tag{9}$$

From Equation 9 we have:

$$\frac{\partial f}{\partial \alpha_i} = \frac{\mu \sum_{j \neq i} \alpha_j}{(\alpha_i + \sum_{j \neq i} \alpha_j)^2} > 0, \qquad i = \overline{1 \ldots N} \tag{10}$$

Since $\sum_{j \neq i} \alpha_j$ is always positive, it follows from Equation 9 that $\frac{\partial f}{\partial \alpha_i} > 0, \forall i$. This implies that given other players' strategies, player i will set α_i as high as possible in order to maximize its payoff. Notice that Equation 9 is valid only if $\sum_{j=1}^{N} \alpha_j < B$. Otherwise, TCP Vegas, according to [17], behaves exactly like TCP Reno. In this case, there are two possibilities [17]:

$$\lambda_i^{Reno} = \begin{cases} \frac{(1-\gamma^2)(\omega+1)^2}{2(1-\gamma)(\omega^2+\omega)+1} \frac{\mu}{N} < \frac{\mu}{N} & \text{if } \omega \geq \frac{\gamma}{1-\gamma} \\ \frac{\mu}{N} & \text{otherwise.} \end{cases} \tag{11}$$

Thus, we have two cases:

Case 1: $w < \frac{\gamma}{1-\gamma}$

It is important to note that in this case, the link is fully utilized both for TCP Vegas and TCP Reno. Furthermore, in TCP Reno style performance, the bandwidth is fairly (equally) shared between flows (because they have the same RTT). Denote $\alpha^* = (\alpha_1^*, \alpha_2^*, ..., \alpha_N^*)$ be the Nash equilibrium of the game in this case. Without losing generality, we can assume that $\alpha_1^* \leq \alpha_2^* \leq ... \leq \alpha_N^*$. Notice that in Nash equilibrium, we must have $\alpha_1^* = \alpha_2^* = ... = \alpha_N^*$. Otherwise, player 1 has the incentive to deviate (i.e. to increase its number of backlogged packets - α_1) in order to get higher throughput, because in Reno style performance, it would get a fairer share of the total bandwidth (i.e. $\frac{\mu}{N}$). As a result, we have the Nash equilibria for this game: $\alpha^* = (\alpha_1^*, .., \alpha_N^*)$ where $\alpha_i^* \geq \lfloor \frac{B}{N} \rfloor$, $\forall i$. This means that, in this case, in Nash equilibrium, the parameter α can be arbitrarily large.

Case 2: $w \geq \frac{\gamma}{1-\gamma}$

In this case, the link is not fully utilized. Following similar reasoning as in Case 1, we have a set of Nash equilibria defined as follows: $\Omega = \{\alpha = (\alpha_1, ..\alpha_N) | \alpha_1 \leq \alpha_2 \leq .. \leq \alpha_N\}$ with the conditions that $\sum_{i=1}^{N} \alpha_i = B - 1$ and $\alpha_1 \geq \frac{(1-\gamma^2)(\omega+1)^2}{2(1-\gamma)(\omega^2+\omega)+1} \frac{B-1}{N}$. The latter expression simply means that even player 1 (who gets the smallest bandwidth) would not deviate, so no other player would deviate. If this condition does not hold, player 1 would deviate to get higher bandwidth share.

Our final comment on this unique Nash equilibrium is that each TCP Vegas flow (player) maintains the number of its own backlogged packets as many as possible. As a result, the buffer is nearly full and the queueing delay is unnecessarily high. A nearly full buffer may cause many difficulties for TCP Vegas (e.g. the estimation of *baseRTT* might be inaccurate if there are already many packets in the queue when the connection starts)

4   Conclusion

We have demonstrated, by using game-theoretic approach, how TCP Vegas' inherent pricing schemes as well as the parameter setting impact on its performance. Our analysis shows that these inherent pricing schemes result in a rate control equilibrium state that is a Nash equilibrium in game-theoretic terms which is also a global optimum of the all-Vegas networks. We also proved that the parameter setting of TCP Vegas is very vulnerable to selfish actions of the users. This poses a serious threat to the possible deployment of FAST TCP in the future Internet.

References

1. A. Akella, R. Karp, C. Papadimitrou, S. Seshan, and S. Schenker, *Selfish behavior and stability of the Internet: A game-theoretic analysis of TCP* ACM SIGCOMM, 2002.
2. Scott Schenker, *Making Greed Work in Networks: A game-theoretic analysis of switch service disciplines*, IEEE/ACM Transactions on Networking, vol. 3, 1995.
3. Y. A. Korilis and A. A. Lazar, *On the existence of equilibria in noncooperative optimal flow control*, Journal of the ACM, vol. 42, no 3 pp. 584-613, 1995.
4. M. J. Osborne and A. Rubenstein, *A course in game theory*, Cambridge, Massachusetts: The MIT Press, 1994.
5. J. B. Rosen, *Existence and uniqueness of equilibrium points for concave n-person games*, Econometrica, vol. 33, pp. 520-534, Jul. 1965.
6. P. Reichl, B. Stiller, *Pricing model for Internet services*, http://www.tik.ee.ethz.ch/cati
7. F. Kelly et al, *Rate control for communication networks: Shadow prices, proportional fairness and stability*, Journal of Operation Research Society, vol 49, no. 3, pp. 237-252, March 1998.
8. Y. Korilis et al, *Incetive-compatible pricing strategies in noncooperative networks*, in Proc. of IEEE Infocom 1998.
9. T. Alpcan, T. Basar, *Distributed algorithms for Nash equilibira of flow control games*, to appear in Annals of Dynamic Games, 2004
10. S. Low and D. Lapsley, *Optimization flow control, I: basic algorithm and convergence*, IEEE/ACM Transactions on Networking, 7(6):861-874, December 1999.
11. S. Low, L. Peterson, and L. Wang, *Understanding Vegas: a duality model*, Journal of ACM, 49(2):207-235, March 2002.
12. S. Low, *A duality model of TCP and queue management algorithms*, IEEE/ACM Transactions on Networking, October 2003.
13. J. MacKie-Mason, H. Varian, *Pricing the Internet*, in B. Kahin and J. Keller, eds., Public Access to the Internet, MIT Press, Cambridge, MA, 1995.
14. J. MacKie-Mason, H. Varian, *Pricing congestible network resources*, Journal on Selected Areas of Communications, pp. 1141-1149, 13(1995).
15. S. Athuraliya, V. Li, S. Low, Q. Yin, *REM: active queue management*, IEEE Network, June 2001.
16. L. Brakmo, S. O'Malley, and L. Peterson, *TCP Vegas: new techniques for congestion detection and avoidance*, IEEE/ACM SIGCOMM 94, London, UK, Sept. 1994.
17. T. Bonald, *Comparison of TCP Reno and TCP Vegas*, Workshop on the modeling of TCP, 1998.
18. Cheng Jin, David X. Wei and Steven H., *Low FAST TCP: motivation, architecture, algorithms, performance*, IEEE Infocom, March 2004.

BarterRoam: A Novel Mobile and Wireless Roaming Settlement Model

Eng Keong Lua[1], Alex Lin[2], Jon Crowcroft[1], and Valerie Tan[3]

[1] University of Cambridge, Computer Laboratory,
William Gates Building,15 JJ Thomson Avenue
Cambridge CB3 0FD, United Kingdom
{eng.keong-lua,jon.crowcroft}@cl.cam.ac.uk
http://www.cl.cam.ac.uk
[2] Bluengine Holdings
alex@bluengine.com
[3] Infocomm Development Authority, Singapore
valerie_tan@ida.gov.sg

Abstract. This paper describes *BarterRoam:* a new novel mobile and wireless roaming settlement model and clearance methodology, based on the concept of sharing and bartering excess capacities for usage in Visiting Wireless Internet Service Provider (WISP) coverage areas with Home WISP. The methodology is not limited to WISPs; it is applicable to virtual WISPs and any Value-Added Services in the mobile and wireless access environments. In the Broadband Public Wireless Local Area Network (PWLAN) environments, every WISP provides its own coverage at various locations or *Hotspots.* The most desirable option to help WISPs to reduce cost in providing wider coverage area is for the Home and Visiting WISPs to collaborate for customer seamless access via bilateral or multilateral agreement and proxy RADIUS authentication [1]. This is termed a roaming agreement. Due to the large number of WISPs desiring to enter the market, the bilateral or multilateral roaming agreements become complex and unmanageable. The traditional settlement model is usually based on customer's usage plus margin. In the broadband PWLAN environment, most WISPs and customers prefer flat-rated services so that they can budget expenses accordingly. The current settlement model will not be able to handle the preferred flat-rated settlement. Hence, a novel flat-rated settlement model and clearance methodology for wireless network environments is proposed to enable multiple service providers to trade their excess capacities and to minimize cash outflow among the service providers via barter trade mode. We are unaware of other comparative work in this area.

1 Introduction

With wireless and mobile networks evolving into Internet networks, allowing users wireless access to email, web browsing and all available Internet services, comes the challenge of charging and billing for the wireless roaming, services, and applications provided. Provision of these services on wireless and mobile networks presents unique challenges to the service providers, especially in the areas of metering and accounting, Quality of Service (QoS), Service Level Agreements (SLAs), authentication, wireless

J. Solé-Pareta et al. (Eds.): QofIS 2004, LNCS 3266, pp. 348–357, 2004.
© Springer-Verlag Berlin Heidelberg 2004

roaming, etc. Wireless Internet Service Provider (WISP) roaming is different from 2G-style roaming because the Internet model is different from the 2G roaming framework. 2G evolved from the Public Switched Telecommunications Networks (PSTN) model, which contains relatively dumb telecommunication user devices, centralized control, strict network hierarchies, and a circuit-switched basic service. The Internet model contains powerful user terminals, end-to-end service deployment, a core network, and best effort TCP/IP packet-forwarding service. In addition, the Internet is exemplified by the flexibility provided by IP subnet and Network Address Translation (NAT) islands. Thus, inflexible 2G-style roaming agreements where each party must provide connectivity to all requesting customers of another party, which could as a result, create imbalance due to roaming traffic not being symmetrical.

In the broadband Public Wireless Local Area Network (PWLAN) environment, every Wireless Internet Service Provider (WISP) provides its own coverage at various locations (Hotspots). Roaming between Hotspots belonging to different WISPs offers mobility and more access points. By signing a bilateral roaming agreement WISPs allow their end-users to use each other's network. An end-user can today travel abroad and only use another WISP network in two cases: A local pre-paid (scratch)-card that allows a time or Kbytes limited access to the visited network [4]. The service is bought from a local WISP to access its covered public Hotspots. The end-user becomes a client of the foreign WISP and is, by definition, not a *roamer*. This is also referred to as *Plastic Roaming*. Secondly, a subscription is bought with a broker or aggregator [1] who will allow the usage of all networks in the alliance. Strictly speaking this is not a roaming scenario as the end-user pays its charges to the aggregator or broker and does not in the strictest sense have a *home* network.

Most of the roaming issues can easily be solved between two WISPs. The authentication and billing processes can simply be implemented as long as both networks agree on the process. However, in a market with a multitude of networks with a variety of backgrounds it is necessary to implement workable standards so that the whole market can benefit from straightforward roaming procedures. Some vendors and players develop Subscriber Identity Module (SIM) [7], [6] card-based authentication modules with the use of the Mobile system Home Location Register (HLR) and Visitor Location Register (VLR) for roaming settlement; it will be difficult to introduce this into the whole market. Although the SIM card authentication is a very valuable and proven method; it will be necessary to agree on certain authorization procedures. This will include workable operating procedures that deal with interoperability between technologies and authentication.

In the situation where the Home WISP's customers enter into another Visiting WISP's coverage area, where the Home WISP do not have coverage, the following scenarios could happen:

- The customer does not have access, although his hardware is capable of gaining access, but he is not authorized.
- The customer can gain access by signing up with the Visiting WISP either via hourly rate or as a new subscriber to Visiting WISP.
- The Home and Visiting WISPs collaborate to allow the customer seamless access via bilateral agreement and proxy RADIUS [2] authentication. This is termed a roaming arrangement.

The third option is most desirable because it helps WISPs to reduce cost in providing wider coverage area by sharing networks [8].

Due to the large number of WISPs desired to enter the market, the bilateral or multilateral roaming arrangement become complicated and unmanageable when the numbers of WISP in the ecosystem increase. A settlement house similar to the concept of a credential center [9], managing the roaming arrangement is needed. The tradition settlement model as mentioned earlier is usually based on customer usages plus margin. In the broadband PWLAN environment, most WISPs and customers prefer flat rated services because they can budget expenses accordingly.

In this paper, we propose *BarterRoam*: a new novel mobile and wireless roaming settlement clearance model, based on the concept of sharing and bartering excess capacities for usage in Visiting WISP coverage areas. A bilateral or multilateral roaming agreement would have been established between the WISPs, so that the proposed roaming settlement system would be implemented. We assume in addition to roaming agreement, some incentive mechanisms [5] to be rewarded or routing protocol that achieves the truthfulness and cost-efficiency in a game theoretic sense [3] exist, to ensure that roaming domains offer the right amount of resources that is economically justifiable for contribution. The methodology is not limited to WISP; it is applicable to virtual WISP and value-added services. The objectives of this settlement methodology are:

- To enable multiple service providers to trade their excess capacities for usage in other service providers' service coverage areas;
- To minimize cash outflow from any service provider by enabling the trade of ex-cess capacities via barter trade mode.

2   Background

2.1   Assumptions

We are assuming that every WISP builds its service with excess capacity or *headroom* to maintain the desirable Quality of Service (QoS). The excess capacity is not used and will be wasted. Therefore, after provisioning sufficient resources with desirable QoS for their own internal customers, it is desirable to provide access to other WISPs' customers to use the excess service and network, thus increasing revenue and utilization. We also assume that bilateral or multilateral roaming agreement among WISPs have to be established for offering the service to other WISPs' customers, and WISPs' customers will have more pervasive coverage of roaming access. This will provide more value-added benefits to customers.

2.2   Concepts and Working Models

The settlement methodology works as follows:

- Every WISP will allow Proxy RADIUS authentication via the settlement clearing server and therefore allow the Home WISP users to roam into Visiting WISP coverage area.

- Every WISP will compute their excess capacities based on number of Hotspots owned, Hotspot capacities, number of customers, average customer usages and operation duration. This excess capacity is then contributed to the settlement ecosystem.
- Based on the capacity computed, the Visiting WISPs in the entire ecosystem allow the Home WISP's users to use up to the same amount of excess capacity Visiting WISPs contributed.
- The settlement is done monthly or upon agreed periodic billing cycle.

If the Home WISP's users use less than the Home WISP's contribution, then, no additional settlement is needed. In this situation, the excess capacity is wasted, but at a lower wastage limit. If the total utilization of the Home WISP's users exceeded the Home WISP's contributed amount, the below settlement computation follows:

- Home WISP's user usage percentage distribution amongst the Visiting WISP area is computed.
- Retail flat rate for each Visiting WISP is used to compute the total due the Visiting WISP.
- Wholesales discount is applied to compute the actual amount due.

3 Proposed Novel Settlement Model and Methodology Flow

3.1 Stage I – Due Diligence

Before a WISP can join the network, the following information is needed to ensure that the new WISP is compatible in pricing and QoS level:

- *Number of Hotspot* ($N_{Hotspot}$) - The number of wireless service coverage areas.
- *Hotspot capacity* ($N_{HotspotUsers}$) - This is the average number of concurrent users that can gain access to the service. The capacity is usually determined by brand and type of access points (AP) used, i.e. AP supports the number of users.
- *Operating hour* (H) - This is the time where user can gain access to the network. Operating hours is usually in sync with the Hotspot opening hour, but not necessary limited to. The maximum operating hour is 24 hours.
- *Operating day* (D) - This is days in a month where the Hotspot opens for business. The norm used in this settlement methodology is up 30 days per month.
- *Number of users* (N_{Users}) - This is the total signed up users that the WISP had acquired at the time of application and in Stage I.
- *Monthly average utilization per user* (μ) - This is computed from RADIUS log at the time of WISP application and in Stage I. For example, if the utilization pattern for a user is 2 hours per day, and there are 20 days per month of utilization, then there are $2 \times 20 = 40$ hours of monthly average utilization per user within Home WISP network.
- *Retail flat rate* (α) - This is the monthly flat rate the Home WISP bill its customers.
- *Wholesales discount* (β) - This is the discount from the flat rated retail hourly price the Visiting WISP sell the service to Home WISP.

- *Average Hotspot rental price* (γ) - This is the property rental price that the venue operator paid for that location, i.e. average property space rental price per sq. ft. of the property market.

The parameters will be revised at each billing cycle, or at pre-agreed periodic cycle, taking into consideration such as Hotspots that are not operational, new Hotspots, additional users, etc. With all the above parameters, the following secondary parameters are computed:

- *Total WISP capacity* - $CP_{WISPtotal} = N_{Hotspot} \times N_{HotspotUsers} \times H \times D$
  Unit: Hours
- *Own utilization* - $CP_{WISPutilization} = N_{Users} \times \mu$
  Unit: Hours
- *Excess capacity* - $CP_{WISPexcess} = CP_{WISPtotal} - CP_{WISPutilization}$
  Unit: Hours
- *WISP hourly rate* - $P_{WISPhourly} = \alpha \div \mu$
  Unit: Dollars
- *WISP contribution* - $\phi_{WISPcontrib} = CP_{WISPexcess} \times P_{WISPhourly}$
  Unit: Dollars
- *Broadband Value Index (BVI)* - This is introduced and used to ensure that the WISP pricing and contribution is equitable amongst the WISP within the same city where the living standard and QoS should not fluctuate too much.   Unit: None
  $BVI = \frac{1}{\gamma} + \{ \frac{(1-\beta) \times \alpha \times N_{Hotspot}}{N_{Users}} \}$

Note that for virtual WISP, the number of Hotspot is 0 (Zero) and contribution amount is $0 (Zero Dollars). There will be a pledge for [*average user utilization*], and this data is available from the WISP whom the virtual WISP collaborate with.

Once the new WISP is confirmed to have excess capacity and its BVI is within allowable range (e.g. within standard deviation and/or mean range of values of the current service providers), it is included into the ecosystem and the contract with the WISP is entered to include the following:

- Confirmed *Wholesales discount* (β)
- Confirmed *QoS commitment*
- Confirmed *QoS commitment*
- Confirmed *settlement period*
- Confirmed *commitment to the ecosystem period*
- Confirmed *per-user flat rate* payable to the settlement organization to facilitate roam-ing authentication
- Confirmed *per transaction fee* payable to the settlement organization to facilitate settlement computation

Once the service contract is entered, the list of Hotspots of new WISP will be published. The system for inclusion of new WISP and QoS reporting are illustrated in Figure 1 and Figure 2.

The following illustrations show the assumptions and Stage I parameters computation, in Table 1 and Table 2. The *Standard Deviation* and *Mean* for BVI are 0.997 and 1.462 respectively, are obtained from the list of WISPs' BVI. Assumptions made for the initial parameters:

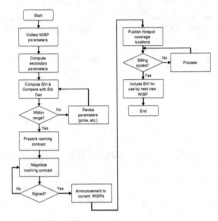

Fig. 1. System Description for inclusion of new WISP

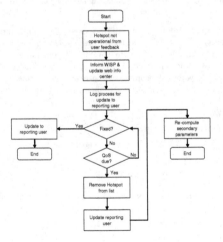

Fig. 2. System Description for QoS Report

Table 1. Assumptions for Parameters

WISPs	$N_{Hotspot}$	$N_{HotspotUsers}$	H (hours)	D	N_{Users}	μ (hours)	α	β	γ ($ per sq. ft.)
A	20	20	24	30	1000	40	$80	30%	$4
B	100	20	12	30	8000	35	$49	20%	$3
C	60	16	12	30	2000	30	$30	30%	$8
D	10	16	12	20	200	40	$80	40%	$2
E (Virtual)	0	0	0	0	2000	41	$67.20	0%	$0

Table 2. Assumptions for Secondary Parameters Computations

WISPs	$CP_{WISPtotal}$	$CP_{WISPutilization}$	$CP_{WISPexcess}$	$P_{WISPhourly}$	$\phi_{WISPcontrib}$	$\%\phi_{WISPcontrib}$	BVI
A	288000	40000	248000	$2.00	$496000	34.01%	1.37
B	720000	280000	440000	$1.40	$616000	42.24%	0.82
C	345600	60000	285600	$1.00	$285600	19.58%	0.76
D	38400	8000	30400	$2.00	$60800	4.17%	2.90
E (Virtual)	0	82000	0	$1.64	$0	0%	Nil

1. All WISPs have excess capacities that they are willing to barter trade.
2. Each Hotspot's access point (AP) can handle up to 20 concurrent users.
3. If each Hotspot has average sitting capacity of 80 patrons, and 20% will be accessing the service concurrently. This worked out to be about 16 concurrent users.
4. All Hotspots operate up to 24 hours a day, with average up to 30 days a month.

3.2   Stage II – Settlement Computation Models

In the situation whereby the Home WISP's users' utilization at the Visiting WISP service area exceeds the Home WISP contribution, the settlement is done by the following computation procedures. As illustrated in Table 3, we assume the utilization in hours at the Visiting WISPs (note that E is a virtual WISP). Firstly, the system tabulates the amount ($Amount_{Visiting}$) of utilization at the Visiting WISPs: $Amount_{Visiting} = CP_{VisitingUtilization} \times P_{WISPhourly}$. The system further checks for over-capacity by computing the exceeded amount ($Amount_{Exceed}$), to ensure the QoS of the ecosystem: $Amount_{Exceed} = \sum(Amount_{Visiting}) - \phi_{WISPcontrib}$. In this example, users of WISP B and WISP D did not exceed their contribution, as illustrated in Table 5. The system further checks for Visiting WISPs' capacity overload to ensure that total utilization by the Visiting users are within the overall available network excess capacity, i.e. $\sum CP_{WISPVisitingtotal} < CP_{WISPexcess}$, as shown in Table 4. With the $Amount_{Visiting}$ computed, the utilization distribution for the Home WISP at the Visiting WISP is then tabulated using the below expression and illustrated in Table 6: $\%Amount_{Visiting} = \frac{Amount_{Visiting}}{\sum(Amount_{Visiting})} \times 100$. This will ensure fair distribution of $Amount_{Exceed}$ collected from the Home WISPs. The final stage of computation of the WISPs' actual amount payable for its users' utilization at the other Visiting WISPs' networks is done as follows: $Amount_{Payable} = \%Amount_{Visiting} \times Amount_{Exceed} \times \beta$.

Table 3. Utilizations in hours at the Visiting WISPs, $CP_{VisitingUtilization}$

	WISP A	WISP B	WISP C	WISP D	WISP E	$\sum CP_{VisitingUtilization}$
A User	Nil	320000	80000	0	Nil	40000
B User	20000	Nil	100000	10000	Nil	130000
C User	150000	10000	Nil	2000	Nil	162000
D User	1000	3000	5000	Nil	Nil	9000
E User	6000	6000	52800	18000	Nil	82800
$\sum CP_{WISPVisitingtotal}$	177000	339000	237800	30000	Nil	

Table 4. Perform check if within overall available network excess capacity, $\sum CP_{WISPVisitingtotal} < CP_{WISPexcess}$

	WISP A	WISP B	WISP C	WISP D	WISP E
$\sum CP_{WISPVisitingtotal}$	177000	339000	237800	30000	Nil
$CP_{WISPexcess}$	248000	440000	285600	30400	Nil
Check!	Okay	Okay	Okay	Okay	Nil

Table 5. Utilization Amount at the Visiting WISPs, $Amount_{Visiting} = CP_{VisitingUtilization} \times P_{WISPhourly}$

	WISP A	WISP B	WISP C	WISP D	$\sum Amount_{Visiting}$	$Amount_{Exceed}$
A User	Nil	$448000	$80000	$0	$528000	$32000
B User	$40000	Nil	$100000	$20000	$160000	$0
C User	$300000	14000	Nil	4000	$318000	$32400
D User	$2000	$4200	$5000	Nil	$11200	$0
E User	$12000	$8400	$52800	$36000	$109200	$109200

Table 6. Utilization distribution for the Home WISP at the Visiting WISP, $\%Amount_{Visiting} = \frac{Amount_{Visiting}}{\sum (Amount_{Visiting})} \times 100$

	Payee				
	WISP A	WISP B	WISP C	WISP D	WISP E
Wholesales Discount (β)	30%	20%	30%	40%	Nil
WISP A	Nil	84.85%	15.15%	0%	Nil
WISP B	25.00%	Nil	62.50%	12.50%	Nil
WISP C	94.34%	4.40%	Nil	1.26%	Nil
WISP D	17.86%	37.50%	44.64%	Nil	Nil
WISP E	10.99%	7.69%	48.35%	32.97%	Nil

Note: Further contra settlement can be achieved by matching the WISP's payment, e.g. WISP C need only pay WISP A: $9170 - $1455 = 7155, instead of the full amount being transferred. Table 7 illustrates the computation, and the WISPs' *total* amount payable is also computed, i.e. $\sum (Amount_{Payable})$.

Table 7. Computation of the actual amount payable, $Amount_{Payable} = \%Amount_{Visiting} \times Amount_{Exceed} \times \beta$

| | Payee | | | | | |
	WISP A	WISP B	WISP C	WISP D	WISP E	Total Payable
Wholesales Discount (β)	30%	20%	30%	40%	Nil	$\sum(Amount_{Payable})$
WISP A	Nil	$5430	$1455	$0	Nil	$6885
WISP B	$0	Nil	$0	$0	Nil	$0
WISP C	$9170	$285	Nil	$163	Nil	$9618
WISP D	$0	$0	$0	Nil	Nil	$0
WISP E	$3600	$1680	$15840	$14400	Nil	$35520

3.3 Stage III – Reconfirmation Check

The settlement organization will further check to see if the WISP's users are consuming too much usage at the Visiting WISPs. This is done by computing the following parameters for each WISP and performing matching with the *average hour/user/month*. From Stage I and II, for each WISP, the *average cost per user* and *total utilization hour* are obtained. Thus, for each WISP, the utilization at the Visiting WISPs can be computed as the *average hour/user/month*. The parameters' computations are illustrated below:

- average cost per user $= \frac{\sum(Amount_{Payable})}{N_{Users}}$
- total utilization hour $= \sum CP_{VisitingUtilization}$
- average hour/user/month $= \frac{\sum(CP_{VisitingUtilisation})}{N_{Users}}$
- margin/user $= \alpha -$ average cost per user

If the utilization at the Visiting WISPs exceeded the Home utilization in terms of *average hour/user/month*, the WISP will have to pay the penalty of such leakage. This could happen due to the leakage or sharing of account. For virtual WISP E, the computed utilization is compared with the previously pledged [*average user utilization*], and this computed data is available from the WISPs whom the virtual WISP E has collaborated with. As illustrated in below tables, for the case of virtual WISP E, the computed utilization of 41.4 average hour/user/month is almost the same as the pledged value at 41

WISP E (Virtual)	
average cost per user $= \frac{\sum(Amount_{Payable})}{N_{Users}}$	$17.76
total utilization hour $= \sum CP_{VisitingUtilization}$	82800
average hour/user/month $= \frac{\sum(CP_{VisitingUtilisation})}{N_{Users}}$	41.4
margin/user $= \alpha -$ average cost per user	$49.44

WISP A Extra Cost	
average cost per user $= \frac{\sum(Amount_{Payable})}{N_{Users}}$	$6.88
total utilization hour $= \sum CP_{VisitingUtilization}$	400000
average hour/user/month $= \frac{\sum(CP_{VisitingUtilisation})}{N_{Users}}$	400
margin/user $= \alpha -$ average cost per user	$73.12

average hour/user/month. In the case of WISP A, the computed utilization at Visiting WISPs of 400 average hour/user/month is far exceeding the Home utilization of 40 average hour/user/month. This could be due to leakage or sharing of account. So, WISP A will pay the penalty of such leakage.

4   Conclusions

We have described the concept of BarterRoam model, a wireless roaming system for the contribution of excess capacity to barter trade for usage in Visiting WISP coverage area as an effective means of providing QoS that will enable the Home WISP to expand its coverage footprint. This is done:

- – Without incurring huge Hotspot setup cost;
- – Without incurring huge roaming settlement cost when Home WISP users roam into Visiting WISP coverage area.

This proposed settlement methodology sheds light on a formal computational approach towards catering for flat-rate charging mechanism for all WISPs (peer-level) in mobile and wireless environments.

References

1. B. Anton, B. Bullock, J. Short. Best Current Practices for Wireless Internet Service Pro-vider Roaming. Wi-Fi Alliance Public Document, 2003. http://www.wi-fi.org.
2. IETF. RADIUS: Remote Authentication Dial In User Service. RFC2138. http://www.ietf.org/rfc/rfc2138.txt.
3. L. Anderegg and S. Eidenbenz. Ad hoc-vcg: a truthful and cost-efficient routing protocol for mobile ad hoc networks with selfish agents. In *Proceedings of the 9th annual international conference on Mobile computing and networking*, San Diego, CA, USA, 2003.
4. G. Camponovo, M. Heitmann, K. Stanoevska-Slabeva, and Y. Pigneur. Exploring the wisp industry: Swiss case study. In *Proceedings of 16th Electronic Commerce Conference*, Bled, Slovenia, 2003.
5. J. Crowcroft, R. Gibbens, F. Kelly, and S. Ostring. Modelling incentives for collaboration in mobile ad hoc networks. In *Proc. of WiOpt'03*, 2003.
6. European Telecommunications Standards Institute. Global Multimedia Mobility (GMM) - A Standardization Framework. ETSI/PACi6-(96) 16 Part B, 1996.
7. K. M. Martin, B. Preneel, C. J. Mitchell, H.-J. Hitz, G. Horn, A. Poliakova, and P. Howard. Secure billing for mobile information services in UMTS. In *Proceedings of Fifth International Conference on Intelligence in Services and Networks Technology for Ubiquitous Telecom Services*, pages 535–548, 1998.
8. B. Patel and J. Crowcroft. Ticket based service access for the mobile user. In *Proceedings of the 3rd annual ACM/IEEE international conference on Mobile computing and networking*, pages 223–233, ACM Press, Budapest, Hungary, 1997.
9. H. Wang, J.-L. Cao, and Y. Zhang. Ticket-based service access scheme for mobile users. *Australian Computer Science Communications*, 24(1):285–292, 2002.

Charging for Web Content Pre-fetching in 3G Networks[*]

David Larrabeiti, Ricardo Romeral, Manuel Urueña, Arturo Azcorra,
and Pablo Serrano

Universidad Carlos III de Madrid, Av. Universidad 30, Leganés 28670, Spain
{dlarra,rromeral,muruenya,azcorra,pablo}@it.uc3m.es
http://www.it.uc3m.es

Abstract. Web pre-fetching is a technique that tries to improve the QoS perceived by a user when surfing the web. Previous studies show that the cost of an effective hit rate is quite high in terms of bandwidth. This may be the reason why pre-fetching has not been commonly deployed in web proxies. Nevertheless, the situation can change in the context of 3G, where the radio access is a shared scarce resource and the operator may find useful to exchange *fixed-network bandwidth* by *perceived QoS* for subscribed customers. Most importantly, in UMTS it is possible to charge for this service even though pre-fetching is provided by a third party. This paper studies this scenario, identifying the conditions where pre-fetching makes sense, describes the way OSA/Parlay could be used to enable charging, presents a tool developed for this purpose and analyses several issues related to charging for this service.

1 Introduction

Web pre-fetching is a well-known technology that has met a singular market niche in so-called internet boosters. The target of this sort of applications is to increase the effective QoS perceived by the end user by making use of spare access bandwidth to pre-fetch and cache those web pages most probable to be visited by the user in his/her next HTTP request. The average performance gain is driven by the ability of the prediction module to foresee the next link selected by the user. The context of this application is dial-up internet access over a low speed modem (e.g. V.34 or V.90). At first sight a reader may think that these internet boosters could be incorporated in 3G terminals and thus enhance the access bandwidth and virtually remove currently high RTTs (Round-Trip Times) from UMTS terminal to the Internet. However the charging schemes applicable in 3G that rate the volume of carried traffic make this option not realistic.

---

[*] This work was partially funded by the IST project Opium (Open Platform for Integration of UMTS Middleware) IST-2001-36063 and the Spanish MCYT under project AURAS TIC2001-1650-C02-01. This paper reflects the view of the authors and not necessarily the view of the referred projects.

J. Solé-Pareta et al. (Eds.): QofIS 2004, LNCS 3266, pp. 358–367, 2004.
© Springer-Verlag Berlin Heidelberg 2004

An alternative to this is the deployment of a proxy cache in the network that features pre-fetching (Fig.1). This way, over-fetching does not happen on the limited-bandwidth radio segment and, consequently, this service can actually be delivered in a cost-effective way by the network operator and even exploited by a third party as discussed in this paper. Therefore pre-fetching can be used as a way to provide differentiated application-specific QoS to a number of users on a subscription basis.

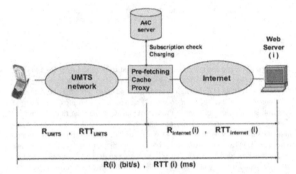

Fig. 1. Network-based pre-fetching in UMTS

The purpose of this paper is to show that this is possible and to identify key issues in its deployment, including the study of charging strategies for this specific QoS provisioning mechanism. On this sense, this paper is organized as follows. Section 2 provides a quick overview of pre-fetching and its practical limits. Once described the nature of the technique to be exploited, section 3 discusses how this service could be exploited externally via OSA/Parlay. Section 4 describes a test scenario deployed in a real UMTS network, that uses a pre-fetcher and charges for its usage. Finally a number of conclusions are drawn from this experience in section 5.

2   Web Pre-fetching

The research carried out in techniques to optimize web caching is very extensive. Therefore we shall try to cite only those works that bring key ideas required to understand the process, and how to charge for it in 3G. A broader survey can be found in [1].

As already defined, web pre-fetching is a technique that tries to improve the quality of service perceived in web browsing. The idea behind pre-fetching is enhancing an HTTP cache with initiative to retrieve in advance those web objects most likely to be downloaded by its users. This way, the hit ratio is increased and the effective latency perceived by the user is reduced, under the premise that there is an excess of band-width available.

An important early key work in pre-fetching techniques is [2] where the authors ana-lyze the latency reduction and network traffic for personal local caches, and introduce the idea of measuring conditional html link access probabilities and compressing this information with Prediction by Partial Matching. With this method and by using web-server-trace-driven simulation they obtain a reduction of 45% in latency at the cost of doubling the traffic. This can be considered a practical bound for prediction based on *client access probabilities*, which is the probability that the user accesses the link from a given page based on the user's personal navigation history.

With regard to pre-fetching with shared network caches, as mentioned above the scenario applicable to 3G, [3] estimates the theoretical limits of perfect caching plus perfect pre-fetching in a reduction of client latency of 60%. More realistic results (40-50%) are documented in other works [4] that study the origin of the limit of im-provement: it comes from the increase of delay and burstiness caused by the extra traffic put on the network by over-fetching. Regarding mobility, [5] analyses the effect of moving from one access technology to another on the effectiveness of pre-fetching.

As a conclusion from this brief survey techniques and bounds, it can be said that:

- The improvement obtained by pre-fetching is not fixed and depends on the pre-dictability of the user behaviour.
- The gain is measured in reduction of retrieval latency. The maximum perform-ance is around 50%. This means that the perceived improvement is actually de-termined by round trip times, size and speed of our context. In a context with low RTTs and high speeds the gain can be negligible.
- Effective pre-fetching is expensive in terms of bandwidth consumption. The optimum is obtained at 100% of over-fetching. Hence different strategies will be required depending on whether the link bandwidth is shared or not, and its cost.

3   Charging for Third-Party Provided Pre-fetching Service in 3G

The conclusions derived from the above observations seem to justify why, nowadays, pre-fetching is not enabled in web proxies. In fact, the main reason for web proxies campus and corporative networks is not just the reduction of latency, but the saving of a fair amount of bandwidth in the shared internet access link. Pre-fetching would reverse the latter positive effect (bandwidth saving) for the sake of the varying no-ticeability former effect (latency). Moreover, in the fixed internet access business model there is not an easy framework to charge the user –not a terminal- according to complex rules.

On the other hand, 3G subscribers are identified individually after they turn on their terminal and type in their PIN, and there is a working billing system available. The following situation may be frequent. The subscriber has paid for a high class ubiquitous wireless Internet access, and finds that the perceived QoS is under the real access capacity, due to bottlenecks and long delays not (only) in the access but somewhere in the Internet. In this 2G-3G context, network-based pre-fetching may be a tool to cover this demand, improve delivered QoS and charge accordingly.

3.1 Charging Schemes for Pre-fetching

Once described the nature of the service, its context and theoretical bounds, several charging policies for pre-fetching can be studied:

- Flat rate. This option is the simplest to deploy as it only requires subscription checking. Its drawback is that it relays strongly on the confidence of the subscriber about the promise performance gain. This is a problem that also suffer some commercial offers for differentiated services. In fact, as already described, pre-fetching can not guarantee a pre-determined QoS due to the random variables involved (mainly, the user will and the conditions of the path to the chosen web servers). Therefore, such charging may be rendered unfair by the user, due to the uncertainty of the obtained QoS.
- Charging per GGSN bandwidth devoted to pre-fetching on behalf of a given subscriber. This is directly not feasible due to the high number of subscribers sharing the Internet access of the 3G network. It is not possible to pre-allocate bandwidth for all. An alternative is to pay for sharing the bandwidth proportionally among all active pre-fetching subscribers. Again the charging procedure is very simple, yet the problem comes from the fact that more bandwidth share simply means more probability of obtaining an uncertain amount of extra QoS.
- Charging per pre-fetched information. The rationale for not recommending this option is the same as the previous one. In both cases, however, there is a direct relation between the extra communication costs caused by each subscriber and the charged amount.
- Charging per saved delay. This is the fairest approach from the subscriber perspective as the charging is directly related, not only to the success of the prediction module in terms of hits, but to the exact gain achieved by each hit. On the other hand, it is the most difficult to implement. The client is charged only for the accumulated amount of saved delay. The service provider must have careful control of incurred costs, but has the flexibility to allocate to the pre-fetching activity just unused resources.
- Fixed quota plus per-saved-delay charging. Given the operation costs, a fixed subscription fee is actually necessary, complementary to charging per saved delay, in order to establish a point where service provider and client viewpoints meet.

The computation of saved delay is not straightforward. It could be estimated by this formula:

$$\Delta\delta(I) = \sum_{i \in I} (sizeof(i)(\frac{1}{\min(R_{internet}(i), R_{UMTS})} - \frac{1}{R_{UMTS}}) + k(RTT_{internet}(i))) \tag{1}$$

Where I is the set of retrieved web objects during a navigation session and function *sizeof()* returns their respective size. Bit rates R and round trip times RTT are defined by Fig. 1. R_{UMTS} is considered constant, and $R_{internet}$ must be estimated and recorded by the pre-fetcher when the object is downloaded (the simplest implementation will log the size and download duration, integrating delay due to connection setup and data transfer). The reader must note that the RTT term is particularly important when retrieving small objects or when the bottleneck in the path is the UMTS access. Factor $k \cdot 1$ models the times the RTT is required for the TCP slow-start mechanism to reach the capacity of the path. A linear charging scheme proposed by the authors by this service would be given by:

$$\alpha = K_{subscription} + K_t \sum_I \Delta\delta(I) \tag{2}$$

Where $K_{subscription}$ and K_t are expressed in *monetary unit* and *monetary unit/ms* respectively.

3.2 Provisioning and Charging by a Third Party

The open network interface promoted by 3GPP[6] for UMTS (Universal Mobile Telecommunication System, the ETSI European standard of the 3G IMT-2000 system launched commercially in Europe in the fourth quarter of 2003) provides a framework for A4C, along with a significant number of network control functions, that makes it possible for third-party service providers the delivery of a wide-range transaction-oriented services upon this telecommunications network. This open network interface is named OSA (Open Service Access) [7] and adopts work from Parlay [8]. In this section we study how OSA can be used to enable the provision of pre-fetching by a third party.

Figure 2 shows the main elements involved in this scenario. The pre-fetcher keeps track of the user navigation and, when a given link access probability threshold is exceeded, the associated object is pre-fetched. Based upon the time and object size log available at the cache, it is possible to estimate the perceived delay gain, and charge the user accordingly through the OSA interface. Note that provisioning of pre-fetching by an external entity implies that all the internet traffic is handled by this entity. This requires low delay operator- service provider and may require NAT (Network Address Translation).

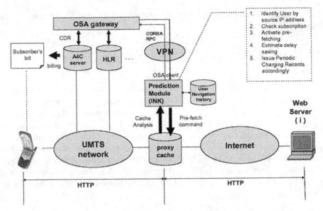

Fig. 2. Charging via OSA/Parlay

The interaction with OSA works as follows. Once the application, in this case the pre-fetching proxy, has been successfully authenticated by the Framework SCF (Service Capability Functions) of the Parlay Gateway, it is able to access to the authorized SCFs. The SCFs required for this service are the Charging SCF and a non-standard Terminal Session SCF which will be described in the next section. In particular, all the communication between the application and the Parlay gateway employs CORBA, although it is modelled as Java classes. For example, in the case of the Charging SCF, the relevant API calls are: from the initial `ChargingManager` object, applications may create several `ChargingSession` objects that refer to a concrete user and merchant. This class implements the interface to perform the most relevant operations of credit/debit to request charging the user for some amount of money or in some volume unit as in bytes, directly (`directCreditUnitReq()`) or towards a reservation (`debitAmountReq()`) for pre-paid services.

A quick overview of some UMTS aspects is required to understand a few implementation issues of charging via OSA of IP-address-identified services.

The UMTS architecture is strongly influenced by compatibility with the 2G digital telephony system (GSM) and the switched packet data service evolved from it GPRS (General Packet Radio Service). Two conceptually new elements have been introduced: the SGSN (Serving GPRS Support Node) and the GGSN (Gateway GPRS Support Node). These devices are in charge of data packet switching. In outline, the SGSN deals with mobility across RNCs (Radio Network Controller), following mobile stations in its service area and with AAA functions; whereas the GGSN is the actual gateway to Internet (see UMTS forum [6] documents for further details).

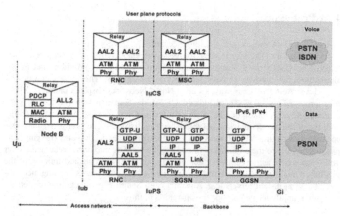

Fig. 3. UMTS protocol architecture

Fig. 3 shows the respective protocol stacks running in each of these elements to transport IP packets. The purpose of this figure is to reflect that the transport of IP packets inside the UMTS network is complex and that the only fixed point in the network where persistent caching is possible is just behind the GGSN, as this is the single internet access for the UMTS subnetwork. Another important issue when trying to charge for pre-fetching is that the "always-on" feature (a fixed global public IP address permanently allocated to the terminal) implies a non-scalable resource consumption in UMTS nodes. Therefore, IP addresses are provided on demand via the dynamic creation of *contexts* for each data session. This means that the implicit authentication given by the source IP address is not valid all the time and the binding <MSISDN , IP address> must be checked whenever the proxy observes a new data flow from an IP address inactive for a period longer than the guard time given by UMTS to reassign the address T_{guard}. Furthermore, since the release of an IP address is not conveyed to the proxy, charging requires periodic charging transactions on the OSA gateway interfacing the UMTS network. Just a leaky bucket whose period is less than the IP address reassignment guard time T_{guard} is enough to guarantee that the identity of the user is still valid and the charge goes to the right bill. This rate for CDR generation also determines the throughput required at the OSA gateway to accomplish a target Grade of Service. This can be determined by well known traffic engineering formulae such as Erlang-B or Engset, where the maximum number of active data sessions that can be charged (the number of servers in that formula) is given by $m = C_{server}/C_{client}$, where C_{server} is the capacity of the server in transactions/minute, and $C_{client} > 1/T_{guard}$ is the CDR generation rate of a single data session.

4 A Prototype

In the context of project [9], a testing scenario for the concepts developed in this paper was set up. The targets were to assess the viability and performance of pre-fetching in UMTS, to evaluate provisioning by a third-party provider, and to test alternatives for charging for this service via OSA/Parlay. The OSA gateway employed was AePONA Causeway, set up at Nortel Networks Hispania premises, Vodafone provided its UMTS network and UC3M developed a web pre-fetcher [10]. Commercial exploitation of UMTS services had not yet started at the time of testing, and, therefore, the system was not tested with real subscribers. The pre-fetcher was connected through a VPN to the OSA gateway for authentication and charging purposes according to the criterium defined in section 3. The maximum delay gain (over 40%) was easily achieved after a number of code optimizations and training of the prediction module. This gain translated into tens of seconds when browsing medium-sized pages at distant servers (e.g. Australia from Spain) even though the pre-fetcher was suboptimally located outside the operator's premises.

Inheriting terminology from the Telephony service, the OSA gateway issued Call Detail Records (CDR) with a flexible arbitrary format that enabled the production of a detailed charging log containing application-specific information about the charged event. As explained before, the experiment required an extension to the OSA API: the Terminal Session SCF. This extension permits to find out what user (i.e. which MSISDN number) is making a given HTTP request by simply checking the packet source IP address. This new functionality is very important in order to make OSA fully transparent to the end user.

A main practical result of the experience is that, regardless of the usage of pre-fetching and caching, the utilisation of a network proxy in UMTS is always advisable. The main reason is the speed of proxies. Proxies have multi-threading retrieval capabilities not usually available at UMTS terminal's web browsers. Furthermore, in the scenario deployed for the trial, the proxy was located 200ms away from the terminal (RTT=400ms) and still the improvement was significantly high in sequential retrievals.

5 Conclusions

Several conclusions can be drawn from the previous discussion and from the practical experience obtained with the test platform.

- Today the mechanics of pre-fetching are well known and it is clear that the performance gain is expensive in terms of bandwidth consumption and added traffic burstiness when performed on behalf of a large population of clients. However, in the context of internet access through 3G networks, where the business model is quite different from the classical accesses, pre-fetching can be an added-value service that can be offered to a certain type of users on a subscription basis.

- Network-based pre-fetching is viable in UMTS. A rough estimation of the maximum performance gain obtained is given by a 100% hit ratio, which, in the typical delay-bandwidth UMTS-Internet scenarios, leads to up to 40% of delay reduction (accessing far away or low speed web servers) at a cost of double bandwidth consumption. This means tens of seconds of saving when downloading a medium complexity web page. As demonstrated in IST project Opium, OSA enables a business model for pre-fetching in UMTS very difficult to deploy for fixed internet access subscribers, and whose low scalability must be controlled by subscription.

- It must be remarked that a network-based prefetcher does not improve the throughput in the radio segment, but the perceived end-to-end bandwidth due to bottlenecks and delay existing in the fixed part of the network. In other words, the only case when this pre-fetching scenario is really cost-effective is when accessing far away servers. Otherwise, it is the optimized multithreading capabilities of the proxy what predominates and still justifies the insertion of a proxy.

- Thanks to OSA/Parlay it is possible to authenticate, check the subscription, and charge subscribers of web content pre-fetching. This worked as expected and all tests passed. The range of tarification models applicable is very wide (charging for effective pre-fetching, for average virtual bandwidth excess granted, for the amount of pre-fetched information, etc), and its granularity is limited by the rate of charging transactions at the OSA/Parlay gateway, which is constrained itself by a lineal communications/computation overhead. CDRs were generated at a maximum rate of 2 CDR/minute in our tests. CDRs were issued only when the user selected a link that had been pre-fetched on his/her behalf, and the amount of money charged was proportional to the real delay saving. Due to the unpredictable effectiveness of pre-fetching this seems to be the fairest cost model applicable to this scenario.

References

1. Jia Wang. A survey of Web caching schemes for the Internet. ACM Computer Communication Review, 25(9):36{46, 1999.
2. J. G. Cleary and I. H. Witten. Data compression using adaptive coding and partial string matching. IEEE Transactions on Communication, 32:396{402, 1984.
3. Tom M. Kroeger, Darrell D. E. Long, and Jerey C. Mogul. Exploring the bounds of web latency reduction from caching and pre-fetching. In USENIX Symposium on Internet Technologies and Systems, 1997.
4. Venkata N. Padmanabhan and Jerey C. Mogul. Using predictive pre-fetching to improve World-Wide Web latency. In Proceedings of the ACM SIGCOMM '96 Conference, Stanford University, CA, 1996.
5. Z. Jiang and L. Kleinrock. Web pre-fetching in a mobile environment. IEEE Personal Communications, 5:25{34, October 1998.
6. 3rd Generation Partnership Project (3GPP). http://www.3gpp.org, July 2003.

7. 3GPP. Open Service Access (OSA) Application Programming Interface (API). Technical Specification 29.198+.
8. Parlay Group. http://www.parlay.org.
9. Opium project site. http://www.ist-opium.org/.
10. INK tool web page. http://matrix.it.uc3m.es/opium

Provider-Level Service Agreements for Inter-domain QoS Delivery

Panos Georgatsos[1], Jason Spencer[2], David Griffin[2], Takis Damilatis[1],
Hamid Asgari[3], Jonas Griem[2], George Pavlou[4], and Pierrick Morand[5]

[1]Algonet SA, Athens, Greece, {pgeorgat, pdamil}@egreta.com
[2]University College London, UK, {dgriffin, jsp, jgriem}@ee.ucl.ac.uk
[3]Thales Research and Technology Ltd., Reading, UK,
Hamid.Asgari@thalesgroup.com
[4]University of Surrey, Guildford, UK, g.pavlou@eim.surrey.ac.uk
[5]France Telecom R&D, Caen, France, pierrick.morand@rd.francetelecom.com

Abstract. In the current Internet, business relationships and agreements between peered ISPs do not usually make specific guarantees on reachability, availability or network performance. However, in the next generation Internet, where a range of Quality of Service (QoS) guarantees are envisaged, new techniques are required to propagate QoS-based agreements among the set of providers involved in the chain of inter-domain service delivery. In this paper we examine how current agreements between ISPs should be enhanced to propagate QoS information between domains, and, in the absence of any form of central control, how these agreements may be used together to guarantee end-to-end QoS levels across all involved domains of control/ownership. Armed with this capability, individual ISPs may build concrete relationships with their peers where responsibilities may be formally agreed in terms of topological scope, timescale, service levels and capacities. We introduce a new concept of QoS-proxy peering agreements and propose a cascade of inter-domain Service Level Specifications (SLSs) between directly attached peers: each ISP meeting the terms of the SLSs agreed with upstream peers by being responsible for its own intra-domain service levels while relying on downstream peers to fulfill their SLSs.

1 Introduction

There is a growing trend towards IP-based services, not only from a technical perspective e.g. VoIP (Voice over IP) services, but also from business perspectives, e.g. the emergence of Internet-based Application Service Providers. As more and more network-performance-sensitive services migrate to IP networks, best-effort networks no longer meet their QoS requirements. To this end, services require differentiation to provide different QoS levels for different applications over the Internet at large.

The issue of provisioning end-to-end QoS in the Internet is currently being investigated by both research and standardisation communities. Requirements from ISP perspectives and proposals targeted at building MPLS-based inter-domain tunnels

J. Solé-Pareta et al. (Eds.): QofIS 2004, LNCS 3266, pp. 368–377, 2004.
© Springer-Verlag Berlin Heidelberg 2004

have been recently submitted to the IETF [9], [10], [11] by key players in the field. As a pure layer 3 (routed) solution, the European-funded IST research project MESCAL [7] is investigating solutions targeted at building and maintaining a QoS-aware IP layer spanning across multiple domains, and considers architecture, implementation and the business models associated with it [4].

To provide end-to-end QoS across the Internet closer co-operation between multiple ISPs is required. The business relationships between ISPs must be considered and this is the main focus of this paper, although, we do not consider accounting and data collection methods, charging, rating and pricing models in this paper. A number of assumptions are made when approaching the co-operation problem. Firstly, we assume that there is no global controlling entity over all ISPs; secondly, that the problem of intra-domain is already solved and that different domains offer a range of *local Quality Classes* (l-QCs) between their edge routers.

With these assumptions in place we describe and analyse various modes of distributed interaction between ISPs and propose suitable business models to provide both loose, qualitative and statistical, quantitative end-to-end QoS guarantees. However, to achieve this ISP interaction, it is seen that a strengthening of business relationships is required with explicit QoS information being part of these agreements. To achieve this we use *pSLSs* (provider-Service-Level-Specifications) to describe QoS attributes to given destinations. These pSLSs are then used to concatenate l-QCs to form *extended Quality Classes* (e-QCs) to remote destinations. We analyse various ways this concatenation can be achieved and the implications on scalability.

Once the pSLSs have been created, the further problem of financial settlement must be considered. Drawing on concepts from the current Internet and the global PSTN (Public Switched Telephone Network) we propose new peering agreements and examine the flow of monies across the new QoS-enabled Internet.

The paper is organised as follows. Section 2 describes the current arrangements between ISPs and describes the business relationships between international PSTN operators. Section 3 analyses source-based and cascaded models for QoS-peering. Section 4 considers business cases for both loose and statistically guaranteed performance levels and the associated financial settlement issues between ISPs. Finally, section 5 presents our conclusions.

2 Business Relationships and Financial Settlements in the Current Internet and PSTN Networks

The global Internet is a collection of independently operated networks whose organisation in retrospect has been modelled by a three-tiered hierarchy [6] (Figure 1). The connectivity and position in the tier model is dependent on the size of the ISP, its geographic reach, capacity (in terms of link speeds and routing capability) and the available reachable prefixes. While this model is not strictly accurate it serves to demonstrate the variety of ISPs and their relationships within the Internet.

Currently, in the best-effort Internet, there exist two forms of distinct relationships between ISPs for traffic exchange, underlined by respective business agreements: *peer-to-peer* and *transit* (customer-provider). A transit relationship is where one provider provides forwarding (transit) to destinations in its routing table (could be to the global Internet) to another ISP for a charge. Usually, this type of business relationship is between ISPs belonging to different tiers of the three-tier Internet model (lower tier ISP being a customer of the upper tier ISP). Peer-to-peer is the business relationship whereby ISPs reciprocally provide only access to each other's customers and it is a non-transitive relationship. It is a kind of 'short-cut' to prevent traffic flowing into the upper tiers and allows for the direct flow of traffic between the peer-to-peer ISPs.

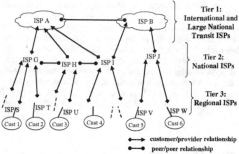

Fig. 1. A post-hoc approximation of a three-tier Internet model with peering/transit agreements.

Financial Settlements

The financial settlements between ISPs primarily depend on their business relationship. In the *service-provider settlement*, a customer (end-customer or ISP) pays a flat rate or a usage-based amount to the provider ISP for reachability to networks, which the provider ISP can reach through its peers, customers or through its own provider ISPs. The customer will always pay whether the traffic is being sent or received.

In the *negotiated-financial settlement*, the traffic volume in each direction is monitored and then payment is made on the net flow of traffic.

In the *settlement-free agreement* (also known as the Sender-Keeps-All, or SKA agreement in the PSTN) neither ISP pays the other for traffic exchange, and they usually split the physical layer costs between them. This settlement is a special case of the negotiated-financial settlement, because either the traffic is symmetric or because the perceived gain to each party is considered worth the agreement.

PSTN Networks

An existing network that requires close business relationships to provide end-to-end better-than-best-effort communication is the PSTN (Public Switched Telephone Network). The inter-connection of international PSTN networks has a number of traits similar to the inter-domain QoS problem: such as resources must be reserved to provide the required level of service but direct interconnection between all networks is not possible and therefore trust relationships are required. When crossing international boundaries the problem of trust becomes more acute, and the transit network topology is a function of political and financial considerations. These peering agreements are overseen by the International Telecommunications Union (ITU) and the World Trade Organisation (WTO) and usually result in a financial settlement like the Accounting Revenue Division Procedure [1]. Here the originating caller at the originating network is charged at a customer rate, and the originating network pays the terminating network to terminate the call at a previously negotiated rate, the settlement rate. The transfer of monies is then usually performed only on the net flow in a negotiated-financial settlement style agreement. If, however, the perceived value to each party is similar, as it may be when the edge operators adjust their customer rates, the inter-operator agreements approach the SKA (sender-keeps-all) agreement.

These ideas could be drawn into a QoS-enabled Internet business model, albeit with the following differences: PSTN has a single QoS traffic unit – the call-minute, whereas a QoS-enabled Internet could have many QoS levels. Also, the PSTN, especially at the international level, is more flat than hierarchical and has organisations in loose control to mediate and arbitrate settlements. Another issue that must be considered is which traffic end-point is the initiator, who gains from the data flow, and who is charged for it [3], as well as methods of charging for multicast traffic. Some of these issues have been considered in the split-edge pricing work in [2].

3   QoS Peering Approaches

In the current Internet, ISPs form business relationships with one another and deploy links between their networks, either directly or through Internet exchange points (IXPs). BGP policies are then deployed to determine which prefixes will be advertised to adjacent providers and subsequently best-effort traffic may be routed across the Internet.

We view that in a QoS-enabled Internet additional agreements are needed to determine the QoS levels, traffic quantities and the destinations to be reached across the pre-existing inter-domain links, together with the agreed financial settlement terms, penalty clauses, etc. The technical aspects of the QoS agreements between ISPs are contained in pSLSs, introduced in section 1. Once the pSLSs are in place, BGP will announce the existence of destinations tagged with the agreed QoS levels, and traffic conforming to the pSLSs may be forwarded to remote destinations and receive the appropriate treatment to meet the agreed performance targets. pSLSs are meant to support aggregate traffic, and they are assumed to be in place prior to any agreements with end customers (via customer-SLSs, or cSLSs) or upstream ISPs (via pSLSs) to use services based on them. They are negotiated according to the business policies of the provider and the outcome of the deployed off-line inter-domain traffic engineering algorithms that use forecasts of customer traffic as input. pSLSs are considered as semi-permanent, only changing when traffic forecasts for aggregate traffic alter significantly, or when business policies are modified. They should not be seen as dynamic entities to be deployed for individual customer flows.

There are many models for the interconnection and service-layer interactions between ISPs. Such models are required not only to establish a complete end-to-end customer service, but also to provide and maintain the hierarchy of related management information flows. Eurescom specified organisational models for the support of inter-operator IP-based services [8]. The models may be grouped into three configurations known as the *cascade model*, the *hub model*, and the *mixture model*: a combination of the two. The type of inter-domain peering impacts the service negotiation procedures, the required signalling protocols, the QoS binding, and path selection. In the following we give an overview of the cascade and source-based model and analyse their pros and cons.

Source-Based Approach
The source-based approach (similar to the hub model from Eurescom) disassociates pSLS negotiations from the existing BGP peering arrangements. The originating domain knows the end-to-end topology of the Internet and establishes pSLSs with a set of adjacent and distant domains in order to reach a set of destinations, with a particular QoS.

As shown in Figure 2, the originating domain (AS1) has the responsibility for managing the overall requested QoS service/connection. To manage customer requests, the provider (AS1) directly requests peering agreement ($pSLS_1$ and $pSLS_2$) with providers AS2 and AS3 and with any other network provider involved in order to create an e-QC (from AS1 to AS3).

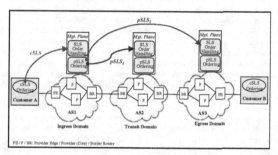

Fig. 2. Source-Based Approach

Cascaded Approach
In the cascaded approach, each ISP established pSLSs only with adjacent ISPs, i.e. those ISPs with whom there are existing BGP peering relationships. Figure 3 gives an overview of the operations in this approach. The domain AS3 supports an intra-domain QoS capability (l-QC_1). AS2 supports an intra-domain QoS capability (l-QC_2) and is a BGP peer of AS3. AS2 and AS3 negotiate a contract ($pSLS_2$) that enables customers of AS2 to reach destinations in AS3 with a QoS (e-QC_1). This process can be repeated recursively to enable AS1 to also reach destinations in AS2 and AS3, but at no point do AS1 and AS3 negotiate directly.

There are no explicit end-to-end agreements in the cascaded approach, each domain may build upon the capabilities of adjacent downstream ASs to form its own e-QCs to the required destinations. This recursive approach results in an approximation of end-to-end agreements with the merit, as discussed in the following section, of being more scalable.

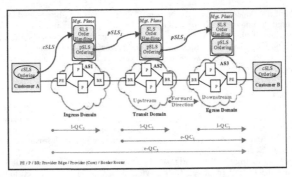

Fig. 3. Cascaded Approach

Strengths and Limitations of the Source-Based and Cascaded Approaches

The originating domain in the source-based approach requires an up-to-date topology of the Internet including the existence and operational status of every physical link between ASs. Whereas, in the cascaded approach, each ISP in the chain only needs to know its adjacent neighbours and the status of related interconnection links.

In both approaches inter-domain routing is pSLS constrained, i.e. traffic will only pass through ASs where pSLS agreements are already in place. Since the originating domain in the source-based approach has topological knowledge of all domains and their interconnections, it is possible to exercise a finer degree of control over the chain of pSLSs through to the destinations. In the cascaded approach each AS participating in the chain does not have all topology data and the initiator is obliged to use the e-QCs previously constructed by the downstream domain. Therefore there is less flexibility and control of the whole IP service path, which may result in sub-optimal paths, although the QoS constraints of the traffic will still be met.

In conclusion, a single point of control for the service instances is the compelling feature of the source-based approach. However it would be difficult to manage for more than a few interconnected ISPs and it is expected that most providers would prefer the cascaded approach, which reflects the loosely coupled structure of the Internet. The cascaded approach makes it possible to build IP QoS services on a global basis while only maintaining contractual relationships with adjacent operators. Hence, this approach is more scalable than the source-based approach. This also reflects the current behaviour of BGP.

4   Inter-domain Business Relationships

Considering a hop-by-hop, cascaded approach for interactions between providers, the following business cases are proposed. Figure 4 depicts the business case, which directly corresponds to the business model of the Internet as it stands today. The business relationships (transit and peer-to-peer) need to be supported by appropriate pSLSs to allow the exchange of QoS traffic. Due to the bi-directional nature of these business relationships and their broad topological scope, only services with loose, qualitative QoS guarantees may be supported in this case.

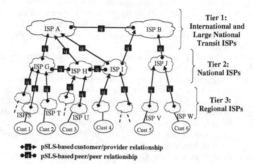

Fig. 4. Case A: Provisioning services with loose QoS guarantees

To provision services with statistical guarantees on quantitative bandwidth and performance metrics a so-called *upstream-QoS-proxy* (or simply *QoS-proxy*) relationship needs to exist between ISPs. Figure 5 illustrates this second business case.

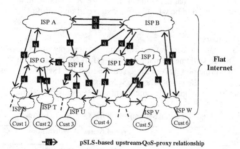

Fig. 5. Case B: Provisioning services with statistical QoS guarantees

In the QoS-proxy relationship, either of the ISPs may agree with the other ISP to provide a transit QoS-based connectivity service to (a subset of) the destinations it can reach with this QoS level. The ISP offering the transit QoS service would have built its QoS reach capabilities based on similar agreements with its directly attached ISPs and so on.

Two things are worth noting about this business relationship: first, its liberal, unidirectional nature, where either ISP can use the other as a QoS proxy; and, second, its strong collaborative and transitive nature, which is built in a cascaded fashion. The

QoS-proxy business relationship differs from the peer-to-peer and customer-provider relationships in the current best-effort Internet in the connotation of the established agreements on traffic exchange and subsequently in the directionality of the traffic flows. In customer-provider relationships, agreements are established for transporting traffic from/to the customer ISPs to/from the provider ISPs and in the peer-to-peer relationships, agreements are established for the ISPs to exchange traffic on a mutual basis; in QoS-proxy relationships, agreements may be established independently in either direction, as each ISP wishes.

The differences between the two business models, as discussed above, are attributed to the diverse types of services each model is set-up to provide. In a best-effort or loose-QoS-based connectivity service Internet, geographical coverage is clearly the strongest selling point; hence, the customer-provider and peer-to-peer relationships. In a statistical-QoS-based service Internet, provided that there is a global demand for related services, the ability to provide such QoS levels becomes clearly an asset. As such, what is required, is to seek for suitable peers to deliver such QoS to desired destinations; hence, the QoS-proxy relationship, which equally applies to all ISPs - regardless of their size.

The business model based on the QoS-proxy relationship resembles the current practice in the traditional telecom service world. The QoS-proxy business agreements could be seen as corresponding to call-termination agreements established between telephony operators/providers. Furthermore, in the telco's world, synergies between operators are built mainly on grounds of reliability and competitiveness, much as the flat Internet business model implies. This resemblance is not surprising; telephony calls and IP services based on statistically guaranteed quantitative QoS metrics, provided that there is a global demand for them, are very similar in that they are both commodities, which need to be widely offered at a certain quality.

Financial Settlements

First, it should be made clear that the financial settlements in a QoS-aware Internet are *in addition* to the settlements made for best-effort connectivity. Broadly speaking, the following two principles govern these settlements: the ISP who requests the pSLS pays the other; and in the cases where pSLSs exist in both directions or the pSLSs have the connotation of mutual agreements, payment reconciliation may take place.

Table 1. Financial Settlements in the QoS-aware Internet

Type of business relationship	Type of financial settlement
customer-provider	service-provider settlement
peer-to-peer	negotiated-financial, or, settlement-free agreement
QoS-proxy	service-provider settlement, if only one ISP requests pSLSs, or, negotiated-financial or settlement-free agreement, if ISPs request pSLSs from each other

5 Conclusions

We propose in this paper that agreements for QoS traffic exchange need to be in place between ISPs in addition to agreements for best-effort connectivity. As with all agreements between ISPs, the establishment of these pSLSs is instigated by the business objectives of the ISP. By augmenting the peer-to-peer and transit (customer-provider) relationships of today's Internet with QoS-based pSLSs we may achieve loose inter-domain QoS guarantees. For provisioning harder QoS guarantees to specific destinations a new QoS-proxy business relationship is proposed. This type of business relationship could be thought as being the QoS Internet counterpart of call-termination agreements in the PSTN or VoIP business world.

The following points are worth mentioning regarding the proposed business relationships. First, they are built on top of existing best-effort agreements and they are not exclusive: e.g. ISP A may act as transit for ISP B, while using ISP C as a QoS-proxy, if this best fits its business objectives. The proposed business models therefore offer an incremental, best-effort compatible migration path towards a QoS-capable Internet.

Second, a key aspect of the proposed business relationships is that they are based on pSLSs, i.e. service agreements, implying legal obligations. On one hand, this offers a tangible lever for ensuring trust between ISPs, which after-all is the bottom-line of any inter-domain solution. On the other hand, this might discourage ISPs. However, given that QoS-based services are offered to end-customers on the basis of Service Level Agreements, requiring the same for ISPs seems reasonable and fair.

Third, the feasibility of the realisation of the proposed business relationships is currently being undertaken by the IST MESCAL project [7]. The MESCAL solutions encompass service management and traffic engineering functions and rely on interactions between adjacent ISPs both at the service layer - for pSLS establishment - and the IP layer - for determining, configuring and maintaining suitable inter-domain QoS routes. The MESCAL validation work covers:

- information models for describing pSLSs under each of the identified business relationships,
- the logic and protocols for pSLS negotiation,
- mechanisms for parsing and extracting the necessary traffic engineering information from the pSLSs,
- extensions to the BGP protocol and associated route selection algorithms to allow the exchange and processing of QoS routing information based on the established pSLS,
- off-line traffic engineering algorithms for determining the required set of pSLSs based on anticipated QoS traffic demand and dimensioning network resources accordingly.

Inevitably, the price to be paid is mainly the increase of the size of the Internet routing tables – growing linearly with the number of the distinct levels of QoS offered. Further simulations and testbed experimentation work [5] for assessing inter-domain routing performance aspects is currently underway.

Acknowledgements. The work described in this paper has been carried out in the IST MESCAL project which is partially funded by the Commission of the European Union. The authors would like to acknowledge the contribution of their colleagues in the MESCAL project for their contributions to the ideas presented in this paper.

References

1. Andreas Thuswaldner, "The international revenue settlement debate", Telecommunication Policy Volume 24 Issue 1 (2000) pp 31-50.
2. Bob Briscoe, "The Direction of Value Flow in Connectionless Networks", Proceedings of Networked Group Communication: First International COST264 Workshop, NGC'99, Pisa, Italy, November 17-20, (1999).
3. David D. Clark, "Combining Sender and Receiver Payments in the Internet", Tele-communications Research Policy Conference (1996).
4. Paris Flegkas, et al., "D1.1: Specification of Business Models and a Functional Architecture for Inter-domain QoS Delivery", http://www.mescal.org/, (2003).
5. Michael Howarth, et al., "D1.2: Initial specification of protocols and algorithms for inter-domain SLS management and traffic engineering for QoS-based IP service delivery and their test requirements", http://www.mescal.org/, (2004).
6. G. Huston, "ISP Survival Guide", John Wiley and Sons 1998. ISBN 201-3-45567-9
7. http://www.mescal.org
8. Eurescom Project P1008, "Inter-operator interfaces for ensuring end-to-end IP QoS", Deliverable 2, Selected Scenarios and requirements for end-to-end IP QoS management, (2001).
9. JP Vasseur, Arthi Ayyangar, "Inter-area and Inter-AS MPLS Traffic Engineering", IETF Internet Draft, (2004).
10. JL Le Roux, JP Vasseur, Jim Boyle, "Requirements for Inter-area MPLS Traffic Engineering", Internet Draft, TEWG Working Group, (2004).
11. Raymond Zhang, JP Vasseur, "MPLS Inter-AS Traffic Engineering requirements", Internet Draft, TEWG Working Group, (2004).

Operations Support System for End-to-End QoS Reporting and SLA Violation Monitoring in Mobile Services Environment

Bharat Bhushan[1], Jane Hall[1], Pascal Kurtansky[2], and Burkhard Stiller[2]

[1] Fraunhofer FOKUS, Kaiserin-August-Allee 31, D-10589 Berlin, Germany
{bhushan, hall}@fokus.fraunhofer.de
[2] ETH Zürich, Computer Engineering and Networks Laboratory TIK Gloriastrasse 35,
CH-8092 Zürich, Switzerland
{kurtansky, stiller}@tik.ee.ethz.ch

Abstract. Delivering end-to-end QoS is emerging as a key requirement for the success of innovative mobile data services. However, overall service quality tends to be susceptible to variations in the performance of mobile networks. SPs (service providers) will need to know what performance and QoS parameters are important in a service delivery chain, and how they can be measured and aggregated to build an overall picture. This is proving to be challenging and placing new requirements on traditional OSSs (Operations Support Systems). For example, OSS process flows involved in SLA (Service Level Agreement) violation monitoring are markedly more complex. Consequently, OSS functions are more interdependent. There is a need to map out and define such interdependencies. This paper presents an OSS architecture that enables QoS reporting and SLA violation monitoring. Key OSS functions are defined and an information model capturing QoS requirements is presented. The results of the implementation and validation of the architecture are also presented.

1 Introduction

QoS is a measure that indicates the collective effect of service performance that determines the degree of satisfaction of a user of the service. The measure is derived from ability of the network and computing resources to provide different levels of services to applications and associated sessions. Customers requiring 3G services for business purposes are demanding guarantees on service quality. Providing such guarantees and maintaining agreed QoS levels is made more complex by end-to-end nature of QoS support in a roaming environment [1]. Service quality can vary over different network technologies and OSSs need to interoperate in order to ensure that the agreed quality of service can be achieved wherever users are located. This requires performance data from each of the service providers involved in delivering the service in order to establish overall quality of service that the customers are using.

Providing 3G services, e.g., location-based services, or corporate intranet access on mobile networks requires a transport connection that transverses many providers and

J. Solé-Pareta et al. (Eds.): QofIS 2004, LNCS 3266, pp. 378–387, 2004.
© Springer-Verlag Berlin Heidelberg 2004

networks [2]. Maintaining QoS in such an environment is going to be complex and will require more versatile and comprehensive approach towards QoS management [3] [4]. Given this situation, there is a strong need for a clear-cut understanding of managed system, management functions, and the management information. The problems that OSS developer face is that QoS area is too wide. This paper presents a coherent view of what is essential for the management of end-to-end QoS and SLA violations.

The rest of the paper is organized as follows. Sect. 2 presents a mobile QoS management environment. Sect. 3 defines the OSS functions required. Sect. 4 presents information model required for SLA violations management. Sect. 5 provides more detailed information on the OSS components implemented to realise the OSS functions presented in this paper. Sect. 6 concludes the paper.

2   Mobile QoS Management Environment

The word *environment* here captures two important aspects of service quality in mobile networks: delivery and measurement (see Fig. 1).

QoS delivery is intertwined with measurement and a broader understanding of this matter requires that these two aspects are seen in a combined view. The environment shown presents QoS and SLA violation management through horizontal and vertical cross-sections. Horizontal cross-section undertakes QoS delivery view, beginning from the UE to the server-side, as well as the transport networks in between. Vertical cross-section, fully described in Sect. 3, undertakes QoS measurement.

There are three provider domains namely MAN (Mobile Access Network), EDN (External Data Network) and CAN (Content/Application Network) from where resources supporting a service are drawn. This configuration clearly demarcates the boundaries of domains that support end-user services from the UE to the server-side

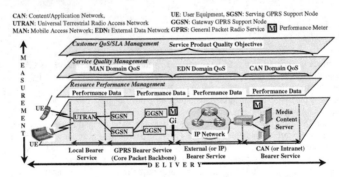

Fig. 1. Mobile QoS management environment

[5]. Measuring end-to-end QoS involves measuring QoS of the three domains or segments shown in Fig. 1. MAN combines Local and GPRS Bearer Services, which implement TCP/UDP/IP and GTP (GPRS Tunnelling Protocol) [6] [7]. The second segment is the EDN, which corresponds to External (or IP) Bearer Service. This is the IP backbone that carries Web applications, FTP, and email traffic and supports RSVP, DiffServ [8]. Lastly, CAN is an intranet belonging to a SP or third-party SP.

The aggregated values of a service product KQI (Key Quality Indicator) [9] are derived from values of lower level service KQIs, which focus on performance of individual services. Services depend on the reliability of network and service resources and KPI (Key Performance Indicator) values [9] are calculated from the performance data provided by the resources. An example of service KQI can be 90% availability of end-to-end network connection. To calculate this value, there can be at least two KPIs, namely resource up-time and resource down-time, whose values are required.

3   OSS Functions and Process Flow

The OSS functions defined in this section concern with the OSS support required by the SPs, mobile network operator, the customer, as well as between SP and third-party SP. To define OSS functions, layers of the eTOM [20] model are used in the Assurance business process. Requirements of these stakeholders on OSS are formulated from the Assurance business processes and incorporated in the definitions.

3.1   OSS Function Definitions

Customer QoS/SLA Management. This function ensures that the performance of a product delivered to customers meets the SLA parameters, or product KQI objectives, as agreed in the contract concluded between the SP and the customer. This function has four specific operations. *SLA specification* specifies SLA parameters for a new product, including soft and hard thresholds. *SLA monitoring* collects service KPI values achieved, aggregates them into product KQI values achieved. *SLA analysis* analyses product KQI values achieved against the SLA parameters agreed. It triggers further action if the service KQI values achieved do not meet the agreed SLA values. *SLA reporting* produces product performance summaries and trend reports of SLA parameters values achieved compared with the agreed SLA values.

Service Quality Management. This function ensures that the performance of a service delivered to customers meets the service KQI objectives. It uses mathematical algorithms that transform KPI values into KQI values. This function has four specific operations. *Service KQI specification* specifies service KQIs and their objectives for a new service, including soft and hard thresholds. *Service KQI monitoring* collects resource KPI values achieved and aggregates them into service KQI values achieved using algorithms. *Service KQI analysis* analyses service KQI values achieved against the service KQI objectives and issues warnings and alarms if the values achieved do not meet the objectives. *Service KQI reporting* produces service performance summaries and trend reports of KQI values achieved compared with the KQI objectives.

Resource Performance Management. This function ensures that the performance of the resources involved in delivering services meets the resource KPI objectives. This function has four specific operations. *Resource KPI specification* specifies resource KPIs and their objectives. *Resource KPI monitoring* collects performance data and aggregates it into KPI values achieved. *Resource KPI analysis* analyses KPI values achieved against the KPI objectives. It issues warnings and alarms if the resource KPI values achieved do not meet the KPI objectives. *Resource KPI reporting* produces performance summaries and trend reports of resource KPI values achieved.

3.2 Process Flow

The process flow shown in Fig. 2 is based on the assumption that customers subscribe to SP for the services they use.

The SP also has appropriate supplier/partner (S/P) relationship with the one or more third-party SPs. The process flows shows how the SP aggregates its own data with the performance and usage data it receive from third-party SPs. The process flow steps shown in Fig. 2 are as follows:

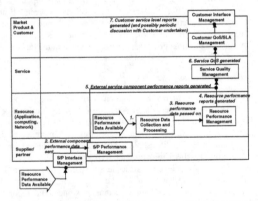

Fig. 2. Service Quality Monitoring and SLA Violation Management Process Flow

During normal operation, performance data that is used for monitoring of service levels as well as for longer-term capacity prediction is collected on an ongoing basis from the service-providing infrastructure by *Resource Data Collection & Processing* (step 1 in Fig. 2). During normal operation, performance data from external service components of third-party SPs is sent on an ongoing basis to *S/P Performance Management* for general monitoring of service levels (step 2). *Resource Data Collection & Processing* sends performance data to *Resource Performance Management* for further analysis (step 3). *Resource Performance Management* sends performance reports (step 4) and *S/P Performance Management* sends external component performance reports (step 5) to *Service Quality Management* for QoS calculations and to maintain statistical data on the supplied service instances. *Service Quality Management* analyses the performance reports received and sends overall service quality reports to *Customer QoS/SLA Management* so that it can monitor and report aggregate technology and service performance (step 6). *Customer QoS/SLA Management* checks the quality reports it receives against the individual Customer SLA and establishes that no SLA violation has occurred (step 7).

4   Information Model

The model shown in Fig. 3 deals with SLA, QoS and the surrounding classes. It is based on input from the TM Forum SID [10] [11] [12] models but simplified and specialised to meet the requirements of the work presented in this paper. The model presented here introduces a new relationship between KeyQualityIndicatorSLSParam and KeyPerformanceIndicatorSLSParam, which indicates that KQI values used to determine compliance with service level objectives are derived from KPI values.

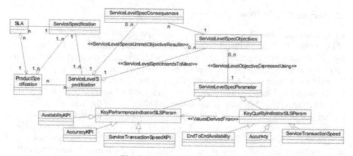

Fig. 3. QoS Information Model

4.1   Customer Contract View

ProductSpecification. This is a collection of services offered to customers. Enterprises create product offerings from product specifications that have additional market-led details applied over a particular period of time. The relationship between ProductSpecification and SLA signifies what services are *agreed* with the customer.
ServiceSpecification. This is an abstract base class for defining the ServiceSpecification hierarchy. It defines the invariant characteristics of a Service and may also exist within groupings, such as within a Product. The relationship between ProductSpecification and ServiceSpecification signifies what services are *offered* to the customer.

ServiceLevelAgreement. A SLA is a formal negotiated agreement between two parties regarding provided Service Level [13]. It is designed to create a common understanding about services, priorities, responsibilities, etc. It is usually written down explicitly and has legal consequences. Its main aim is to achieve and maintain a specified QoS in accordance with ITU-T and ITU-R Recommendations [13]. These procedures and targets are related to specific service availability or performance.

ServiceLevelSpecification. This is a group of service level objectives, such as thresholds, metrics, and tolerances, along with consequences that result from not meeting the objectives. Service level specifications are also characterised by exclusions associated with the specification.

ServiceLevelSpecObjectives. These objectives can be seen as a quality goal for a ServiceLevelSpecification defined in terms of metrics and thresholds associated with the parameters. The *conformanceTarget* attribute determines whether ServiceLevelSpecObjectives are met. Attribute *conformanceComparator* specifies whether ServiceLevelSpecObjectives are violated above or below the conformanceTarget.

ServiceLevelSpecConsequences. It is an action that takes place in the event that a ServiceLevelObjective is not met. Violations to SLA usually result in some consequence for the SP.

4.2 Service Quality View

KeyQualityIndicatorSLSParam. ServiceLevelSpecification parameters can be one of two types: KQIs and KPIs. A KQI provides a measurement of a specific aspect of the performance of a Product or a Service (i.e. ServiceSpecification). A KQI draws its data from a number of sources, including KPIs. This subsection deals with KQIs and the next subsection with KPIs. KeyQualityIndicatorSLSParam is the superclass of EndToEndAvailability, Accuracy and ServiceTransactionSpeed [14].

EndToEndAvailability KQI is the likelihood with which the relevant product or service can be accessed at customer's request. To determine the value of end-to-end availability, aggregated values MAN, EDN and CAN availability are measured and then combined in a single KQI value. The main attributes of this class are *manAvailability, ednAvailabilityPercentage, canAvailability* [5]. *KQITransformationAlgorithm* is a procedure used to calculate the KQI value. Accuracy defines the fidelity and completeness in carrying out the communication function. This KQI informs *about* the degree to which networks and services are dependable and error free. The attributes of this class are *networkingAccuracy, serviceTransactionAccuracy*, and *contentInfoReceptionAccuracy* [5]. ServiceTransactionSpeed defines the promptness with which network and service complete end user transactions. The main attributes of this class are *serviceSessionSetupRate, serviceResponseTime*, and *contentInfoDeliveryRate* [5].

4.3 Network Performance View

KeyPerformanceIndicatorSLSParam. KPIs provide a measurement of a specific aspect of the performance of a service resource or group of service resources of the

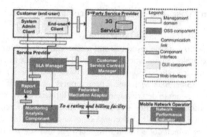

Fig. 4. OSS Component Architecture

same type. A KPI is restricted to a specific resource type. This is the superclass of three subclasses: AvailabilityKPI, AccuracyKPI and ServiceTransactionSpeedKPI.

AvailabilityKPI represents the KPIs whose values are required for the calculation of the EndToEndAvailability KQI value. The attributes of this subclass are *pdpContextSet-upTime* [7], *httpRoundTripDelay* [15], *up-time*, and *down-time* [5]. AccuracyKPI represents the KPIs whose values are required for the calculation of the Accuracy KQI value. The main attributes of this subclass are *sduLossRatio* [6], *httpRoundTripDelay*, *BER* (Bit Error Rate) [6] and *jitter*. ServiceTransactionSpeedKPI represents the KPIs whose values are required for the ServiceTransactionSpeed KQI value. The main attributes are *congestion* [5], *serviceResourceCapacity* [16], *serviceRequestLoad, pdpContextSet-upTime,* and *contentInfoDeliveryRate* [5].

SID models are not very clear about combining values of two or more KQIs to form a single combined KQI value, as proposed in WSMH [9]. To remedy this, the information model presented derives subclasses from KeyQualityIndicatorSLSParam, which represent the combined KQI whereas subclasses' attributes represent the KQIs.

5   OSS Components and Validation Scenario

OSS functions have been validated through development of the components that provide those functions and then by testing them on 3G network and with a location-based service in a validation scenario.

5.1   Components Architecture

The component architecture has undertaken process flows described in Sect. 3. The process flows and the OSS functions are used as a blueprint to assemble a set of OSS components in an integrated management system (Fig. 4).

SLA Manager. This component maintains customer SLAs and for ensuring that the delivered QoS meets the QoS specified in customer contracts or other product specifications. It ensures that the provisions in the contract are being met and issues warnings and alarms if this is not the case. This means that it must be able to monitor, analyse and report on the actual QoS achieved compared with the QoS specified in the SLA.

Monitoring Analysis Component (MAC). MAC collects and aggregates values of KPIs from network and service resources and transforms them into KQI values using algorithms, for example, mathematical equations. The MAC uses the value of one or more KQIs and service quality objectives as a basis and issues reports containing warnings or alarms to the SLA Manager. These reports inform the SLA Manager that the service quality is about to deteriorate or has deteriorated below a threshold value.

Network Performance Evaluator (NPE). NPE measures resource performance in KPI value. The NPE is a rule-based component and operates in conjunction with performance meters, which collect data from network and service resources. Meters can be programmed by means of filter specifications, which are simple instruction rules.

Report Log. The Report Log component receives service quality reports from the MAC and stores them in a log. It also allows the service provider to view and browse through the report log via a web browser.

Supporting Components. There are two more components namely FMA (Federated Mediation Adaptor) and CSC (Customer Service Contract) Manager that also take part in SLA management. The CSC Manager provides methods for the management of SLA data. The FMA component collects SLA violations reports and forwards them to a rating and billing facility.

5.2 Validation Scenario

The main objective of validation scenario is to demonstrate that end-to-end service quality of 3G services can be measured and reported to monitor SLA violation. The SP, shown in Fig 4., integrates a location-based service and virtual home environment service, both of which are provided by third-party SPs. The end-use client is the single point of access and authentication for the customer. SP obtains performance data from each of the third-party SPs and Mobile Network Provider, and aggregates and analyses the data to establish overall service quality and also to produce overall QoS reports. If a SLA violation occurs, violation reports are sent to the Billing process. An algorithm is used to measure service availability as a percentage value, which is compared against a service quality objective. In the scenario, objective for overall service availability is set to 90%, and an alarm is raised if service availability value reaches below 90%. A software system showing this validation scenario has been successfully demonstrated in the trials of an EU IST project called AlbatrOSS [17] and fuller details on validation can be found in [18]. The concepts presented in this paper are also being modified and further developed in another IST project called Daidalos [19].

5.3 Component Operations

This section uses UML Message Sequence Diagrams to illustrate in Fig. 5 how OSS components interact in order to achieve the goal of validation scenario.

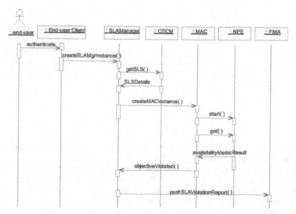

Fig. 5. Service Quality Meaurement and SLA Violation Reporting

The sequence begins when the end-user logs on and begins to use the service. This is the SLA initialisation phase in which the SLA Manager obtains the SLS details for the service being used by the end user and creates an instance of the MAC. MAC continually interacts with the NPE and obtains the values of KPIs from web-based servers running over UMTS network, measures the value of the KQIs, and compares it against the service quality objective set in the SLS. As the KQI values go below the objective, the MAC informs the SLA Manager that the service quality reached below the objective, which may have a consequence and an action need to be taken. The action may state that the end user is entitled to receive a discount. In this case, a report stating the action to be taken is sent to the FMA, which forwards it to a Rating facility.

6 Summary and Conclusions

The main theme of this paper is measurement and aggregation of service quality in a mobile service provision chain, and analysis of service quality against certain objectives. In this scenario, interdependencies among OSS functions are more pronounced compared to traditional OSSs. This paper tries to map out such interdependencies.

This paper presents an OSS that is based on requirements of the personalised mobile data services. The focus is mainly on the specific functionality in an OSS for 3G and beyond mobile networks, although traditional OSS functions that may need to be extended for mobile data environment have also been considered. These functions define the scope of the system; their specification lays the foundation for the component and interface design in management system architecture. The requirements of the specified functions for the QoS information have also been considered and comprehensive information model has also been produced. It has been designed as a light and flexible structure able to support the key OSS functions defined industrial consortium such as 3GPP and TMForum. This work provides practical insight into the way the TMForum SID models can be applied to develop QoS management system. The paper also presents an OSS component architecture that implements the OSS functions defined in the paper. The validation scenario has provided the opportunity to test out the OSS functions and information model.

References

1. Fawaz, et al: "Service Level Agreement and Provisioning in Optical Network", IEEE Communication Magazine, January 2004.
2. Choi: "Location-based Service Provisioning for Next Generation Wireless Network", International Journal of Wireless Information Network, Vol. 10, No. 3, July 2003.
3. Mahajan: "Managing QoS for Multimedia Applications in the Differentiated Services Environment", Journal of Network and Systems Management, Vol. 11, No. 4, Dec 2003.
4. Duan, Zhang, and Hou: "Service Overlay Networks: SLAs, QoS and Bandwidth Provisioning", IEEE/ACM Transactions on Networking, Vol. 11, No. 6, Dec 2003.
5. Bhushan, et al: "Measuring Performance of Third-Generation Wireless Networks for the End-to-End QoS Analysis", IEEE/Comsoc ICN 04 proceedings, Guadeloupe, Feb 2004.
6. 3GPP: "Technical Specification Group Services and System Aspects: Quality of Service (QoS) concept and architecture", 3GPP 23.107 V5.10.0 (2003-09) (Release 5)
7. 3GPP: "Technical Specification Group Services and System Aspects: End-to-End Quality of Service (QoS) concept and architecture", 3GPP TS 23.207 V6.0.0 (2003-09) (Release 6).
8. ITU-T Y.1541, Global Information Infrastructure and IP Aspects, Quality of service and network performance, Network Performance Objectives for IP-based services, May 2002.
9. GB923, *Wireless Service Measurements Handbook*, V3.0, www.tmforum.org, March 2004.
10. GB922, SID Model. Concepts, Principles, and Domains, Member Evaluation Version 3.1, www.tmforum.org, July 2003
11. GB922, SID Model. Addendum 4SP – Service Overview Business Entity Definitions, Addendum-4SO, Member Evaluation V 2.0, www.tmforum.org, July 2003
12. GB922, SID Model: Addendum 1A -Common Business Entity Definitions – Agreement (including SLA), Addendum-1A, Member Evaluation V 3.1, www.tmforum.org, July 2003.
13. GB917, *SLA Management Handbook*, V1.5, www.tmforum.org, June 2001.
14. Ooden, Ward, Mullee: "Quality of Service in Telecommunications", IEE Telecommunication Series 39, Published by IEE UK, 1997, ISBN: 0 85296 919 8.
15. RFC 2681, A Round-trip Delay Metric for IPPM, IETF, Sept 1999.
16. RFC 3148, A Framework for Defining Bulk Transfer Capacity Metrics, IETF, July 2001.
17. www.ist-albatross.org
18. Evaluation of Trial System Version 2, IST-2001-34780 AlbatrOSS Project Deliverable 11, December 2003. www.ist-albatross.org
19. www.ist-diadalos.org
20. GB921, *enhanced Telecom Operations Map*™ (eTOM). The Business Process Framework, TM Forum Approved Version 4.0, www.tmforum.org, March 2004.

Author Index